T CELL SUBSETS IN INFECTIOUS AND AUTOIMMUNE DISEASES

The Ciba Foundation is an international scientific and educational charity (Registered Charity No. 313574). It was established in 1947 by the Swiss chemical and pharmaceutical company of CIBA Limited—now Ciba-Geigy Limited. The Foundation operates independently in London under English trust law.

The Ciba Foundation exists to promote international cooperation in biological, medical and chemical research. It organizes about eight international multidisciplinary symposia each year on topics that seem ready for discussion by a small group of research workers. The papers and discussions are published in the Ciba Foundation symposium series. The Foundation also holds many shorter meetings (not published), organized by the Foundation itself or by outside scientific organizations. The staff always welcome suggestions for future meetings.

The Foundation's house at 41 Portland Place, London W1N 4BN, provides facilities for meetings of all kinds. Its Media Resource Service supplies information to journalists on all scientific and technological topics. The library, open five days a week to any graduate in science or medicine, also provides information on scientific meetings throughout the world and answers general enquiries on biomedical and chemical subjects. Scientists from any part of the world may stay in the house during working visits to London.

Ciba Foundation Symposium 195

T CELL SUBSETS IN INFECTIOUS AND AUTOIMMUNE DISEASES

1995

JOHN WILEY & SONS

Chichester · New York · Brisbane · Toronto · Singapore

Published in 1995 by John Wiley & Sons Ltd,
Baffins Lane, Chichester,
West Sussex PO19 1UD, England
Telephone: National (01243) 779777
International (+44) 1243 779777

Other Wiley Editorial Offices

John Wiley & Sons, Inc., 605 Third Avenue,
New York, NY 10158-0012, USA

Jacaranda Wiley Ltd, 33 Park Road, Milton,
Queensland 4064, Australia

John Wiley & Sons (Canada) Ltd, 22 Worcester Road,
Rexdale, Ontario M9W 1L1, Canada

John Wiley & Sons (SEA) Pte Ltd, 37 Jalan Pemimpin #05-04,
Block B, Union Industrial Building, Singapore 2057

Suggested series entry for library catalogues:
Ciba Foundation Symposia

Ciba Foundation Symposium 195
ix + 262 pages, 22 figures, 7 tables

Library of Congress Cataloging-in-Publication Data

T cell subsets in infectious and autoimmune diseases.
p. cm. — (Ciba foundation symposium ; 195)
Editors: Derek Chadwick (Organizer) and Gail Cardew.
Based on a symposium held at the Ciba Foundation, London, March, 1995.
Includes bibliographical references and index.
ISBN 0 471 95720 8 (alk. paper)
1. T cells—Congresses. 2. Immunopathology—Congresses.
3. Communicable diseases—Immunological aspects—Congresses.
4. Autoimmune diseases—Immunological aspects—Congresses.
I. Chadwick, Derek. II. Cardew, Gail. III. Series.
[DNLM; 1. T-Lymphocyte Subsets—congresses. 2. Autoimmune Diseases—congresses. 3. Communicable Diseases—congresses. W3 C161F v. 195 1995 / QW 568 T1144 1995]
QR185.8.T2T16 1995
616.9'0479—dc20
DNLM/DLC
for Library of Congress 95-33408
CIP

British Library Cataloguing in Publication Data

A catalogue record for this book is available from the British Library

ISBN 0 471 95720 8

Typeset in 10/12pt Times by Dobbie Typesetting Limited, Tavistock, Devon.
Printed and bound in Great Britain by Biddles Ltd, Guildford.
This book is printed on acid-free paper responsibly manufactured from sustainable forestation, for which at least two trees are planted for each one used for paper production.

Contents

Symposium on T cell subsets in infectious and autoimmune diseases, held at the Ciba Foundation, London 7–9 March 1995

This symposium is based on a proposal made by F. Y. Liew

Editors: Derek Chadwick (Organizer) and Gail Cardew

Participants

A. K. Abbas Immunology Research Division, Departments of Pathology, Brigham and Women's Hospital and Harvard Medical School, 221 Longwood Avenue, Boston, MA 02115, USA

P. M. Allen Department of Pathology, Washington University School of Medicine, Box 8118, 660 South Euclid Avenue, St Louis, MO 63110-1093, USA

D. A. Cantrell Lymphocyte Activation Laboratory, Imperial Cancer Research Fund, PO Box 123, 44 Lincoln's Inn Fields, London WC2A 3PX, UK

R. L. Coffman Department of Immunology, The DNAX Research Institute of Molecular and Cellular Biology, 901 California Avenue, Palo Alto, CA 94304-1104, USA

D. Cosman Immunex Research and Development Corporation, 51 University Street, Seattle, WA 98101, USA

R. W. Dutton Department of Biology, University of California at San Diego, Cancer Center, 9500 Gilman Drive, La Jolla, CA 92093-0063, USA

F. W. Fitch Department of Pathology and the Ben May Institute, University of Chicago, 5841 South Maryland Avenue, Chicago, IL 60637, USA

R. Flavell Yale University School of Medicine, 310 Cedar Street, New Haven, CT 06520-8011, USA

S. H. E. Kaufmann Max-Planck-Institute for Infectious Biology, Monbijoustrasse 2, 10117 Berlin, Germany

P. J. Lachmann MRC Molecular Immunopathology Unit, MRC Centre, Hills Road, Cambridge CB2 2QH, UK

J. R. Lamb Department of Immunology, St Mary's Hospital Medical School, Norfolk Place, London W2 1PG, UK

F. Y. Liew Department of Immunology, University of Glasgow, Western Infirmary, Glasgow G11 6NT, UK

R. M. Locksley Departments of Medicine and Microbiology/Immunology, University of California, Box 0654, C-443, San Francisco, CA 94143, USA

M. T. Lotze Department of Surgery , University of Pittsburgh, 200 Lothrop Street, Room W1543, Promedical Science Tower, Pittsburgh, PA 15261, USA

A. J. McMichael Molecular Immunology Group, Institute of Molecular Medicine, John Radcliffe Hospital, Headington, Oxford OX3 9DU, UK

D. W. Mason MRC Cellular Immunology Unit, Sir William Dunn School of Pathology, University of Oxford, Oxford OX1 3RE, UK

N. A. Mitchison (*Chairman*) Deutsches Rheuma-Forschungszentrum Berlin, Nordufer 20, D-13353 Berlin, Germany

T. R. Mosmann Department of Immunology, 865 Medical Sciences Building, University of Alberta, Edmonton, Alberta T6G 2H7, Canada

V. Navikas (*Bursar*) Department of Neurology, Huddinge Hospital, S-14186 Huddinge, Sweden

I. A. Ramshaw Division of Cell Biology, John Curtin School of Medical Research, Australia National University, Canberra, ACT 2601, Australia

M. Röllinghoff Institut für Klinische Mikrobiologie, Friedrich-Alexander-Universität Erlangen-Nürnberg, Wasserturmstrasse 3, D-91054 Erlangen, Germany

S. Romagnani Division of Clinical Immunology and Allergy, Institute of Clinical Medicine 3, University of Florence, Viale Morgagni 85, I-50134 Florence, Italy

G. M. Shearer Experimental Immunology Branch, National Cancer Institute, National Institutes of Health, Bethesda, MD 20892, USA

A. Sher Immunobiology Section, Laboratory of Parasitic Diseases, National Institute of Allergy and Infectious Diseases, Building 4, Room 126, National Institutes of Health, Bethesda, MD 20892, USA

S. L. Swain Department of Biology - 0063, University of California, 9500 Gilman Drive, San Diego, CA 92093-0063, USA

G. Trinchieri The Wistar Institute for Anatomy and Biology, 3601 Spruce Street, Philadelphia, PA 19104, USA

Introduction

N. A. Mitchison

Deutsches Rheuma-Forschungszentrum Berlin, Nordufer 20, D-13353 Berlin, Germany

The study of T cell subsets has made great progress in the past few years. The audience of the International Congress of Immunology in Toronto in 1986 was thrilled by Tim Mosmann's new concept of two subsets of T helper cells, Th1 and Th2 cells. Some of us will recall a much earlier meeting at the Rockefeller Villa, Lake Garda, Italy, which was held to mark an equally decisive turning point: the introduction of Th cells and T cytotoxic (Tc) cells. I remember Harvey Cantor, then deeply involved with the Ly1, Ly2 and Ly3 markers, accepting the new functional terminology only when it was pointed out to him that it would immortalize his own initials. We all like to rewrite history in our own image: my own presentation at the Toronto Congress also dealt with splitting the $CD4^+$ T cell subset, but with major histocompatibility complex (MHC) restriction, idiotype linkage and suppression playing as large a part as cytokines. Tim Mosmann found that many of the differences between H-2A- and H-2E-restricted T cell subsets, and also most of the help-versus-suppression differences, boiled down to Th1 versus Th2 cytokines. Our task at this symposium is to find out where the Th1/Th2 revolution has taken us so far, and where it is likely to lead us in the future.

The two key discoveries on which the revolution is built can be summarized as follows. (1) T cell clones are divided into Th1 cells, which tend to express γ-interferon (IFN-γ), and Th2 cells, which tend to express interleukin 4 (IL-4). (2) Naive T cells stimulated in the presence of IL-12 or IFN-γ differentiate into Th1 cells (or Tc1 cells). In contrast, if they are stimulated in the presence of IL-4, they differentiate into Th2 cells (or Tc2 cells). These discoveries are applicable to both mice and humans. The experiments which underpin them are documented in three admirable chapters in recent editions of *Annual Review of Immunology* (Fitch et al 1993, Seder & Paul 1994, Romagnani 1994).

Two lines of enquiry stem from these foundations. One addresses the underlying molecular cell biology. How exactly does the differentiation between T cell subsets take place? Should we be looking at the promoter regions of the three key cytokines or is there something still deeper in the differentiated state of the cell that permits one or other cytokine to be expressed? Is the differentiation irreversible? None of these questions have been answered. However, one can be optimistic because the answers lie within the normal reductionist programme in cell biology, and they are, therefore, within the reach of contemporary molecular methods.

The second line of enquiry, which runs upwards into the integrated activity of the immune system, is more challenging. How does the immune system maintain a proper balance between the T cell subsets? Is this balance defective in immunological disease? If so, how can we rectify the imbalance? This symposium will show that we have many animal models which mimic the cytokine perturbations evident in human immunological disease, and that we are learning how to manipulate the balance in these models. All this is good news, but we are still a long way from understanding why the balance is impaired in disease states and how to manipulate it by methods that may be used in humans.

These difficulties were brought home to me during the 15th European Rheumatology Research Workshop for Rheumatology Research, Erlangen/Bamberg, Germany, March 16–19 1995 (see Clin Rheumatol 14[2] 1995), which was devoted to the question of whether T cell cytokine subsets control rheumatoid arthritis. Several groups have observed T cells making cytokines within the rheumatoid synovium. This is an achievement because older methods applied to this problem yielded negative results. However, the step from simply observing these cytokines to deciding whether or not they are important still seems enormous.

In my opinion, some of the best evidence for the importance of T cell cytokine balance in human disease comes from HLA associations. Susceptibility to rheumatoid arthritis is influenced by class II HLA genes. It is well known that the presence of certain of these genes increases susceptibility to the disease and its subsequent severity. Significantly, other class II HLA genes decrease this susceptibility. Furthermore, these protective alleles are effective when heterozygous, indicating that they have an active protective function, rather than merely failing to do something. The same is true for type 1 diabetes, in which the protective effect is even larger. Indeed, it is likely that other, rarer immunological diseases will be similarly controlled, although for these it will take longer to collect the results. The function of class II HLA genes is to guide, or restrict, the activation of a subset of T cells. When some of these genes push the disease in one direction, and others push it in the opposite direction, it is difficult to imagine what they could be doing, other than activating two groups of counteractive T cells. Also, how could this counteractivity operate, other than through the production of mutually inhibitory cytokines? As ever in biology, other possibilities cannot be entirely dismissed, and the hypothesis has to be tested experimentally. Monika Brunner and I have been doing just this in a mouse model of rheumatoid arthritis and also with other antibody responses, in all of which the presence of an H-2A^{b} allele is associated with suppression. In the one case in which we examined the

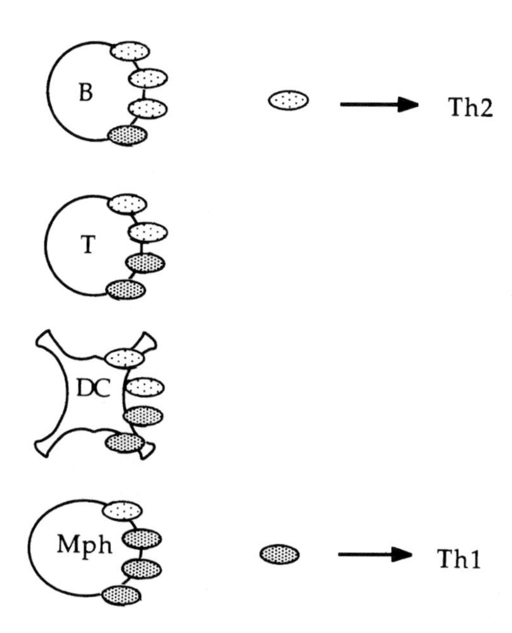

FIG.1. How selective expression of class II major histocompatibility complex (MHC) molecules might affect T cell function. B, B cell; DC, dendritic cell; Mph, macrophage; T, T cell; Th, T helper. The light stippled MHC molecule is over-represented on B cells and may, therefore, preferentially activate Th2 cells; the heavily stippled one is over-represented on macrophages and may, therefore, preferentially activate Th1 cells.

cytokine balance, the H-2A^{b} allele inhibited the production of IL-4 (Mitchison & Brunner 1995, A. Mitchison & M. Brunner, unpublished results 1995). There remains the question of how the HLA genes act so selectively. Our idea is that promoter polymorphisms may be responsible, by regulating differential expression in different types of antigen-presenting cell (Mitchison & Brunner 1995). This possibility is summarized in Fig. 1, which depicts two allelic class II MHC molecules, one of which is over-expressed in B cells and, consequently, tends to activate Th2 cells. The other is over-expressed in macrophages and has the reverse effect. One striking feature is the high degree of polymorphism in the X and Y boxes controlling the expression of class II MHC genes both in humans (Louis et al 1993) and in the mouse haplotypes H-2^{b}, H-2^{d}, H-2^{k} and H-2^{q} (data courtesy of R. Lauster, accession numbers X86147–X86156 in the EBI nucleotide sequence bank). J. F. Eliaou has pointed out (personal communication, Louis et al 1993) that this seems like a consequence of natural selection.

The self-stabilizing nature of the cytokine balance makes it an attractive target for therapy because any restoration would prove long-lasting. Accordingly, a wide range of therapeutic strategies are being explored. Old drugs, such as thalidomide (McHugh et al 1995) and rolipram (Raine 1995), are being re-evaluated for their cytokine effects. A more direct approach of injecting IL-4 into the synovial cavity (Miossec 1993) has been advocated. The pharmacokinetics of the cytokines themselves are unimpressive, except perhaps for IL-12 with its relatively large, two-chain molecule. My guess is that the best, or at least the most revealing, approach will be directed towards the genes encoding cytokines, and that the best place to implant them will be into T cells themselves.

There is no shortage of ideas. The only cloud on the horizon for us immuno-tinkerers is the possibility that in the future these diseases will be treated by auto-transplantation of stem cells. Alan Tyndall (Rheumatologische Universitätsklinik, Basel, personal communication) suggests that under certain circumstances (i.e. in the worst-prognosis cases, not too late in the disease course) this procedure could be justified: after all, stem cell transplantation is an established medical procedure in other diseases. Perhaps we do not need to tinker with the niceties of the immune system, but instead just blast it away. Consider the humbler problem of infertility. In the past, sterile couples were examined in great detail by skilful endocrinologists and immunologists. Most of that has gone by the board, superseded by *in vitro* fertilization.

As science moves from exploration to engineering, we are confronted with an unlimited number of openings to explore. By concentrating on the exploration rather than the engineering, this symposium keeps within manageable bounds. The reader should bear in mind that, although the ideas discussed here are attractive in the abstract, their ultimate test will come in the dusty arena of therapy.

References

Fitch FW, McKisik MD, Lancki DW, Gajewski TF 1993 Differential regulation of murine T lymphocyte subsets. Annu Rev Immunol 11:29–48

Louis P, Eliaou JF, Kerlon-Candon S, Pinet V, Vincent R, Clot J 1993 Polymorphism in the regulatory region of HLA-DRB genes correlating with haplotype evolution. Immunogenetics 38:21–26

McHugh SM, Rifkin IR, Deighton J et al 1995 The immunosuppressive drug thalidomide induces T helper cell type 2 (Th2) and concomitantly inhibits Th1 cytokine production in mitogen-stimulated and antigen-stimulated human peripheral blood mononuclear cell cultures. Clin Exp Immunol 99:160–167

Miossec P 1993 Proprietes anti-inflammatorires de l'interleukine 4. Rev Rhum Mal Osteo-Artic 60:119–124

Mitchison NA, Brunner C 1995 Association of $H\text{-}2A^b$ with resistance to collagen-induced arthritis in *H2*-recombinant strains: an allele associated with reduction of several apparently unrelated responses. Immunogenetics 42:239–245

Raine CS 1995 Multiple sclerosis: TNF revisited, with promise. Nat Med 1: 211–214
Romagnani S 1994 Lymphokine production by human T cells in disease states. Annu Rev Immunol 12:227–258
Seder RA, Paul WE 1994 Acquisition of cytokine-producing phenotype by $CD4^+$ T cells. Annu Rev Immunol 12:635–674

DISCUSSION

Abbas: Can you please expand on your comment that the presence of the H-$2A^b$ allele inhibits the production of interleukin 4 (IL-4)?

Mitchison: With other alleles, for example H-$2A^k$, a conspicuous burst of IL-4 production is observed within less than 24 h in response to allo-4-hydroxyphenylpyruvate dioxygenase (allo-HPPD). This is first detected 11 h after immunization. However, this IL-4 burst is missing when the H-$2A^b$ allele is present, even in individuals heterozygous for the H-$2A^b$ allele, because these suppressive effects are always dominant or very nearly dominant.

Abbas: Is this study a comparison between BALB/c and H-2 mice?

Mitchison: No, it's a comparison between H-2-recombinant strains on the B10 background. Therefore, non-major histocompatibility complex (MHC) differences don't confuse the issue.

Liew: Is this a general phenomenon?

Mitchison: The cytokine burst has only been studied for allo-HPPD, but other responses, such as to bovine type II collagen and probably also to allo-CDw90, are under the same genetic control.

Liew: We have tried to induce arthritis in IL-4 knockout H-2^b mice, but we did not observe an effect (I.B. McInnes & F.Y. Liew, unpublished results). The collagen-induced arthritis can only be induced in DBA/1 (H-2^q) mice. Is it possible that IL-4 is involved in the regulatory mechanism in this model?

Mitchison: No, because the MHC exon product is necessary for operation of this control mechanism.

Flavell: One prediction of this concept is that information in the regulatory part of the promoter will confer a particular phenotypic property. Consequently, one might find haplotypes where the information has been transferred by some intergenic process, such as gene conversion. Have you looked at whether there are any haplotypes that carry the transcription regulatory unit and, therefore, the peptide-binding component into a different gene?

Mitchison: Not yet. We are planning experiments in which we would reshuffle exons and introns. Ricchiardi's antigen-presenting cell lines will provide a way of testing them (Lutz et al 1994).

Abbas: Do the promoters influence the level of expression of the MHC class II molecules? This would be one way of interpreting your results, i.e. that you are really influencing the density of peptide MHC complexes.

Mitchison: Yes that's a popular notion. However, we think that more 'intelligent' mechanisms are involved, i.e. differential expression between different antigen-presenting cells.

Mason: There are differences in the level of expression of MHC class II molecules on the kidney proximal tubule of different rat strains (D. Mason, unpublished results). For instance, there is a 100-fold difference between the rat strains DA and PVG. I don't know what the significance of that result is, but the difference is striking.

Mitchison: That must involve an additional factor because that observation is unlikely to result from promoter differences. Can you compare the alleles in the same rat?

Mason: They are actually different strains.

References

Lutz MB, Granucci F, Winzler C et al 1994 Retroviral immortalization of phagocytic and dendritic cell clones as a tool to investigate functional heterogeneity. J Immunol Methods 174:1–2

Differentiation and tolerance of $CD4^+$ T lymphocytes

Abul K. Abbas, Victor L. Perez, Luk van Parijs and Richard C. K. Wong

Immunology Research Division, Departments of Pathology, Brigham & Women's Hospital and Harvard Medical School, 221 Longwood Avenue, Boston, MA 02115, USA

Abstract. The development of effector and memory populations of T lymphocytes is determined by antigen-induced growth and differentiation of naive T cells, and it is regulated by antigen-induced functional tolerance and cell death. $CD4^+$ helper T lymphocytes that vary in their profiles of cytokine production and in effector functions also show distinct responses to antigens and co-stimulatory signals, and they differ in their sensitivity to tolerance induction. Thus, stimuli that trigger T cell growth and differentiation, as well as mechanisms that inhibit T cell expansion, determine both the magnitude and the nature of T cell-dependent immune responses to protein antigens.

1995 T cell subsets in infectious and autoimmune diseases. Wiley, Chichester (Ciba Foundation Symposium 195) p 7–19

The nature and magnitude of immune responses to all protein antigens are determined by the types and numbers of $CD4^+$ helper and regulatory T lymphocytes that are induced following exposure to the antigens. Naive $CD4^+$ T cells recognize antigens in the form of processed peptides presented by major histocompatibility complex (MHC)-encoded molecules on specialized antigen-presenting cells (APCs). As a result of antigenic stimulation, antigen-specific T cells proliferate and differentiate into effector populations that secrete cytokines and perform various functions which eliminate the antigens. The best defined functionally distinct subsets of helper T cells are Th1 and Th2 cells (Mosmann & Coffman 1989, Romagnani 1994, Seder & Paul 1994). Th1 cells secrete interleukin 2 (IL-2), which serves as their autocrine growth factor, and γ-interferon (IFN-γ), which activates macrophages and stimulates the production of opsonizing and complement-fixing antibodies. Thus, the Th1 subset is responsible for phagocyte-mediated host defence against infectious microbes, and it is capable of causing inflammation and tissue injury in infectious and autoimmune diseases. Th2 cells produce IL-4, IL-5 and IL-10, and they are responsible for phagocyte-independent defence reactions, e.g. against helminthic parasites. It is possible that a major physiological function of Th2 cells is to control Th1-dependent and macrophage-mediated inflammation because IL-10 is the prototypic anti-macrophage cytokine and IL-4 antagonizes many of the activities of IFN-γ. Other subpopulations of effector CD4$^+$ T cells that produce various combinations of cytokines may also exist, and effector populations of CD8$^+$ T cells may show similar functional heterogeneity (this volume: Mosmann et al 1995, Fitch et al 1995). Some antigen-stimulated T cells develop into long-lived and recirculating, but functionally quiescent, memory cells. The mechanisms that determine the decision of an individual T lymphocyte to differentiate into an effector or a memory cell are not well understood. It is also not established whether a memory T cell is permanently committed to a particular pattern of cytokine production or whether it can be induced upon restimulation with antigen to develop into Th1, Th2 and other effector populations (Swain 1994).

Elucidating the mechanisms that control the development of effector and memory T cells is central to understanding both protective anti-microbial immunity and pathological autoimmune reactions against self-protein antigens. The net result of antigenic stimulation is a balance between lymphocyte growth and differentiation on the one hand, and tolerance on the other. In this chapter we review the stimuli that regulate the development of Th1 and Th2 subsets, and the mechanisms that induce tolerance in these T cell subpopulations.

Differentiation of CD4$^+$ T lymphocytes into Th1 and Th2 subsets

The activation of naive T lymphocytes is initiated by recognition of MHC-associated peptide antigen, and it usually requires additional stimuli provided by co-stimulators expressed on APCs. In addition, cytokines produced at the site of antigen stimulation influence the pattern of T cell differentiation. Numerous experimental models have been employed to analyse the roles of antigen, co-stimulators and cytokines in the development of Th1 and Th2 cells. Perhaps the most powerful of these models uses T cells from transgenic mice expressing T cell receptors (TCRs) of known peptide and MHC specificities (Hsieh et al 1992, Seder et al 1992). A consensus view that has emerged from these experimental systems is that cytokines are powerful inducers of Th1/Th2 differentiation, with IL-4 being an obligatory Th2-inducing cytokine, and both IL-12 and IFN-γ promoting the development of Th1 cells (this volume: Coffman et al 1995, Trinchieri 1995). These conclusions are supported by *in vivo* results using model protein antigens and infectious agents. The roles of co-stimulators and antigens in regulating the patterns of T cell differentiation are not as well defined. Recent experiments suggest that blocking either of two co-stimulators, B7-1 and B7-2, has distinct effects on T cell responses to a myelin antigen (Kuchroo et al 1995), but it has proved difficult to establish conclusively distinct Th1/Th2-inducing effects of these co-stimulators in simpler *in vitro* or *in vivo* models. Both the concentration of antigen and the nature of the TCR: antigen interactions (Sloan-Lancaster & Allen 1995, this volume) may also influence the relative development of Th1 and Th2 subsets, but the mechanisms of these effects are poorly understood.

The molecular mechanisms that are responsible for selective cytokine gene expression in effector T cell subsets are the focus of active investigation in many laboratories. Cytokines such as IL-12 and IL-4 may stimulate the production of transcription factors, called STAT (signal transducer and activator of transcription) proteins, which bind to specific target sequences in the promoters of cytokine-responsive genes. IL-4 induces a STAT that binds to and presumably activates the IL-4 promoter, and this may be the key mechanism responsible for IL-4-induced Th2 differentiation. Finally, effector T cell subsets generated *in vitro* vary in their functional stability. Thus, IL-12-induced Th1 populations can be readily converted into IL-4 producers by exposure to IL-4, but IL-4-induced Th2 populations appear irreversibly committed (Perez et al 1995). The stability of Th2 cells may limit one's ability to control pathological Th2-mediated immune reactions therapeutically.

Mechanisms of peripheral T cell tolerance

Protein antigens can be administered in ways that induce tolerance in lymphocytes instead of activation. The phenomenon of tolerance is important because it ensures that the immune system does not respond to self-antigens, and because it is an important determinant of the response to foreign antigens. In addition, the induction of tolerance is being actively investigated as a therapeutic approach for inhibiting deleterious immune responses.

Peripheral tolerance results from the encounter of mature lymphocytes with tolerogenic forms of antigens in peripheral lymphoid tissues. (In contrast, central tolerance is due to the interaction of immature lymphocytes with antigens in the generative lymphoid organs, i.e. the bone marrow and thymus.) There are several mechanisms of peripheral T cell tolerance. Each is induced in distinct ways, is mediated by different biochemical pathways and affects different T cell subpopulations.

T cell clonal anergy

The phenomenon of anergy was first described in $CD4^+$ T cell clones cultured with peptide antigens that were presented by other T cells or by chemically treated APCs (Schwartz 1990). It is generally believed that anergy develops when a T cell recognizes its antigen in the absence of co-stimulation (Schwartz 1992). However, it has proved difficult to demonstrate anergy as a consequence of antigen recognition in the presence of antagonists that block co-stimulators of the B7 family. There are also no clear examples of anergy induced in uncloned populations of T cells. Anergy may be induced *in vivo* by the administration of antigens without adjuvants (Kearney et al 1994). However, the contribution of clonal anergy to T cell tolerance *in vivo* remains difficult to establish, mainly because there is no reliable way of identifying anergic T cells *in vivo*. Anergy may also be induced by the recognition of peptide antigens carrying mutations in their TCR contact residues (Sloan-Lancaster & Allen 1995, this volume).

T cell clonal anergy is due to a block in the transcription of the gene encoding IL-2, resulting in the failure of T cells to proliferate (Schwartz 1992). Transcription of the gene encoding IL-4 is relatively co-stimulator independent (McKnight et al 1994), so it is difficult to anergize IL-4-producing Th2 clones. Thus, anergy may result in Th2-dominant immune responses.

Activation-induced cell death

Repeated or chronic stimulation with high doses of antigens results in apoptotic death of specific T cells. This activation-induced cell death is a form of suicide. It is due to the induction of Fas (CD95) and Fas ligand on activated T cells because it does not occur in T cells of mice homozygous at the *lpr* (lymphoproliferative) locus, which is due to a mutation in the gene encoding Fas, or in T cells of *gld* (generalized lymphoproliferative disease), which have a point mutation in the gene encoding Fas ligand (Singer & Abbas 1994), and it can be blocked by a chimeric Fas–Ig fusion protein (e.g. Ju et al 1995). Fas-dependent cell death is the mechanism responsible for deletion of specific T cells in mice given superantigens and in TCR transgenic mice injected with high concentrations of specific protein antigens. Unlike clonal anergy, activation-induced cell death requires stimulation of T cells by competent APCs and it is dependent on IL-2 production, presumably because such stimulation is needed for the expression of the Fas ligand (Table 1).

The importance of activation-induced cell death for the maintenance of self-tolerance is most dramatically illustrated by the *lpr/lpr* and *gld/gld* mouse

TABLE 1 Mechanisms of peripheral T cell tolerance: postulated differences

	Anergy	*Deletion*
Inducers	antigen recognition without co-stimulation	chronic or repeated antigenic stimulation and activation
Mechanisms	IL-2 gene transcription blocked	Fas ligand-mediated, activation-induced cell death. Augmented by IL-2 exposure
Inhibitors	anti-CD28 antibody (provides co-stimulation)	Fas–Ig protein (competitive antagonist of Fas ligand)

models of systemic autoimmunity (Cohen & Eisenberg 1991). Presumably, the same suicide mechanism is responsible for at least some situations in which tolerance is induced by supra-optimal doses of foreign protein antigens. Among cloned cell lines, Th1 clones are higher expressors of Fas ligand than Th2 clones, and are, therefore, more susceptible to activation-induced cell death. Recently activated bulk populations of both Th1 and Th2 cells are known to undergo apoptosis upon cross-linking of antigen receptors, e.g. by high concentrations of anti-TCR antibody (Singer & Abbas 1994). The basis for this difference in cloned cell lines and uncloned, activated T cells is not yet established.

T cell-mediated suppression

The concept that some T cells can inhibit the responses of other T cells is an old one, but it is being redefined with increasing precision. The best examples of T cell-mediated suppression are situations in which antigen-specific Th2 cells produce cytokines such as IL-4 and IL-10 that inhibit the effector functions of specific Th1 cells. This phenomenon has been demonstrated in numerous parasitic infections (Locksley et al 1995, this volume). The administration of high doses of aqueous protein antigens leads to T cell tolerance, with selective inhibition of Th1 responses. That this apparent tolerance is at least partly due to the inhibitory actions of Th2 cells is shown by the finding that many of the manifestations of 'high-dose tolerance' can be prevented if the development of Th2 cells is blocked (Burstein & Abbas 1993). Aqueous antigens may be presented by co-stimulator-deficient APCs, and they may induce Th2 differentiation because IL-4 production is relatively co-stimulator independent. Suppression by Th2-like cells may also contribute to oral tolerance (Weiner et al 1994).

Thus, peripheral tolerance in mature T cells may be induced under different conditions of antigen exposure and by distinct mechanisms, including functional anergy, apoptotic cell death and active suppression. In most of these situations, Th2 cells are less tolerance sensitive. Not surprisingly, therefore, tolerogenic forms of antigens tend to inhibit Th1 responses preferentially.

Conclusions

The stimuli that induce differentiation and tolerance in CD4+ T lymphocytes and the biochemical mechanisms of these processes are being elucidated in a variety of experimental systems. The challenge that faces immunologists is to relate the experimental models to infectious and autoimmune diseases. Only then will the pathogenesis of these diseases be understood in mechanistic terms, enabling the development of rational therapeutic strategies.

References

Burstein HJ, Abbas AK 1993 *In vivo* role of interleukin 4 in T cell tolerance induced by aqueous protein antigens. J Exp Med 177:457–463

Coffman RL, Correa-Oliviera R, Mocci S 1995 Reversal of polarized T helper 1 and T helper 2 cell populations in murine leishmaniasis. In: T cell subsets in infectious and autoimmune diseases. Wiley, Chichester (Ciba Found Symp 195) p 20–33

Cohen PL, Eisenberg RA 1991 *lpr* and *gld*—single gene models of systemic autoimmunity and lymphoproliferative disease. Annu Rev Immunol 9:243–269

Fitch FW, Stack R, Fields P, Lancki DW, Cronin DC 1995 Regulation of T lymphocyte subsets. In: T cell subsets in infectious and autoimmune diseases. Wiley, Chichester (Ciba Found Symp 195) p 68–85

Hsieh C-S, Heimburger AB, Gold JS et al 1992 Differential regulation of T helper phenotype development by interleukins 4 and 10 in an $\alpha\beta$ T cell receptor transgenic system. Proc Natl Acad Sci USA 89:6065–6071

Ju S-T, Panka DJ, Cui H et al 1995 Fas (CD95)/FasL interactions required for programmed cell death after T-cell activation. Nature 373:444–448

Kearney ER, Pape KA, Loh DY, Jenkins MK 1994 Visualization of peptide-specific T-cell immunity and peripheral tolerance induction *in vivo*. Immunity 1:327–339

Kuchroo VK, Das MP, Brown JA et al 1995 B7-1 and B7-2 costimulatory molecules activate differentially the Th1/Th2 development pathways: application to autoimmune disease therapy. Cell 80:707–718

Locksley RM, Wakil AE, Corry DB, Pingel S, Bix M, Fowell DJ 1995 The development of effector T cell subsets in murine *Leishmania major* infection. In: T cell subsets in infectious and autoimmune diseases. Wiley, Chichester (Ciba Found Symp 195) p 110–122

McKnight AJ, Perez VL, Chea CM, Gray GS, Abbas AK 1994 Costimulator dependence of lymphokine secretion by naive and activated $CD4^+$ T lymphokines from T cell receptor transgenic mice. J Immunol 153:5220–5225

Mosmann TR, Coffman RL 1989 Th1 and Th2 cells: different patterns of lymphokine secretion lead to different functional properties. Annu Rev Immunol 7:145–173

Mosmann TR, Sad S, Krishnan L, Wegmann TG, Guilbert LJ, Belosevic M 1995 Differentiation of subsets of $CD4^+$ and $CD8^+$ T cells. In: T cell subsets in infectious and autoimmune diseases. Wiley, Chichester (Ciba Found Symp 195) p 42–54

Perez VL, Lederer JA, Lichtman AH, Abbas AK 1995 Stability of Th1 and Th2 populations. Int Immunol 7:869–875

Romagnani S 1994 Lymphokine production by human T cells in disease states. Annu Rev Immunol 12:227–257

Schwartz RH 1990 A cell culture model for T lymphocyte clonal anergy. Science 248:1349–1356

Schwartz RH 1992 Costimulation of T lymphocytes: the role of CD28, CTLA-4, and BB1/B27 in interleukin-2 production and immunotherapy. Cell 71:1065–1068

Seder RA, Paul WE, Davis MM, de StGroth BF 1992 The presence of interleukin 4 during *in vitro* priming determines the lymphokine-producing potential of $CD4^+$ T cells from T cell receptor transgenic mice. J Exp Med 176:1091–1098

Seder RA, Paul WE 1994 Acquisition of lymphokine-producing phenotype by $CD4^+$ T cells. Annu Rev Immunol 12:635–673

Singer GG, Abbas AK 1994 The Fas antigen is involved in peripheral but not thymic deletion of T lymphocytes in T cell receptor transgenic mice. Immunity 1:365–371

Sloan-Lancaster J, Allen PM 1995 Signalling events in the anergy induction of T helper 1 cells. In: T cell subsets in infectious and autoimmune diseases. Wiley, Chichester (Ciba Found Symp 195) p 189–202

Swain SL 1994 Generation and *in vivo* persistence of polarized Th1 and Th2 memory cells. Immunity 1:534–552

Trinchieri G 1995 The two faces of interleukin 12: a pro-inflammatory cytokine and a key immunoregulatory molecule produced by antigen-presenting cells. In: T cell subsets in infectious and autoimmune diseases. Wiley, Chichester (Ciba Found Symp 195) p 203–220

Weiner HL, Friedman A, Miller A et al 1994 Oral tolerance: immunologic mechanisms and treatment of animal and human organ-specific autoimmune diseases by oral administration of autoantigens. Annu Rev Immunol 12:809–837

DISCUSSION

Mosmann: You made the provocative suggestion that a T helper 2 (Th2) response might be more important as a regulator than a Th1 response. Although it is difficult to assign relative importances, is it possible that the reason for this is that the Th2 response is more important for multicellular parasites (which are much less important than many other parts of the immune system in Europe and North America)?

Abbas: You're correct, it is subjective to assign relative importances. However, T cell immunodeficiency states that are associated with severe infections are all deficiencies of Th1 responses. One has to search the literature to find examples of helminthic infections in which Th2 cells are protective (Urban et al 1992). The published reports, although important, are few enough to be virtually anecdotal. In contrast, there are numerous examples of Th1 cell deficiencies or Th2 over-stimulations that lead to increased infections (this volume: Coffman et al 1995, Locksley et al 1995, Shearer et al 1995). Therefore, although interleukin 4 (IL-4) production may be important in helminthic infections, if one assigned a quantitative estimate of the importance of protection against infections, it is probably less important than a Th1-mediated response.

Coffman: One could probably make a viable argument to support the suggestion that the only useful function of a Th2 response is to regulate Th1-like responses. However, this role does not account for the effector functions mediated by Th2 cells. For example, IgE, eosinophils and mast cells are potent effector mechanisms, but they do not have such obvious regulatory roles.

Abbas: But what's the point of having an immediate hypersensitivity reaction? I've never been able to answer this question convincingly. It has been suggested that it both mediates gut peristalsis and eliminates helminths. But it is difficult to design the experimental models to determine whether Th2 cells are beneficial because helminth immunity and survival models are difficult to study.

Mitchison: I would like to ask Abul Abbas a question about prior priming. Even with transgenic mice kept in a clean animal house, many of the apparent differences between Th1 and Th2 cells may reflect prior priming. Isn't it difficult, even in your elegant culture system, to sort out the effects of prior priming from intrinsic differences between Th1 and Th2 cells?

Abbas: Surface markers may be used as indicators of prior priming. T cells from transgenic mice are routinely sorted on the basis of T cell receptor (TCR) or CD4 expression, and either L-selectin or low levels of CD44. One can argue about how definitive those markers are but they are the best we've got.

Mitchison: Were your experiments performed with sorted cells?

Abbas: Yes.

Allen: I would like to mention the problems of defining Th2 cell tolerance. In our system, we found that the cells made wild-type levels of IL-4 but they failed to proliferate (Sloan-Lancaster et al 1994), which is in agreement with your results. The definition of Th2 cell tolerance may be operational in that the expansion of the cell is limited, although it can still produce cytokines.

Abbas: We studied co-stimulator dependence with bulk Th1 and Th2 cell populations, and we found that *in vitro* differentiated Th2 cell populations do proliferate. Proliferation is more rapid in response to live antigen-presenting cells (APCs) but they still proliferate in response to fixed APCs (McKnight et al 1994). In non-clonal systems we can't be completely sure that every cell in the population is a Th2 cell, so how heavily should we rely on clonal systems versus systems that are closer to normal? The latter have obvious limitations; for example, problems with the degree of purity, so that not all the so-called Th2 cells in an IL-4-producing population of cells are actually committed Th2 cells.

Lamb: Do you see any differences in susceptibility between Th1 and Th2 cells to the induction of anergy when superantigens are used?

Abbas: No. Because superantigen tolerance, in my mind, is not clonal anergy but is a model of activation-induced cell death. It occurs when large numbers of T cells are switched on to make IL-2, and when peptides are injected into TCR transgenic T cells. These are all models of activation-induced cell death, and they are all Fas dependent because superantigen-induced deletion doesn't occur in mice homozygous at the *lpr* locus. Th1 cell clones express higher levels of Fas than Th2 cell clones (Ramsdell et al 1994). It is possible, therefore, that superantigen-induced deletion would preferentially affect differentiated Th1 cells. However, this result is not quite as clear with bulk populations of cells (Singer & Abbas 1994), which again highlights the problem of whether we are looking at pure populations of cells or not.

Flavell: You described the effect of fixed APCs on what you call Th1 and Th2 cells. You differentiated the Th cells for a week, restimulated them and then you measured IL-2 production in Th1 cells and IL-4 production in Th2 cells. It is possible that the IL-2 is produced by leftover naive cells and not by Th1 cells. The predominant product of that restimulation is actually γ-interferon (IFN-γ), so you could look at the levels of IFN-γ instead. Have you done this?

Abbas: We observed about a 70% reduction in the levels of IFN-γ in Th1 cells stimulated with fixed APCs. This is not as drastic as the decrease in the levels of IL-2, although the levels of both IL-2 and IFN-γ can be co-stimulated to normal levels.

Flavell: We have looked at transcription factor activation requirements in the effector cell population versus naive T cells, and we have found that these requirements change drastically as a consequence of the various stages of activation (Nakamura et al 1995, this volume). My working hypothesis is that the effector cell is completely transformed in terms of what it takes to switch on its transcription factors.

Swain: We've looked at effector Th1 and Th2 cell populations, and we find that IL-2 production is extremely co-stimulation dependent, regardless of the differentiation state of the cell (C. Dubey, M. Croft & S. Swain, unpublished observations). The production of other cytokines is much less co-stimulator dependent than IL-2. The dependency profile of each cytokine varies a little—IL-5 and IFN-γ are intermediate between IL-4, which is co-stimulator independent, and IL-2. Therefore, it is possible that co-stimulation dependency has more to do with the cytokine than with the type of cell.

Mitchison: You detected a low level of IL-2 transcripts. Is it possible that IL-2 is not involved?

Abbas: The bulk of the available evidence suggests that IL-2 is required to initiate the process of Th1/Th2 differentiation (Seder et al 1994). This is probably because IL-2 is a critical growth factor, but it is not a differentiation factor.

Flavell: How much IL-2 is made by those Th1 cells compared with the initial naive cells?

Abbas: About 10-fold more.

Locksley: Is it possible that you're not achieving a B7-2 signal in your CHO (Chinese hamster ovary) cell transfectants because the machinery is not present in a CHO cell? If you used APCs and natural B7-2 you might activate the signal transduction pathways necessary for the production of other co-stimulatory models, for instance.

Abbas: No. Because in our hands the effects of B7-2 are not easily distinguishable from those of B7-1.

Locksley: But you're using CHO cells, which may not have the machinery to transduce a signal to B7-2. Therefore, B7-2 might not induce IL-12 in CHO cells, for instance. Have you looked at this in a system in which the transduction machinery potential through B7-2 is intact?

Abbas: We have looked at live (i.e. mitomycin treated) APCs in the presence of antibodies against B7-1 and B7-2, which is similar to what you're asking. We do not see a clear difference in the effects of these two co-stimulators in mediating the differentiation of T cells (A. N. Schweitzer, R. C. K. Wong, A. K. Abbas & A. H. Sharpe, unpublished results). The problem is that this is a negative result, so one cannot be sure that the blocking effects of the antibodies are comparable or complete.

Dutton: B7-1 is not the only co-stimulatory pathway, which may explain why the blocking experiments fail.

Abbas: Yes. We are using CHO transfectants precisely because none of the other pathways are present, so we can introduce CD54 (intercellular adhesion molecule 1), for example, if we wished to.

Lotze: But you may observe a different end result if none of the signal transduction pathways are available. You may also observe a different response using Fab fragments than if you used whole antibodies.

Abbas: The signal is coming to the APC from either the natural ligand or from the anti-B7-2 antibody, and the function of the APC is altered. This is not the conventional view of co-stimulation, but it is one that we're having to pay attention to.

Mosmann: The secondary effects of the cytokines are also important in these co-stimulation experiments. You speak of IL-2 synthesis being blocked and IFN-γ synthesis being partially suppressed. This may be a secondary effect simply because IL-2 stimulates IFN-γ production. Have you added exogenous IL-2 to the culture to see if it restores IFN-γ synthesis?

Abbas: No, we haven't done that experiment.

Liew: We have blocked IL-2 production by adding a nitric oxide donor to cultured Th1 cell clones. We observed that the levels of IFN-γ can be partially restored when exogenous IL-2 is added (A. W. Taylor-Robinson & F. Y. Liew, unpublished results).

Is it possible to bias the expansion of Th2 cells *in vitro* by adding a high dose of antigen?

Abbas: We have titrated antigen over a wide dose in two different TCR transgenic systems. In the pigeon cytochrome c system, Th1 responses are dominant in the absence of any exogenous cytokines. This is because these mice are on a B10 background, which has a propensity to develop Th1 responses. Decreasing the concentration of antigen results in a gradual loss of response until, eventually, cells cannot be recovered from the priming culture, but we do not see a change to Th2 responses. Also, when we administer a large amount of aqueous antigen *in vivo*, we observe an 'immune deviation' towards Th2 dominance (Burstein et al 1992).

Mitchison: But the claim that decreasing ligand density favours a Th2 response is important.

Abbas: I don't believe that such a claim has been published.

Liew: In 1972 Chris Parish and I reported an experiment which showed that the induction of Th cells mediating delayed-type hypersensitivity (Th1 cells) and those helping antibody synthesis (Th2 cells) in rats can be differentially induced by injecting different amounts of antigen (Parish & Liew 1972).

However, the conditions required for a similar preferential induction of Th1 and Th2 cells have not yet been reproduced *in vitro*.

Swain: We have performed similar experiments with decreasing amounts of antigen (and sometimes we can get away with not using any antigen) and then adding exogenous IL-2. We observe, even in the transgenic model where we cannot detect anything except resting naive cells, that some cells grow out and make IL-4 or IFN-γ. This response is not observed in the absence of exogenous IL-2 or if percoll gradients are performed. It is possible that as these cells grow out, a few have sufficient IL-2 receptors to proliferate and become cytokine-producing cells. This may explain why some people have observed an apparent switching of cytokines *in vitro* in the transgenic mouse. Therefore, this observation may simply be artefactual.

Flavell: We have similar results (T. Nakamura & R. A. Flavell, unpublished results), but when an IL-12-secreting population is induced to make IL-4, we don't know whether the same cells are switching to make IL-4 or whether different cells are differentiating from an uncommitted precursor population.

Abbas: We've tried to answer that question by doing limiting dilution assays, but the results of these experiments have not been definitive. We have observed that after activation, the phenotype of the entire population changes (Perez et al 1995), but this doesn't answer definitively whether uncommitted cells are present in that population.

Mitchison: Can you switch the phenotype of the clones?

Abbas: No.

Mosmann: This seems to be a fairly general experience because we have found that even a fairly short-term clone seems to be unswitchable.

Mitchison: Doesn't that answer the question?

Abbas: No, because if the same T cell population is restimulated repeatedly in the presence of IL-12, for example, one would predict that at some stage a population of cells would arise that would not be convertible. In other words, one would get closer and closer to producing a cloned cell line. We've never done that experiment. It's a difficult experiment to do.

Mitchison: But why bother, if the result with the clones is clear?

Flavell: A clone may not be relevant to the population of cells that Abul Abbas is looking at.

Mitchison: But when the cells are stimulated, each stimulator cell turns itself into a clone.

Flavell: No, a clone is something that you have induced *in vitro* in order that it may self-propagate under certain circumstances.

Mosmann: The question that we're skirting around is, what is the number of definable stages of differentiation for T cells? Because we know that there are initial cells making IL-2, and there are Th1 and Th2 clones. However, the intermediates stages may be quite fuzzy. Richard Flavell has some results that support the existence of multiple cytokine-producing cells, which are not necessarily Th0 cells, along that pathway (Nakamura et al 1995, this volume). We don't know how many different states there are. Also, this population could be a mixture of uncommitted and Th1 cells, or it could be a cell type that we have not yet isolated because it is not stable enough to isolate as a clone.

Abbas: Is a clone just the progeny of a single cell taken at any time, or are the clones that most of us now study qualitatively different from the bulk cell populations that we started with? I would argue that clones are probably qualitatively different because they've gone through many more cycles of activation and differentiation.

Liew: Nevertheless, there is a difference between Th1 and Th2 cell populations because the Th2 population cannot be switched back to Th0 or converted to Th1.

Abbas: Yes, this is the result in our bulk cell populations. But Richard Flavell would argue that the Th2 cells have far fewer uncommitted precursors. There's no way of getting around that.

Flavell: We're using the ablation approach to get around it.

Abbas: But only if the ablation is lineage specific.

Flavell: Yes, but it's quite clear that a population of Th1 differentiated cells cannot be killed, so ablation of effector cells is clearly lineage specific (Nakamura et al 1995, this volume).

Coffman: I would like to change the subject for the last question. You observed a loss of IL-4 signal transduction, at least as measured by a loss of activation of the signal transducer and activator of transcription (STAT) protein for IL-4. This is surprising because Th1 clones retain their proliferative response to IL-4.

Abbas: There is no difference between Th1 and Th2 clones because the IL-4 STAT protein is not clone specific, it's IL-4 inducible, so the expression of the IL-4 STAT protein is turned on whenever IL-4 is present.

References

Burstein HJ, Shea CM, Abbas AK 1992 Aqueous antigens induce *in vivo* tolerance selectively in IL-2 and IFN-gamma-producing (Th1) cells. J Immunol 148:3687–3691

Coffman RL, Correa-Oliviera R, Mocci S 1995 Reversal of polarized T helper 1 and T helper 2 cell populations in murine leishmaniasis. In: T cell subsets in infectious and autoimmune diseases. Wiley, Chichester (Ciba Found Symp 195) p 20–33

Locksley RM, Wakil AE, Corry DB, Pingel S, Bix M, Fowell DJ 1995 The development of effector T cell subsets in murine *Leishmania major* infection. In: T cell subsets in infectious and autoimmune diseases. Wiley, Chichester (Ciba Found Symp 195) p 110–122

McKnight AJ, Perez VL, Shea CM, Gray GS, Abbas AK 1994 Costimulator dependence of lymphokine secretion by naive and activated CD4+ T lymphocytes from T cell receptor transgenic mice. J Immunol 153:5220–5225

Nakamura T, Rincón M, Kamogawa Y, Flavell RA 1995 Regulation of $CD4^+$ T cell differentiation. In: T cell subsets in infectious and autoimmune diseases. Wiley, Chichester (Ciba Found Symp 195) p 154–172

Parish CR, Liew FY 1972 Immune response to chemically modified flagellin. III. Enhanced cell-mediated immunity during high and low zone antibody tolerance to flagellin. J Exp Med 135:298–311
Perez VL, Lederer JA, Lichtman AH, Abbas AK 1995 Stability of Th1 and Th2 population. Int Immunol 7:869–875
Ramsdell F, Seaman MS, Miller RE, Picha KS, Kennedy MK, Lynch DH 1994 Differential ability of Th1 and Th2 T cells to express Fas ligand and to undergo activation-induced cell death. Int Immunol 6:1545–1553
Seder RA, Germain RN, Linsley PS, Paul WE 1994 CD28-mediated costimulation of interleukin 2 (IL-2) production plays a critical role in T cell priming for IL-4 and interferon gamma production. J Exp Med 179:299–304
Shearer GM, Clerici M, Sarin A, Berzofsky JA, Henkart PA 1995 Cytokines in immune regulation/pathogenesis in HIV infection. In: T cell subsets in infectious and autoimmune diseases. Wiley, Chichester (Ciba Found Symp 195) p 142–153
Singer GG Abbas AK 1994 The Fas antigen is involved in peripheral but not thymic deletion of T lymphocytes in T cell receptor transgenic mice. Immunity 1:365–371
Sloan-Lancaster J, Evavold BD, Allen PM 1994 Th2 cell clonal anergy as a consequence of partial activation. J Exp Med 180:1195–1205
Urban JF, Madden KB, Svetic A et al 1992 The importance of Th2 cytokines in protective immunity to nematodes. Immunol Rev 127:205–220

Reversal of polarized T helper 1 and T helper 2 cell populations in murine leishmaniasis

Robert L. Coffman*, Rodrigo Correa-Olivieira† and Simonetta Mocci*

**Department of Immunology, The DNAX Research Institute of Molecular and Cellular Biology, 901 California Avenue, Palo Alto, CA 94304-1104, USA and †Laboratorio de Immunologica Celular e Molecular de Parasitas, Cento de Pesquisas Rene Rachou-FIOCRUZ, AV. Augusto de LIMA 1715, Belo Horizonte, MG 30190-002, Brazil*

Abstract. T helper 1 (Th1) and Th2 cells are the major subsets of fully differentiated $CD4^+$ T cells in the mouse. The spectrum of cytokines characteristic of each subset determines the distinctive regulatory and effector functions mediated by each subset. We have used the murine model of *Leishmania major* infection to study the question of whether highly polarized populations of normal T cells are as stable in their cytokine phenotype as Th clones or whether the phenotype can be altered with regulatory cytokines. Interleukin 4 (IL-4) appears to be a key cytokine for Th2 responses as it is necessary for both the initial differentiation of Th responses to *L. major* and the stability of ongoing responses. Furthermore, IL-4 is capable of converting highly polarized Th1 responses to Th2 responses either *in vitro* or when adoptively transferred to severe combined immunodeficiency mice.

1995 T cell subsets in infectious and autoimmune diseases. Wiley, Chichester (Ciba Foundation Symposium 195) p 20–33

A wide variety of infectious and autoimmune diseases are characterized by $CD4^+$ T cell responses that are strongly polarized to either the T helper 1 (Th1) or Th2 patterns of cytokine production (Sher & Coffman 1992, Liblau et al 1995). In some infectious diseases these responses are not especially effective at controlling and clearing the pathogen, but they cause significant pathology and can even inhibit responses that would be much more effective (Powrie et al 1994). In most organ-specific autoimmune diseases $CD4^+$ T cells (usually Th1-like cells) are key elements of immunopathology. Clearly, one therapeutic approach to such diseases would be to modify or even, in some cases, reverse the polarity of the T cell response to the pathogen or autoantigen (Powrie & Coffman 1993). That this may be difficult is suggested by the absence of published experiments in which cloned Th1 lines are induced to switch to a Th2 pattern of cytokine production or vice versa.

We have addressed this question in a mouse model of *Leishmania major* infection, in which highly polarized Th1 or Th2 responses can be obtained. The infection of most mouse strains with *L. major* leads to the development of a localized cutaneous lesion that resolves spontaneously within about six weeks. Once healed, the mouse is relatively resistant to secondary infection. In a few strains, however, such as BALB/c, the initial infection is not controlled: the lesion continues to grow and ulcerate, the infection spreads to the visceral organs and the mouse usually dies. Both types of mice develop strong T cell responses, predominantly of the $CD4^+$ subset, but the nature of this response is quite different. The healing response represents a highly polarized Th1 response, whereas the progressive, non-healing response, although just as intense, is dominated by Th2 cells (Locksley & Scott 1991).

The Th1 responses of resistant mice can be readily changed to Th2 responses by treatment with anti-γ-interferon (IFN-γ) antibody, provided that the treatment is done approximately coincident with infection (Belosevic et al 1989). Similarly, anti-interleukin 4 (anti-IL-4) antibody will convert non-healing BALB/c mice to Th1 responders if it is given close to the time of infection (Sadick et al 1990, Chatelain et al 1992). Such studies have been important in defining the key roles played by IFN-γ and IL-4 in inducing the differentiation of naive T cells to Th1 or Th2 cells, respectively. However, it has been consistently observed that, after one to two weeks of infection, both Th1 and Th2 responses become resistant to such interventions (Coffman et al 1991, Chatelain et al 1992). This suggests that populations under chronic antigenic stimulation may be terminally differentiated, as appears to be the case with tissue culture-adapted Th clones. There are other possible explanations, however, for the stability of the *L. major*-specific responses *in vivo*, including regulatory influences from a variety of cell types and effects of the parasites themselves.

We have taken two approaches to determine whether polarized populations of normal $CD4^+$ T cells have a fixed cytokine secretion pattern. The first is to transfer unfractionated or CD4-enriched cells from the lesion-draining lymph nodes of four to six week post-infection BALB/c mice into *L. major*-infected CB-17 severe combined immunodeficiency (SCID) mice. A non-healing Th2 response could be transferred to the SCID recipients with as few as 10^6 lymph node cells or the equivalent number of $CD4^+$ T cells. Transfer of the Th2 response was dependent, however, on IL-4 because one dose of anti-IL-4 monoclonal antibody (mAb) given at the time of reconstitution led to the development of a healing Th1 response (R. Correa-Oliveira, K. Varkila, R. L. Coffman, submitted). This well-established Th2 population also gave rise to a healing Th1 response when lower doses of cells were transferred (1×10^5 $CD4^+$ T cells to as low as 1×10^4 $CD4^+$ T cells). The preferential development of Th1 responses at low donor cell numbers has also been observed when naive BALB/c spleen cells were transferred to syngeneic CB-17 SCID mice (Varkila et al 1993). In both cases the cytokine environment is dominated not by IL-4 from the transferred cells, but by IFN-γ and IL-12 from the SCID recipient, and the switch from Th2 to Th1 dominance is prevented by treating the mice with anti-IFN-γ antibody. Similarly, T cells from healed resistant mice developed a Th2 response to *L. major* if transferred to infected SCID mice together with anti-IFN-γ antibody. In summary, highly polarized populations of either Th1 or Th2 cells could be readily transformed into populations with the opposite cytokine phenotype by altering the relative levels of IL-4 and IFN-γ during the first several weeks of adoptive transfer. However, the same populations were quite resistant to the effects of such manipulations as long as they remained in the primary host. What was clear from these experiments, however, was that these polarized $CD4^+$ T cell populations specific for *L. major* retained the capacity to develop quite different cytokine profiles under the appropriate conditions.

The second experimental approach was to determine whether the cytokine pattern of T cell populations polarized *in vivo* could be altered by short-term culture *in vitro* with candidate regulatory cytokines or antibodies. To do this with the *L. major* response, we followed the strategy taken by Maggi et al (1992), who first showed the conversion of cytokine patterns in antigen-specific T cells from allergic or immunized patients. $CD4^+$ T cells were again isolated four to six weeks after infection from lesion-draining nodes of BALB/c mice or BALB/c mice that had been treated prior to infection with anti-CD4 antibody. This transient reduction in $CD4^+$ T cell numbers is an effective way to change the initial Th response of these mice from Th2 to Th1 (Titus et al 1985), thus providing both types of polarized population on the same strain background. By the time the lymph nodes were isolated for these experiments, $CD4^+$ T cell numbers had recovered to normal levels. The isolated $CD4^+$ T cells were stimulated with *Leishmania* antigen (a freeze-thawed preparation of *L. major* promastigotes), irradiated splenic antigen-presenting cells (APCs) and IL-2, and they were cultured with or without the addition of cytokines or anti-cytokine antibodies to attempt modification of the populations. After one week of culture, the T cells were harvested and restimulated with fresh APCs and antigen, but with no other additions. The supernatants of these restimulated cultures were harvested at 60–72 h and assayed for cytokine production by sensitive cytokine ELISAs (Mocci & Coffman 1995).

The most striking results were obtained when a Th1 cell population was cultured with IL-4 for seven days. Upon restimulation, this population produced a typical *L. major*-specific Th2 response, with greatly reduced levels of IFN-γ and high levels of the Th2 cytokines, IL-4 and IL-10 (Fig. 1). In contrast, parallel cultures lacking only IL-4 gave a typical Th1 pattern, with high IFN-γ and little or no detectable IL-4 or IL-10. IL-4 appeared to be both

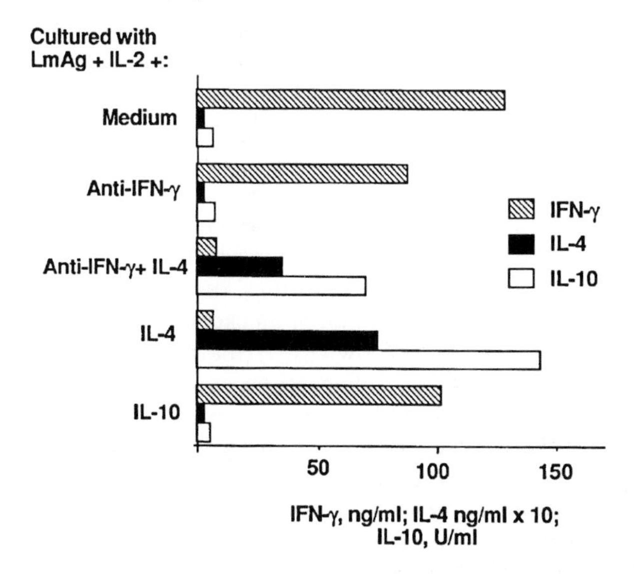

FIG. 1. Conversion of a T helper 1 (Th1) cell population into a Th2 cell population with interleukin 4 (IL-4) *in vitro*. $CD4^+$ T cells were isolated from the popliteal lymph nodes of BALB/c mice that had been induced to make a healing Th1 response by pretreatment with anti-CD4 antibody. Lymph nodes were harvested four weeks after infection. The $CD4^+$ T cells were cultured for one week with splenic antigen-presenting cells and the indicated additions, harvested and restimulated with antigen. The supernatants from the second stimulation were collected at 72 h and assayed for cytokines by ELISA. Further details are found in Mocci & Coffman (1995). IFN-γ, γ-interferon; LmAg, *Leishmania* antigen.

necessary and sufficient to induce this rapid Th1 to Th2 conversion. A variety of other cytokines, including IL-10 and IL-13, were tested, as were Th2 supernatants depleted only of IL-4, but none could stimulate even a partial shift, nor could any enhance the effect of IL-4 (Mocci & Coffman 1995). Surprisingly, it was not necessary to neutralize the high level of IFN-γ produced by the cultures during the first two days of the stimulation in order for IL-4 to have its maximum effect.

The shift from a dominant Th2 cell population was somewhat less efficient, despite the fact that the non-healing response in BALB/c mice contains a small Th1 component that is actively suppressed *in vivo* by IL-4 and IL-10 (Powrie et al 1994). Both IFN-γ and IL-12 induced some IFN-γ production upon restimulation, but they worked much more effectively if anti-IL-4 antibody was also added to the cultures. Even with all three components, however, significant levels of IL-4 and IL-10 were produced upon recall. Only after two to three cycles of culture with antigen, APCs, IL-2 and inducers of a Th1 response were *L. major*-specific populations produced that resembled the highly polarized responses obtained with cells from healing mice.

Further studies on the mechanism of this population shift were done primarily with the IL-4-induced Th1 to Th2 shift (Mocci & Coffman 1995). The induction of this shift was quite rapid, with lowered IFN-γ production and significant induction of IL-4, IL-5 and IL-10 present in cultures restimulated after only five days of primary culture with IL-4. The critical time for exposure to IL-4 was defined in a series of experiments in which IL-4 was added either at various times after the initiation of the cultures, or at the start of culture and removed at various times by the addition of an excess of neutralizing anti-IL-4 antibody. Both types of analysis show clearly that IL-4 exposure during the first 24 h of the cultures has no influence on the differentiation to a Th2 population. Rather, the critical period is from Days 1 to 4 of culture and, for maximum conversion, IL-4 must be present for the entire period. Thus, the first 24 h of stimulation *in vitro* of the population, when many of the key signals are being delivered to the T cell via the T cell receptor and a variety of co-stimulatory molecules, is not the period when the cytokine producing potential of the population can be modified.

It is possible that culture with IL-4 causes a transient decrease in the production of IFN-γ and induces acute production of several Th2 cytokines, but does not permanently alter the predominantly Th1-like differentiation state of the population. By analogy, IL-12 has been shown to induce production of IFN-γ in Th2 clones, but the production is transient and ceases after the IL-12 is removed (Manetti et al 1994). The stability of the induced Th2 state was tested by culture of Th1 cells for one week with or without IL-4 followed by transfer to *L. major*-infected CB-17 SCID mice. Recipients of the induced Th2 population developed a non-healing lesion with a strong and stable Th2 state response, whereas the population that was not shifted transferred a stable, healing Th1 response to the recipients.

It is important to point out that these experiments document a clear shift at the level of the *L. major*-specific cell population, but they do not show whether individual clones of Th1 cells switch their cytokine production profile or whether they are simply overgrown by Th2 cells arising from either a few pre-existing Th2 cells or cells not yet terminally differentiated to either type. We are currently trying to resolve the cellular mechanism by testing whether the cells that switch in culture can be separated phenotypically from Th1 cells. If the precursors of the induced Th2 cell population are distinct from the dominant Th1 cell population, this would rule out a clonal shift in cytokine production as the principal mechanism of this Th1 to Th2 cell population conversion.

Acknowledgement

The DNAX Research Institute is supported by the Schering–Plough Corporation.

References

Belosevic M, Finbloom DS, van der Meide PH, Slayter MV, Nacy CA 1989 Administration of monoclonal anti-IFN-gamma antibodies *in vivo* abrogates natural resistance of C3H/HeN mice to infection with *Leishmania major*. J Immunol 143:266–274

Chatelain R, Varkila K, Coffman RL 1992 IL-4 induces a Th2 response in *Leishmania major*-infected mice. J Immunol 148:1182–1187

Coffman RL, Varkila K, Scott P, Chatelain R 1991 Role of cytokines in the differentiation of $CD4^+$ T cell subsets *in vivo*. Immunol Rev 123:189–207

Liblau RS, Singer SM, McDevitt HO 1995 Th1 and Th2 $CD4^+$ T cells in the pathogenesis of organ-specific autoimmune diseases. Immunol Today 16:34–38

Locksley RM, Scott P 1991 Helper T-cell subsets in mouse leishmaniasis—induction, expansion and effector function. Immunol Today 12:58A–61A

Maggi E, Parronchi P, Manetti R et al 1992 Reciprocal regulatory effects of IFN-gamma and IL-4 on the *in vitro* development of human Th1 and Th2 clones. J Immunol 148:2142–2147

Manetti R, Gerosa F, Guidizi MG et al 1994 Interleukin-12 induces stable priming for interferon-γ (IFN-γ) production during differentiation of human T helper (Th) cells and transient IFN-γ production in established Th2 cell clones. J Exp Med 179:1273–1283

Mocci S, Coffman RL 1995 Induction of a Th2 population from a polarized *Leishmania*-specific Th1 population by *in vitro* culture with IL-4. J Immunol 154:3779–3787

Powrie F, Coffman RL 1993 Cytokine regulation of T-cell function: potential for therapeutic intervention. Immunol Today 14:270–274

Powrie F, Correa-Oliveira R, Mauze S, Coffman RL 1994 Regulatory interactions between $CD45RB^{high}$ and $CD45RB^{low}$ $CD4^+$ T cells are important for the balance between protective and pathogenic cell-mediated immunity. J Exp Med 179:589–600

Sadick MD, Heinzel FP, Holaday BJ, Pu RT, Dawkins RS, Locksley RM 1990 Cure of murine leishmaniasis with anti-interleukin 4 monoclonal antibody. Evidence for a T cell-dependent, interferon gamma-independent mechanism. J Exp Med 171:115–127

Sher A, Coffman RL 1992 Regulation of immunity to parasites by T cells and T cell-derived cytokines. Annu Rev Immunol 10:385–409

Titus RG, Ceredig R, Cerottini JC, Louis JA 1985 Therapeutic effect of anti-L3T4 monoclonal antibody GK1.5 on cutaneous leishmaniasis in genetically susceptible BALB/c mice. J Immunol 135:2108–2114

Varkila K, Chatelain R, Leal LMCC, Coffman RL 1993 Reconstitution of C.B-17 *scid* mice with BALB/c T cells initiates a T helper type-1 response and renders them capable of healing *Leishmania major* infection. Eur J Immunol 23:262–268

DISCUSSION

Sher: One might imagine that in most systems T cell responses progress from a T helper 1 (Th1) response to a Th2 response, but not vice versa. Therefore, your inability to revert Th2-dominated responses makes sense in terms of this established pattern of T cell differentiation.

Coffman: The *Leishmania major* system is simple to work with in some respects because there is no evidence which suggests that a natural progression of Th1 cells to Th2 cells, or vice versa, is occurring. In some of these experiments we manipulated the cells for a week *in vitro*, put them back into SCID (severe combined immunodeficiency) mice, and found that they stayed as either Th1 or Th2 cells, i.e. the same type as when they were injected, for the life of the mouse.

Abbas: In simpler systems, such as in the T cell receptor (TCR) transgenic models, repeated stimulation results in the production of interleukin 4 (IL-4). As soon as a trace of IL-4 is produced, T cells are converted to Th2 cells and a Th2 response ensues (Croft et al 1995). Teleologically, if Th1 cells are thought of as cells that protect against infection and Th2 cells as cells that control inflammatory consequences, then that sequence makes complete sense.

Coffman: An experimental test of this would be that in a stable Th1 response, such as in most healing responses to intracellular parasites, one should be able to induce a Th2 response by adding exogenous IL-4 over a period of a week or two.

Abbas: It's difficult to do experiments with IL-4 *in vivo* because of problems with stability.

Coffman: But you're proposing that only a trace of IL-4 is necessary for a Th2 response. It is certainly possible to maintain a significant level of IL-4 over a period of one to two weeks, for example with IL-4/anti-IL-4 antibody complexes.

Liew: Have you looked at the effect of adding a combination of anti-IL-4 antibody and recombinant IL-12?

Coffman: Yes. We have observed a shift from a Th2 to a Th1 response, although the shift is not as complete as the Th1 to Th2 response shift. γ-Interferon (IFN-γ), IL-12 and anti-IL-4 antibody are all required for the best results in a one week culture (Mocci & Coffman 1995).

Trinchieri: IL-12 is effective only if it is added in the first week. However, Phil Scott has shown that if both IL-12 and pentostan, an antimonial antiparasite drug, are given within the first few weeks of infection, the cells completely revert and give Th1 responses within three to four weeks (Nabors et al 1995).

Liew: What is the result if the infection is ongoing?

Trinchieri: If the infection continues throughout the three to four weeks, the pentostan treatment decreases the parasite load, probably allowing IL-12 to rescue a Th1 response.

Coffman: But you are introducing numerous changes simply by reducing the parasite load.

Flavell: We have to distinguish between experiments performed in SCID mice and in immune-competent mice that have a thymus. In the latter case there is a continuous production of new T cells, which causes confusion because these naive cells continue to expand in the presence of whatever is being introduced. Bob Coffman, are your switching experiments performed in SCID mice or in normal mice?

Coffman: The experiments that I was referring to have been carried out in normal mice with normal thymuses, so presumably there is some replenishment of T cells (S. Mocci, S. Mauze & R. L. Coffman, unpublished results).

Flavell: This is fine as long as there isn't a switch, but if a switch occurs in a non-thymectomized mouse then one cannot distinguish the contribution of newly produced thymic emigrants from the peripheral cells.

Coffman: Yes, but despite the continuous generation of new T cells, responses to *L. major* are very difficult to change after two to three weeks *in vivo*.

Swain: Chronically re-stimulated effectors often undergo apoptosis (Zhang et al 1995). If this is occurring, then it gives other populations of cells a chance to grow. If apoptosis is not occurring and the cells are expanding instead, then competition between cells will occur. Therefore, there are numerous factors that can have an effect on the *in vivo* situation.

Ramshaw: I would like to describe an experiment in which a switch back to a Th1 response from a Th2 response is not observed (Sharma et al 1995). It's a model system that we've been using in which we engineer a cytokine gene into a recombinant virus. Therefore, when the virus replicates, the cytokine is produced at high levels at the site of antigenic stimulation. We found that without a cytokine gene, the virus generates a good cytotoxic T lymphocyte (CTL) response. However, if the gene encoding IL-4 is introduced, the CTL response is completely suppressed and a Th2-like situation is created. Two to four weeks later, when IL-4 is no longer expressed, CTLs can be detected.

Lotze: A similar situation occurs with IL-4 transfections in tumour models, where CTLs are not present early on but are detected later (Golumbek et al 1991).

Ramshaw: Also, if the gene encoding IFN-γ is introduced into these viruses, cell-mediated immunity occurs, and the production of IgG is completely suppressed. Therefore, the expression of IFN-γ at sites of antigenic stimulation can suppress antibody responses (Leong et al 1994).

Mitchison: In my opinion, the present technologies are not going to resolve the question of whether switching in clones resembles switching *in vivo*, and that we need new technologies in order to answer that question. It is possible that an approach involving the double staining of cells will identify cells that are making particular cytokines. One promising method involves covering a T cell with a coat of jelly so that it is possible to see what it has been secreting because it sticks in the jelly coat (Manz et al 1995). It is then possible to see what it is now making by looking inside the cell. Are there other novel approaches that address this issue?

Locksley: Yes, one approach involves putting cytokine promoters upstream of heterologous genes that can be expressed in intact cells. For example, we used the human CD2 gene as a heterologous marker and put the mouse IL-4 promoter upstream (R. Locksley, A. Wakil & N. Killeen, unpublished results). Hopefully, signals that switch on the IL-4 promoter will drive human CD2 expression and tag the cell for fluorescent-activated cell sorter analysis. Diane Mathis (personal communication) is working with a similar system using modified CD4 marker genes driven by either the IFN-γ promoter or the IL-4 promoter in order to delineate these lineages.

Flavell: I've used the green fluorescent protein as a marker for the IL-4 gene (unpublished results), which allows the analysis of cells without antibody staining. Cells can then be detected by fluorescence microscopy. I am also using cell surface staining with CD2, by directing expression of the CD2 gene with the IFN-γ promoter. There is also the issue of promoter specificity, so the ultimate method would be to recombine the gene into the chromosome using the gene targeting. I am now doing these experiments.

Mitchison: Are these methods sensitive enough to pick up the initial tickling of the cytokine promoter?

Flavell: Yes, and the length of time that the protein can be detected will depend on its stability, although we have not measured this yet.

Mitchison: Can this method be used to address the switching question? Presumably, you will be able to combine this method with more conventional ones that detect the bulk of protein which the cell is making.

Flavell: In general yes, but there are several limitations, such as the specificity of the promoter and the stability of the reporter protein. If the protein has a half-life of 10 days, the result will be completely different than if it has a half-life of one day. These issues will have to be addressed, so this method will probably not replace more conventional approaches, but it will be a very useful new trick that will give some quick answers.

Abbas: It seems like we are trying to develop these sophisticated methods because we are giving up the search for surface markers for differentiated cell populations. Is CD45RB a marker that can be used to identify Th1 cells?

Coffman: So far, we have only obtained empirical separations between functional populations, based on the expression of CD45RB under well-controlled situations. In general, this marker cannot be used to distinguish Th1 cells from Th2 cells unequivocally. However, this strategy could be used to resolve whether the cells can switch to Th2 cells or vice versa. We may be able to answer the question if we could show that, in the situation where we can experimentally shift from Th1 to Th2 dominance, for example, the initial population is a mixture of two kinds of cells, i.e. those that will generate Th1 cells and a separate, phenotypically distinguishable population that will generate Th2 cells.

Mosmann: But we should still not abandon the approaches that Richard Flavell and Richard Locksley were describing. These are elegant methods that are generating valuable information. IL-4 and IFN-γ are probably the right cytokines to focus on first because they're the most specific for those two lineages, but looking at the other cytokines will also generate valuable information. For example, it's possible that a cell which produces IL-2, IL-4 and IL-5 but not IFN-γ might be a Th2 cell precursor. Other variations on the Th2 theme may exist, because IL-4 and IL-5 don't always show coordinate regulation within populations. The same sort of methodology using different markers would generate a wealth of information on the other cell types that may exist.

Fitch: Tom Gajewski and colleagues have some interesting observations on the conversion of Th0 cells into cells that have lost the ability to make IL-2 (Gajewski et al 1994). They found that if clonal populations of cells which make IL-2, IL-4 and IFN-γ are stimulated under anergizing conditions, IL-2 production disappears. The anergized Th0 cells, upon acquiring the Th2-like phenotype, change their appearance and become large, they increase their basal levels of intracellular Ca^{2+} and they lose the pattern of tyrosine phosphorylation found in Th1 cells. This altered pattern of secretion is maintained if cells are stimulated with antigen in the absence of co-stimulation.

Recently, they have also found that established Th2 clones can be coaxed into making IL-2 by growing them in the presence of IL-2 for several weeks (Gajewski et al 1995). All of the Th2-like characteristics revert back to a Th0-like state when the cells resume IL-2 production. These observations raise the question of what is the relationship between a cell that makes all three of these cytokines (IL-2, IL-4 and IFN-γ) to one that makes only IL-4 but not IL-2 or IFN-γ? If another population of cells exists, in addition to Th1 cells, Th2 cells and Th0 cells, how does it arise and what is its significance?

Swain: I also have some results which show that memory versions of Th2 cells also make IL-2 (Swain 1994). Therefore, it is possible that IL-2 is regulated in a completely different fashion during differentiation than IL-4 and IFN-γ.

Allen: Bob Coffman, how much is known about the role of TCR ligation versus the role of the cytokines in Th cell differentiation? Can you commit a Th1 cell to become a Th2 cell just by giving it IL-4 without TCR ligation?

Coffman: We have not performed experiments, either *in vitro* or *in vivo*, in the absence of TCR ligation.

Allen: What happens if you take Th1 cells, drive them with IL-4 and look at their phenotype seven days later?

Coffman: Nothing happens in culture if we don't do a TCR ligation.

Allen: Do the signals that are required for proliferation and IL-4 production have to be coordinated for differentiation to occur?

Coffman: They do persist for a short while. One could incubate this population for 48 h in the presence of IL-4, remove the IL-4 and then stimulate with antigen. I have not done this experiment.

Mitchison: Bob Coffman, in a recent paper you separated BALB/c cells by CD45 phenotype and you showed that the CD45RB^{high} cells, which were presumably resting cells, were potential Th1 producers; whereas the CD45RB^{low}

cells, which were presumably activated cells, were Th2 producers (Powrie et al 1994). This is exactly what one would expect in a BALB/c mouse while it is generating a Th2 response. However, if you look in a BALB/c mouse that has been manipulated so that it produces a protective response, or if you look in a different strain of mouse that is resistant, do you see the reverse?

Coffman: In either manipulated BALB/c mice or naturally resistant strains, a minor Th2 cell population cannot be detected, so it is not possible to answer that question. However, I do not understand why you would expect the minor Th1 cell population in a BALB/c mouse to be in the non-activated state. It's not the dominant population, but why should it be non-activated?

Mitchison: Perhaps it's being suppressed.

Coffman: But is that the same as being not activated by antigen? The expression of CD45RB^{high} may be a marker for this suppression, but we cannot equate this with lack of antigen exposure.

Mitchison: But, generally speaking, the higher isoforms are markers of quiescence.

Coffman: I'm not sure that these cells are actually quiescent, but we haven't formally addressed this.

Mitchison: It seems as though the CD45RB isoforms may only reflect an individual's previous immunological exposure.

Mosmann: There was an interesting difference between IL-4 and IL-10 in some of your experiments. Sometimes, IL-10 was more important as an inhibitor of IFN-γ production, and sometimes IL-4 was more important. Could you resolve that apparent difference by separating the process into the differentiation of cells and the activation of cells that have already differentiated? IL-4 is clearly more important for the process of differentiation, whereas IL-10 may be required for the activation of cells once they have differentiated. Does this explain your results?

Coffman: No, because I am assuming that the differentiation of naive cells is no longer a major process in a mouse at four to six weeks post-infection. The underlying assumption has always been that we are studying the regulation of an established population. It is possible that there is a small population of undifferentiated cells hiding within the differentiated population.

Mosmann: Is it possible that natural killer (NK) cells are present? Because they are inhibited by IL-4, which contrasts with T cells.

Coffman: I doubt it. We have shown that the IFN-γ produced in response to our antigen is almost entirely derived from T cells and not from NK cells (K. Varkila & R. L. Coffman, unpublished results).

Lotze: Are the differences between IL-4 and IL-10 occurring at the level of the T cell or at the level of the antigen-presenting cell (APC)?

Coffman: Probably at the level of the APC. Macrophages are clearly one of the primary targets for IL-10-mediated suppression, more so than T cells themselves. We don't know what the targets are for IL-4-mediated suppression.

Abbas: We have addressed this question of the targets of these cytokines, although we have not published the experiments because it's difficult to obtain clear-cut results (C. M. Shea & A. K. Abbas, unpublished results). We performed two-stage culture experiments in which we incubated either T cells from transgenic mice or APCs with a particular cytokine for two to three days, mixed the cell populations in the presence of a neutralizing antibody against the cytokine to block carry over, and stimulated with antigen to study the development of Th1 and Th2 cells. The majority of the evidence suggests that: (1) both IL-4 and IL-12 act directly on T cells; (2) the major, if not exclusive, action of IFN-γ is on macrophages; and (3) the exclusive action of IL-10 is on the APCs.

Lotze: In contrast, we found that both IL-4 and IL-12 can have profound effects on APCs. IL-12 has profound effects on dendritic cells and IL-4 has major proliferative effects on macrophages (W. J. Storkus & M. T. Lotze, unpublished observations). This also contrasts with some of the published results in terms of IL-12 potentially being produced by dendritic cells (Macatonia et al 1995).

Abbas: It may have profound effects on APCs, but I'm asking a more limited question. Is the ability of IL-4 or IL-12 to drive Th2 cell or Th1 cell differentiation due to the effects on APCs or the effects on T cells?

Lotze: It depends on where you look. Most of our systems involve culturing dendritic cells under various cytokine conditions and then giving them peptides. Obviously we're going to bias what we see, based on the fact that we're starting with the APC, i.e. we're not administering the individual cytokine *in vivo*, where they would have the opportunity to act both on APCs as well as on T cells.

Coffman: We may be mixing two processes together which we should try to separate. One is the initial differentiation from a naive bipotential precursor cell, and the other is the regulation of established Th1 and Th2 cells. These are different processes. There's no question that IL-4 acts directly on T cells during differentiation. In contrast, in systems where one can actually see an effect of IL-10 on differentiation, IL-10 acts indirectly via the control of IL-12 and IFN-γ production. However, during a recall response in an established population, it is not known how IL-4 is acting and IL-10 actions are primarily, but not exclusively, aimed at the APC.

Ramshaw: We have administered IL-10 *in vivo*, using recombinant viruses, and we have found that IL-10 *in vivo* has no effect on the immune response.

Abbas: I object strongly to this generalization. Using IL-10-transfected tumours, Bob Tepper's lab has shown that IL-10 inhibits CTL development (Wang et al 1994). Another inaccurate generalization is that IFN-γ inhibits antibody production. We showed that Th1 cell clones block antibody production (Boom et al 1988). If enough IFN-γ is produced, it will shut off B cell proliferation, yet IFN-γ is required for the production of some antibody isotypes. The ability of Th1 clones to inhibit antibody production could be due to Fas ligand-mediated killing of B cells.

Ramshaw: But the quantities of IL-10 and IL-4 in our experiments are the same because they're under the control of virus promoters.

Lotze: How high are the levels?

Ramshaw: You can't measure them *in vivo*.

Mitchison: The Sarvetnick lab also showed striking effects of locally produced IL-10 in insulitis (Wogensen et al 1994).

Mosmann: We have got to be careful not to underestimate the diversity of the immune system. There are numerous different mechanisms that can be directed towards a particular infection or tumour. The question of which cytokine will affect a particular response cannot be predicted on the basis of whether it is a Th1 or Th2 response. There are many other factors involved. Therefore, there's not necessarily any conflict between these different experiments showing different effects of IL-10 on similar responses. These responses are similar but they're not identical.

References

Boom WH, Liano D, Abbas AK 1988 Heterogeneity of helper/inducer T lymphocytes. II. Effects of interleukin 4- and interleukin 2-producing T cell clones on resting B lymphocytes. J Exp Med 167:1350–1363

Croft M, Swain SL 1995 Recently activated naive CD4 T cells can help resting B cells, and can produce sufficient autocrine IL-4 to drive differentiation to secretion of T helper 2-type cytokines. J Immunol 154:4269–4282

Gajewski TF, Lancki DW, Stack R, Fitch FW 1994 "Anergy" of TH0 helper T lymphocytes induces downregulation of TH1 characteristics and a transition to a TH2 like phenotype. J Exp Med 179:481–491

Gajewski TF, Alegre M-L, Fitch FW 1995 Th2 cells can be induced to revert to an IL-2-producing, Th0 phenotype, submitted

Golumbek PT, Lazenby AJ, Levitsky HI, Jaffee LM, Baker M, Pardoll DM 1991 Treatment of established renal cancer tumour cells engineered to secrete interleukin-4. Science 254:713–716

Leong KH, Ramsay AJ, Boyle DB, Ramshaw IA 1994 Selective induction of immune responses by cytokines co-expressed in recombinant fowlpox virus. J Virol 68:8125–8130

Macatonia SE, Hosken NA, Litton M et al 1995 Dendritic cells produce IL-12 and direct the development of Th1 cells from naive $CD4^+$ T cells. J Immunol 154:5071–5079

Manz R, Assenmacher M, Pflüger E, Miltenyi S, Radbruch A 1995 Analysis and sorting of live cells according to secreted molecules, relocated to a cell surface affinity matrix. Proc Natl Acad Sci USA 92:1921–1925

Mocci S, Coffman RL 1995 Induction of a Th2 population from a polarized *Leishmania*-specific Th1 population by *in vitro* culture with IL-4. J Immunol 154:3779–3787

Nabors GS, Afonso LCC, Farrell JP, Scott P 1995 Switch from a type 2 to a type 1 T helper cell response and cure of established *Leishmania major* infection in mice is induced by combined therapy with interleukin-12 and pentostan. Proc Natl Acad Sci USA 92:3142–3146

Powrie F, Correa-Oliveira R, Mauze S, Coffman RL 1994 Regulatory interactions between $CD45RB^{high}$ and $CD45RB^{low}$ $CD4^{+}$ T cells are important for the balance between protective and pathogenic cell-mediated immunity. J Exp Med 179:589–600

Sharma DP, Ramsay AJ, Maguire D, Rolph M, Ramshaw IA 1995 Interleukin 4 expression enhances the pathogenicity of vaccinia virus and suppresses cytotoxic T cell responses, submitted

Swain SL 1994 Generation and in vivo persistence of polarized Th1 and Th2 memory cells. Immunity 1:543–552

Wang L, Goillot E, Tepper RI 1994 IL-10 inhibits alloreactive cytotoxic T lymphocyte generation *in vivo*. Cell Immunol 159:152–169

Wogensen L, Lee MS, Sarvetnick N 1994 Production of interleukin 10 by islet cells accelerates immune-mediated destruction of beta cells in nonobese diabetic mice. J Exp Med 179:1379–1384

Zhang X, Giangreco L, Broome HE, Dargon C, Swain SL 1995 Control of CD4 effector fate. J Exp Med 182:699–710

General discussion I

Memory

Swain: I would like to state a couple of points about memory and the pathway from effectors to memory. We were interested in the length of the polarized response, so we made polarized subsets of effectors from transgenic mice and transferred them to an adopted host. We waited for memory cells to develop, and we found that they were still highly polarized from six weeks to one year later. Also, if the memory cells initially made interleukin 2 (IL-2) and γ-interferon (IFN-γ), or IL-4 and IL-5, they continued to make IL-2 and IFN-γ, or IL-4 and IL-5, for up to a year later (Swain 1994). Another important point about these memory cells is that they produce higher levels of cytokines and they are more effective at driving T helper (Th) responses than if we transferred naive transgenic cells. This brings up the issue of where memory comes from.

We were also interested in what happens to the effectors. When we transferred these effectors to mice, they turned into memory cells. We wondered whether the effector cells were undergoing apoptosis, so we examined what happens to effectors *in vitro*. If we didn't add anything to the effectors but restimulated them with antigen (peptide fragments) and antigen-presenting cells, we observed a high rate of apoptosis (Zhang et al 1995, X. Zhang, L. Carter & S. L. Swain, unpublished results). This rate was faster for polarized Th1 cells than for Th2 cells. However, it is possible to inhibit apoptosis of both the Th1 cells and the Th2 cells with certain combinations of cytokines. Apoptosis of Th1 cells was inhibited by transforming growth factor β (TGF-β) (Zhang et al 1995), whereas inhibition of Th2 cell apoptosis required both IL-2 and TGF-β. This inhibition has a big impact on the cell population as a whole because we observe an expansion of the effector population if we add IL-2 and TGF-β. This may be important when effector populations have to expand to deal with bacterial or viral infections. In the absence of added cytokines, antigen stimulation gives rise to a stable population through a combination of death and proliferation, but expansion of the population occurs when IL-2 alone or IL-2 and TGF-β are added. We also observe significant expansion with these cytokines in the absence of antigen. It is also possible that TGF-β may be involved in the generation of memory through its role of blocking apoptosis.

Allen: In long term memory have you tried to favour a Th2 response by infecting the mice or giving them IL-4?

Swain: I haven't tried that.

Vaccine development

Mitchison: Let us discuss the development of vaccines and the possibilities of therapeutic, post-infection vaccines, such as those that have been claimed to be effective in treating *Leishmania* infection (Convit et al 1989). Their effect may reflect control of the Th1/Th2 balance. Would it be therapeutically effective in *Leishmania* infections to push the balance in favour of a Th1 cell response by direct cytokine manipulation?

Sher: Giorgio Trinchieri has presented some evidence that IL-12 can have remarkable effects in pushing protective immunity towards a Th1 response (Trinchieri 1995, this volume). This is one clear-cut cytokine manipulation that has proven to work, at least in mouse models, both as an adjuvant and also in changing the course of infection itself. It may be possible to use this as a therapy for an established infection.

Trinchieri: IL-12 alone will not change an established Th2-type response into a Th1-type response in the *Leishmania major* infection model in BALB/c mice. It is necessary to treat the mice with the antimonial compound pentostan in addition to IL-12. This was recently shown by Phil Scott (Nabors et al 1995). Pentostan probably reduces the parasite load and increases the levels of secreted antigen, which may result in the rescue of a suppressed Th1 response. One important experiment would be to eliminate the Th2 cell negative feedback loop by knocking out IL-4, IL-10 and possibly also TGF-β, at the same time as giving IL-12. This would determine whether, in the absence of a Th2 response, once an initial Th1 response is induced, there is enough positive feedback to maintain it and to cure the mice.

Mitchison: What is the best way of doing that? Surely, it is not by administering cytokines because the pharmacokinetics are so bad.

Trinchieri: IL-12 treatment is particularly favourable because its half-life is long—it's about 3–4 h in the mouse.

Lotze: In humans the half-life of IL-12 is even longer (M. Atkus, M. Robertson, J. Ritz et al, unpublished observations).

Coffman: I don't agree with the statement that the use of cytokines is limited by their half-life. Our experience suggests that, although this can be a limitation, it does not prevent cytokines from having biologically relevant effects when they are administered as infrequently as once per day.

Mitchison: Also, in the early stages of clinical trials it may actually be an advantage to have an agent that is eliminated rapidly, so that any side effects would be short-lived.

Lachmann: There is another school of thought which takes the view that where the administration of the agonists is too difficult, it may be preferable to use agents that act directly on the receptor, i.e. agonist or antagonist antibodies, or small molecular weight compounds that have the same effects.

Ramshaw: Many of these cytokines are highly concentrated at the site of immune reactivity, which creates a problem. We will be carrying out trials in HIV-infected patients later on this year with an avipoxvirus, which is a non-replicating virus that contains the gag/pol gene of HIV and either the gene encoding IFN-γ or IL-12. When the vaccine antigens (gag/pol) are expressed at the site of immunoreactivity, the cytokines are also co-expressed. In the case of IFN-γ, this results in the switching off of the antibody response and the stimulation of a cell-mediated immune response. This may be a good approach for inducing the switch to a Th1 response (Leong et al 1994).

Liew: I would like to mention some recent results with BRD509, which is an *aroA⁻aroD⁻* vaccine strain of *Salmonella typhimurium* that can be used to deliver the cytokines IFN-γ, tumour necrosis factor α (TNF-α) and migration inhibiting factor (MIF) (D. Xu, S. McSorley & F. Y. Liew, unpublished results). We infected the highly susceptible BALB/c mice with *L. major* and seven to 10 days later, we administered orally a cocktail of BRD509–IFN-γ, BRD509–TNF-α and BRD509–MIF. Control mice, which were administered with BRD509 alone, developed the familiar lethal disease, whereas mice given the cytokines delivered by BRD509 did not develop lesions. The injection of individual cytokines was not as effective. Therefore, this *Salmonella* system can be used to deliver cytokines therapeutically.

Abbas: Were antigens present as well as the cytokines?

Liew: No. Only the cytokines were present. The BALB/c mice were infected with *Leishmania* before the cytokines were introduced with *Salmonella* as a carrier.

Mosmann: Are the results of the *Salmonella* experiment due to an ongoing or very recent response that overlaps with the *Leishmania* response, or do *Salmonella* vectors protect against a much later challenge?

Liew: That is one of the problems. *Salmonella* vectors do not persist for very long. They are present in the liver and spleen only for about three weeks before they are destroyed. Therefore, the protective response is short-lived.

Mitchison: Did you try other approaches to implant the genes encoding the cytokines?

Liew: We have also used the so-called 'naked DNA' technique. This involves the injection of an expression vector containing the gene encoding IFN-γ, for example, directly into the skeletal muscle of mice. These mice then develop significant resistance against *L. major* infection compared to those injected with the empty vector (D. Xu & F. Y. Liew, unpublished results). However, this approach was not as effective as the administration of the *Salmonella* constructs.

Mitchison: There are many strategies for introducing DNA using viral vectors. Cytokine genes have been implanted using naked DNA techniques for the treatment of autoimmunity (Raz et al 1993). Also, genes have been introduced into cells *in vitro* and then implanted *in vivo* (Bessis et al 1995).

Ramshaw: Another technique useful for naked DNA immunization is the gene gun approach. In these experiments the naked DNA of the plasmid is coated onto gold beads. It is not necessary to construct the cytokines in the same plasmid because it is possible to precipitate the plasmid containing the vaccine antigen onto the gold bead with another plasmid that encodes the cytokine. One can fire both of these plasmids into the same cell, and the cytokines are expressed with the antigen and can actually alter the immune response to the vaccine antigen.

Lotze: The Indiana Jones approach to gene therapy!

Mitchison: These naked DNA experiments may shortly enter into clinical trial stages. But on the other hand, these approaches may be unnecessary, and it may be possible to achieve the same effect simply with adjuvants. A process using particular adjuvants, in terms of the cytokines that they illicit, may be the only practical way of applying cytokine manipulation to vaccines on a population scale.

Lotze: Do you mean by picking the appropriate adjuvant to elicit a certain cytokine?

Mitchison: Yes.

Röllinghoff: Do you have a particular example?

Mitchison: It is likely that the Convit–Bloom therapeutic vaccination against leishmaniasis (Convit et al 1989) belongs to this category.

Kaufmann: Mycobacterium bovis bacillus Calmette–Guérin (BCG) has also been widely used for tumour therapy.

Shearer: It is not thought that adjuvants favour a Th1 response, although some are thought to favour a Th2 response. In terms of cytokine regulation, we have compared the effects of adding IL-12 and/or anti-IL-10 antibody to clones generated from HIV^+ patients (Clerici et al 1994). We found that adding both was not as effective as either alone, and that the anti-IL-10 antibody worked at least as well if not better than IL-12. Therefore, an antagonistic approach may be just as effective.

Ramshaw: We have some new results using a different vaccine strategy. We have primed mice with naked DNA encoding a foreign antigen, then boosted with a recombinant avipoxvirus which contains the same gene. The avipoxvirus does not replicate in mammalian cells. The naked DNA immunization produces about 10 μg/ml of specific antibody. One observes the same response if one immunizes with the avipoxvirus alone, which is a reasonable antibody response. If mice are first primed with the DNA, followed by the recombinant avipoxvirus containing the same gene, over 1 mg/ml of antibody to the foreign gene is produced. This response is composed of IgG2a, which is associated with a Th1-type response (Ramshaw et al 1994).

Navikas: Are there any other immunoregulatory substances that can down-regulate Th1 or Th2 responses selectively? We have already heard that some cytokines may influence this balance. However, the phosphodiesterase inhibitor pentoxifylline, for example, can selectively down-regulate a Th1 response and prevent the induction of experimental autoimmune encephalomyelitis (Rott et al 1993). This may be important for developing new autoimmune therapies, and many people believe that it could be a better approach than using cytokines themselves.

Lachmann: Thalidomide has been shown to change Th1 to Th2 responses. This works quite impressively *in vitro* (McHugh et al 1995) and *in vivo* (I. R. Rifkin & S. M. McHugh, unpublished results). The effect is apparently not entirely due to thalidomide's ability to inhibit TNF.

Mitchison: If one could get rid of the side effects, which quite rightly prevent thalidomide from being used on a large scale, then interest in it would be restored. However, the universal comment from clinicians is that there is no way of stopping the supermarket effect, i.e. when one patient says to another while out shopping that a particular drug cured their ulcers and suggests that they try it.

Lachmann: Perhaps a greater problem than the supermarket effect is to persuade any drug company to put it back on the market.

Receptor constructs and cytokine/antibody complexes

Röllinghoff: It is possible to use IL-4 receptor constructs to shift a Th2-type response towards a Th1-type response. This has been shown in the *Leishmania* system by Gessner et al (1994). They used a soluble recombinant mouse IL-4 receptor fused to the hinge CH2 and CH3 domains of a murine IgG2b molecule. This protein persists in the circulation for longer than the original IL-4 receptor. They administered this construct to *L. major*-infected BALB/c mice at Day 0, Day 2 and Day 3 postinfection, and they took out the lymph node cells six weeks later. When they looked for the production of cytokines, they found that a Th1-type response had been generated, with an increased IFN-γ and a decreased IL-4 production.

Lotze: Fred Finkelman talks about the effect of anti-IL-4 antibody in sustaining the action of IL-4. I have never really understood this (Finkelman et al 1993). He says that a particular stoichiometry is required for stabilization, which presumably mimics your genetic constructs. This situation also reminds me of neutralization studies with anti-IL-4 antibody. Is it possible that some of the effects might be driven by IL-4?

Coffman: It is certainly possible. We've done some titrations for longer-lived phenomena than those that Finkelman has reported. We have used IL-4 knockout mice, where the only source of IL-4 is the IL-4 that we introduce (B. W. P. Seymour and R. L. Coffman, unpublished results). The association of IL-4 with anti-IL-4 antibody results in a substantial extension of its functional half-life. This makes sense if one imagines that the half-life is partly due to specific degradation mechanisms which are being blocked when the cytokine is bound to antibody. When IL-4 and anti-IL-4 antibody are present in equimolar ratios, the antibody acts essentially as a reservoir for the cytokine. However, when the antibody is used for neutralization, as in many experiments in my lab, it is added in a large excess, and it blocks the activity of endogenous IL-4. Indeed, Finkelman has shown that an excess of 11B11 antibody will also block the activity of IL-4/anti-IL-4 antibody complexes (Finkelman et al 1993).

Mitchison: Martin Röllinghoff, you found an effect with the IL-4 receptor constructs, as Harald Renz has in his asthma model in mice (Renz et al 1993, 1995), but have you titrated those constructs against an anti-IL-4 antibody?

Röllinghoff: We have compared the anti-IL-4 antibody with three or four different concentrations of IL-4 receptor. We did not do a complete titration curve, so I cannot tell you exactly whether the receptor or the anti-IL-4 antibody works better. In the beginning there were problems with the constructs because of glycosylation.

Mitchison: But which is better so far, constructs or antibodies?

Röllinghoff: It looks as though they have similar potencies.

Cosman: We have some results which suggest that it depends on the system that you use. In some models the antibody is more effective, whereas in others the soluble receptor is more effective (Fanslow et al 1991, Sato et al 1993). It is possible that differences in biodistribution, which depend on the size of the cytokine antagonist, may also be involved. Another complication is that with either antibodies or soluble receptors one may prolong the physical half-life of the cytokine; however, one may or may not prolong the biological half-life of the cytokine.

Lotze: Exogenous cytokines, cytokines plus Fc receptors and cytokines plus antibodies are big molecules. Can they get across into tissues? We presume that they can go everywhere throughout the body but this may not be the case.

Coffman: In the case of the IL-4/anti-IL-4 antibody complex, the answer is probably yes. We can induce wild-type IgE responses with low levels of IL-4/anti-IL-4 antibody complex in IL-4 knockout mice. Therefore, these complexes must be getting into the spleen and lymph nodes.

Mosmann: But most of the responses we try to interfere with involve some local inflammatory response. Vascular leakage associated with inflammation may allow all of these reagents to reach the site that is actually being influenced, so we don't have to assume that they penetrate tissues in general.

References

Bessis N, Boissier M-C, Caput D, Fradelizi D, Fournier C 1995 IL-4 or IL-5 transfected xenogenic fibroblasts in the treatment of collagen-induced arthritis in mice. Clin Rheumatol 14:261

Clerici M, Sarin A, Coffman RL et al 1994 Type 1/type 2 cytokine modulation of T cell programmed cell death as a model for human immunodeficiency virus pathogenesis. Proc Natl Acad Sci USA 91:11811–11815

Convit J, Castellanos PL, Ulrich M et al 1989 Immunotherapy of localized, intermediate, and diffuse forms of American cutaneous leishmaniasis. J Infect Dis 160:104–115

Fanslow WC, Clifford KN, Park LS et al 1991 Regulation of alloreactivity *in vivo* by IL-4 and the soluble IL-4 receptor. J Immunol 147:535–540

Finkelman FD, Madden KB, Morris SC et al 1993 Anti-cytokine antibodies as carrier proteins. Prolongation of *in vivo* effects of exogenous cytokines by injection of cytokine-anti-cytokine antibody complexes. J Immunol 151:1235–1244

Gessner A, Schröppell K, Will A, Enssle KH, Lauffer L, Röllinghoff M 1994 Recombinant soluble interleukin-4 (IL-4) receptor acts as an antagonist of IL-4 in murine cutaneous leishmaniasis. Infect Immun 62:4112–4117

Leong KH, Ramsay AJ, Boyle DB, Ramshaw IA 1994 Selective induction of immune responses by cytokines co-expressed in recombinant fowlpox virus. J Virol 68:8125–8130

McHugh SM, Rifkin IR, Deighton J et al 1995 The immunosuppressive drug thalidomide induces T helper cell type 2 (Th2) and concomitantly inhibits Th1 cytokine production in mitogen-stimulated and antigen-stimulated human peripheral blood mononuclear cell cultures. Clin Exp Immunol 99:160–167

Nabors GS, Afonso LCC, Farrell JP, Scott P 1995 Switch from a type 2 to a type 1 T helper cell immune response and cure of established *Leishmania major* infection in mice is induced by combined therapy with interleukin-12 and pentostan. Proc Natl Acad Sci USA 92:3142–3146

Ramshaw IA, Leong KH, Ruby JC, Ramsay AJ, Boyle D 1994 Recombinant viral vaccines expressing cytokine genes. In: Norrby E, Brown F, Chanock R, Ginsberg H (eds) Vaccines 94: modern approaches to new vaccines including prevention of AIDS. Cold Spring Harbor Laboratory Press, p 29–33

Raz E, Watanabe A, Baird SM et al 1993 Systemic immunological effects of cytokine genes injected into skeletal muscle. Proc Natl Acad Sci USA 90:4523–4527

Renz H, Enssle K, Bradley K, Loader J, Larsen G, Gelfand EW 1993 Prevention of *in vivo* immediate hypersensitivity responses by local and systemic treatment with soluble IL-4 receptor (sIL-4R). J Allergy Clin Immunol 93:261

Renz H, Enssle K, Lauffer L, Kurrle R, Gelfand EW 1995 Inhibition of allergen-induced IgE and IgG1 production by soluble IL-4 receptor. Int Arch Allergy Immunol 106:46–54

Rott O, Cash E, Fleischer B 1993 Phosphodiesterase inhibitor pentoxyfylline, a selective suppressor of T helper type 1- but not type 2-associated lymphokine production, prevents induction of experimental allergic encephalomyelitis in Lewis rats. Eur J Immunol 23:1745–1751

Sato TA, Widmer MB, Finkleman FD et al 1993 Recombinant soluble murine IL-4 receptor can inhibit or enhance IgE responses *in vivo*. J Immunol 150:2717–2723

Swain SL 1994 Generation and *in vivo* persistence of polarized Th1 and Th2 memory cells. Immunity 1:543–552

Trinchieri G 1995 The two faces of interleukin 12: a pro-inflammatory cytokine and a key immunoregulatory molecule produced by antigen-presenting cells. In: T cells in infectious and autoimmune diseases. Wiley, Chichester (Ciba Found Symp 195) p 203–220

Zhang X, Giangreco L, Broome HE, Dargon C, Swain SL 1995 Control of CD4 effector fate. J Exp Med 182:699–710

Differentiation of subsets of CD4+ and CD8+ T cells

Tim R. Mosmann, Subash Sad, Lakshmi Krishnan, Tom G. Wegmann, Larry J. Guilbert and Mike Belosevic*

*Departments of Immunology and *Biological Sciences, University of Alberta, Edmonton, Alberta T6G 2H7, Canada*

Abstract. Our knowledge of the cytokine secretion patterns of T cells and other cells is clearly becoming more complex. The T helper 1 (Th1) and Th2 patterns may represent the extremes of a spectrum of cytokine regulatory patterns controlled by several cell types. $CD8^+$ T cells can also secrete either Th1-like or Th2-like cytokine patterns, and they can contribute to bystander B cell activation. Interactions occur between immune cytokine regulatory networks and other systems, and pregnancy and responses against infection can profoundly influence each other.

1995 T cell subsets in infectious and autoimmune diseases. Wiley, Chichester (Ciba Foundation Symposium 195) p 42–54

Since the discovery that $CD4^+$ T cell subsets secrete different patterns of cytokines, considerable evidence has accumulated that shows the importance of the T helper 1 (Th1) and Th2 subsets in the success of immune responses against infectious agents. However, the Th1/Th2 dichotomy may just represent the beginning of our understanding of an increasingly complex network of cytokine regulatory patterns. In this article, we will discuss recent information on the differentiation pathways of $CD4^+$ T cells, the existence of parallel subsets of $CD8^+$ T cells and the interactions between cytokine regulatory networks in pregnancy and infection.

The Th1 and Th2 cytokine patterns

The characteristic cytokines of both mouse and human Th1 cells are interleukin 2 (IL-2), γ-interferon (IFN-γ) and lymphotoxin (Mosmann et al 1986, Cherwinski et al 1987, Del Prete et al 1991). Preferential expression of these cytokines is observed in a number of immune responses that are biased towards cell-mediated immunity, such as delayed type hypersensitivity (DTH) reactions (Heinzel et al 1991, Tsicopoulos et al 1992). Th1-oriented responses are thought to be more effective against intracellular pathogens because of the induction of a number of cytotoxic and local inflammatory mechanisms (Janeway et al 1988, Mosmann & Coffman 1989, Sher & Coffman 1992).

Th2 cells produce IL-4, IL-5, IL-6, IL-9, IL-10 and IL-13, although the expression of some of these cytokines is not as strictly associated with the Th2 phenotype in human cells as in mouse clones (Mosmann et al 1986, Cherwinski et al 1987, Fiorentino et al 1989). The Th2 cytokine pattern is associated with help for antibody production by B cells, and strong Th2 responses are associated with allergic reactions (Del Prete et al 1991, Kay et al 1991). This is consistent with the individual functions of the Th2 cytokines because IL-4 induces switching to IgE production (Coffman et al 1986), and IL-5 is the major eosinophil growth and differentiation factor (Sanderson et al 1986).

Both Th1 and Th2 responses can be useful for combatting particular infections. Infection by the intracellular macrophage parasite *Leishmania major* can be cured by a Th1 response but not a Th2 response, and the converse is true for infection by the gut parasite *Trichuris muris.* In both cases, manipulations of Th1 and Th2 cytokines can alter both the immune response and the outcome of the infection. For example, treatment with anti-IFN-γ antibody *in vivo* switches a Th1-like curative response against *L. major* into an ineffective Th2 response, whereas anti-IFN-γ switches a Th1-like anti-*T. muris* response into a curative Th2 response (Belosevic et al 1989, Else et al 1994). Thus, there is good evidence that these cytokine patterns can be closely associated with the success or failure of immune responses, and that each response is appropriate for different pathogens.

The Th1/Th2 dichotomy is a simplified model of complex immune responses

Although the Th1 and Th2 patterns are strikingly different, and their relevance in some infections has been clearly documented, there are several reasons to believe that the regulation of immune responses is considerably more complex than these two patterns. Additional CD4+ T cell phenotypes have been described, such as the Th0 phenotype that secretes both Th1 and Th2 cytokines (Firestein et al 1989, Street et al 1990). Other patterns have been described among both mouse and human CD4+ clones, e.g. IL-2, IL-4 and IL-5, but it is not currently known how many of these represent genuine *in vivo* phenotypes.

Another complexity is apparent when the progression of an immune response with time is considered. For some infections, different effector functions may be appropriate at different stages of infection, such as in a mouse malaria model in which Th1 and Th2 responses may be important early and late in infection, respectively (von der Weid & Langhorne 1993, Taylor-Robinson et al 1993).

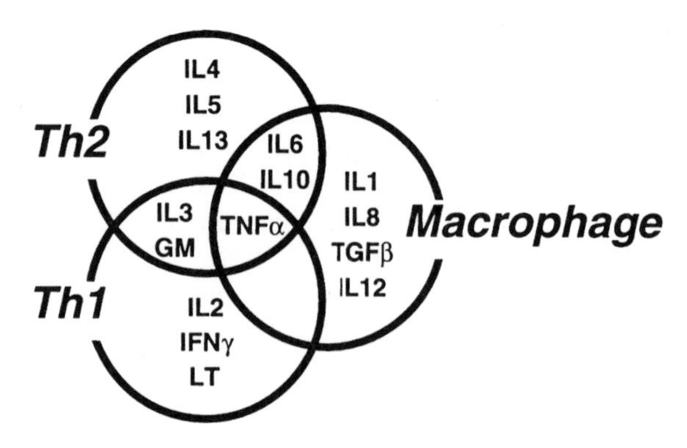

FIG. 1. Overlapping cytokine patterns. GM, granulocyte macrophage; IL, interleukin; LT, lymphotoxin; TGFβ, transforming growth factor β; Th, T helper; TNFα, tumour necrosis factor α.

The location of the immune response is also important because many infections and hence immune responses may be localized to the site of infection and the draining lymph nodes. Successful responses against concurrent infections with different pathogens in different locations would clearly require compartmentalization of the appropriate responses.

Finally, the cytokines characteristic of Th1 and Th2 cells are also produced by many other cell types. For example: IFN-γ is also secreted by natural killer (NK) cells, and some CD8$^+$ T cells and $\gamma\delta$ T cells; IL-4 is also produced by mast cells, and some CD8$^+$ and $\gamma\delta$ T cells; and IL-10 is also produced by macrophages, mast cells and some CD8$^+$ T cells (Kelso & Glasebrook 1984, Brown et al 1987, Sad et al 1995, Ferrick et al 1995). Thus, an overall Th1-like or Th2-like cytokine pattern may be due to the combined contributions of a variety of cells.

Instead of considering the Th1 and Th2 patterns as separate patterns, it may be more informative to consider overlapping patterns of cytokine synthesis, as represented for just three cell types in Figure 1. Cytokines that are generally pro-inflammatory are contributed mainly by macrophages and Th1 cells, although some cytokines common to more than one cell type are involved. Cytokines generally inhibitory for inflammatory reactions are mainly produced by Th2 cells or macrophages. B cell helper cytokines are mainly produced by Th2 cells, although some of these are shared with macrophages, and the Th1-specific cytokine IL-2 is also a strong B cell helper factor. The partial redundancy of cytokine functions also contributes to the complexity of different possible immune effector states. Taken together, all of these complexities suggest that the Th1 and Th2 responses represent the extremes of a spectrum of immune responses that are precisely tailored to particular pathogens.

CD4$^+$ T cell differentiation

It has been known for some time that naive T cells secrete mainly IL-2 when first stimulated, and cells from this population then develop Th1 and Th2 phenotypes under the influence of different cytokines (Le Gros et al 1990, Swain et al 1991, Hsieh et al 1993). IL-4 induces differentiation into Th2 cells, whereas IFN-γ, transforming growth factor β (TGF-β) and IL-12 drive Th1 differentiation. We have found that TGF-β and anti-IFN-γ antibody prevent differentiation into Th2 or Th1 cells, respectively, resulting in the maintenance of CD4$^+$ T cells in a state secreting only IL-2. When the TGF-β/anti-IFN-γ antibody treatment is removed, these cells differentiate rapidly into Th1 or Th2 cells in the presence of IFN-γ or IL-4, respectively. These conditions have been used to establish IL-2-secreting clones. Aliquots of individual clones are able to differentiate into either Th1 or Th2 cells, depending on the cytokine influence (Sad & Mosmann 1994). Thus, the initial IL-2-secreting cells are uncommitted, and they can develop into either of the two extreme effector phenotypes. The *in vivo* equivalent of the rapidly proliferating, IL-2-secreting cell is not clear, but its cell-surface markers and cytokine secretion pattern are similar to those of long-term memory cells.

Cytokine secretion subsets of CD8$^+$ cells: the Tc1 and Tc2 subsets

Although many CD8$^+$ T cell clones and normal CD8$^+$ cell populations secrete mainly Th1 cytokines when stimulated, recent evidence has indicated that CD8$^+$ T cells can also secrete Th2 cytokines (Fong & Mosmann 1990, Erard et al 1993, Croft et al 1994, Seder et al 1992, Maggi et al 1994). We have found that treatment with anti-IFN-γ antibody during initial stimulation leads to the reduction of the subsequent secretion of IFN-γ, and treatment with IL-4 induces the appearance of cells secreting Th2 cytokines. Clones of CD8$^+$ cells were isolated in the presence of IFN-γ plus anti-IL-4 antibody, or anti-IFN-γ antibody plus IL-4. These clones secreted cytokine patterns that were strikingly similar to the Th1 and Th2 patterns, respectively, and so we have proposed the names Tc1 (T cytotoxic 1) and Tc2 for the two subsets (Sad et al 1995). As with Th1 and Th2 cells, the Tc1 and Tc2 phenotypes appear to be relatively stable, and we have not yet been able to convert either subset into the other. Both subsets remain CD8$\alpha^+\beta^+$, CD4$^-$ and NK1.1$^-$.

Functions of Tc1 and Tc2 subsets

Both subsets are strongly cytotoxic, and they can kill more than 50% of allogeneic target cells at an effector : target ratio of 1 : 1. Preliminary results suggest that both subsets are restricted to major histocompatibility complex (MHC) class I and should, therefore, recognize viral antigens presented on

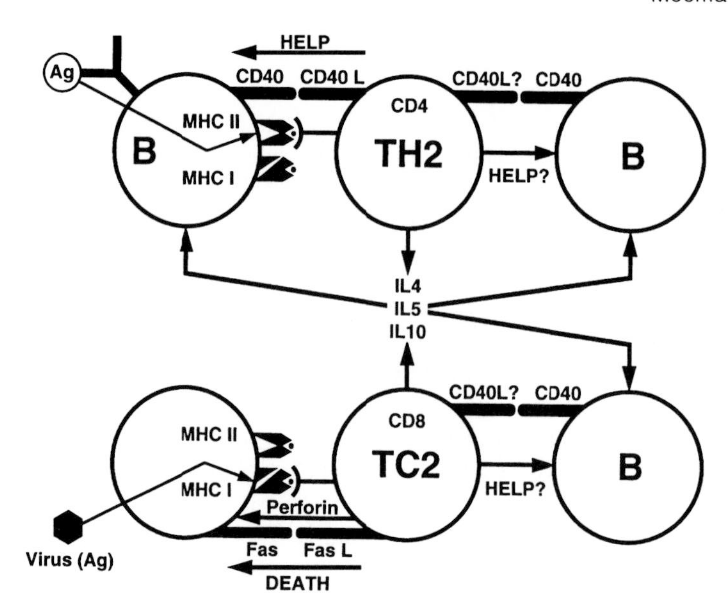

FIG. 2. Do Tc2 (CD8) cells help B cells? Ag, antigen; L, ligand; MHC, major histocompatibility complex; IL, interleukin; TC, T cytotoxic; TH, T helper.

MHC class I as an indication of infection of the target cell. Both subsets are therefore expected to kill virus-infected cells.

Because of the strong helper function of Th2 cells, we have also tested for a potential role of Tc2 cells in B cell help. Th1 cells and especially Th2 cells can provide cognate help for B cells if the T cells are specific for MHC alloantigens on the B cell surface. In contrast, both subsets of $CD8^+$ T cells failed to help alloantigenic B cells, consistent with the strong cytotoxic ability of the $CD8^+$ T cells that should pre-empt any potential helper function by killing the target B cell.

Interestingly, when the Tc2 cells were triggered with plate-bound anti-CD3 antibody, the cells provided help to B cells present in the same culture, as also found by Croft et al (1994), Erard et al (1993) and Maggi et al (1994). This help did not require antigenic recognition of the B cells. Tc2 cells may therefore be able to contribute to bystander help during an immune response, but they are unlikely to provide cognate help.

The possibility of cognate help by Tc2 cells is even less likely when the B cell antigen-processing pathways are considered. Antigen captured by antibody on the surface of an antigen-specific B cell is processed and presented on MHC class II molecules. In contrast, MHC class I molecules, which are recognized by $CD8^+$ T cells, are thought not to contain peptides of antibody-captured antigens, but are instead thought to carry antigen peptides that have been derived from endogenously synthesized proteins, normally indicating infection of the cell. Thus, it is important that recognition by $CD4^+$ and $CD8^+$ T cells should have opposite consequences: $CD4^+$ T cells should help useful B cells that are specific for foreign antigens, whereas $CD8^+$ T cells should kill infected B cells. Figure 2 illustrates this model, and suggests that Tc2 activation may enhance bystander help by secretion of cytokines and, more speculatively, by expressing activating molecules such as CD40 ligand (Armitage et al 1992).

Cytokine cross-regulation between infection and pregnancy

During pregnancy, there is a placental and systemic bias of the immune system towards a Th2 response (Wegmann et al 1993). Cell-mediated responses are often weaker, whereas antibody responses may be exaggerated. IL-10 is constitutively synthesized in substantial quantities in the placenta, which may account for part of the immunomodulatory effect. We have speculated previously (Wegmann et al 1993) that the Th2 bias may be protective for the placenta and fetus because pregnancy can be compromised by cell-mediated responses involving IFN-γ and activated NK cells. The Th2 bias may be useful, for example, in the amelioration of rheumatoid arthritis during pregnancy, but it may also contribute to the weakened resistance of pregnant women against intracellular pathogens that require a Th1 response.

We have recently examined the interactions between pregnancy and infection using a model system of *Leishmania* infection in a normally resistant mouse strain, C57BL/6. We expected that pregnancy might weaken the normal curative Th1 response, and at the same time, a strong Th1 response against the parasite might increase the risk of fetal resorptions.

When pregnancy and infection were initiated at the same time, pregnant mice showed a diminished ability to resolve the infection, as indicated both by sustained footpad swelling and by an increased parasite load in the feet. With two cycles of pregnancy, the failure to resolve infection continued for at least 70 days. Spleen and lymph node cells from normal infected mice secreted high levels of IFN-γ in response to *Leishmania* antigens *in vitro*, whereas cells from pregnant mice produced significantly lower levels. Thus, pregnancy weakened the protective response against *Leishmania* infection, and this correlated with reduced *in vitro* synthesis of IFN-γ in response to antigen.

We compared pregnancy outcomes between infected and control mice, and we found that infection was clearly deleterious to successful pregnancy. The number of resorptions, assessed as uterine scars, was significantly increased in infected mice. Interestingly, we also observed an unexpected increase in the number of mice that appeared to undergo a preimplantation block, suggesting that immune cytokine patterns not only influence placental maintenance, but also that immune mediators may influence fertilization and/or implantation. There appeared to be a short period early in infection that was particularly effective at inducing resorptions and preimplantation blocks, followed by a period with lesser effects and a later period that induced higher resorption rates. These two periods during which the infection interfered with pregnancy correlated approximately with the pattern of increased IFN-γ expression *in vivo*.

References

Armitage RJ, Fanslow WC, Strockbine L et al 1992 Molecular and biological characterization of a murine ligand for CD40. Nature 357:80–82

Belosevic M, Finbloom DS, van der Meide PH, Slayter MV, Nacy CA 1989 Administration of monoclonal anti-IFN-gamma antibodies *in vivo* abrogates natural resistance of C3H/HeN mice to infection with *Leishmania major*. J Immunol 143:266–274

Brown MA, Pierce JH, Watson CJ, Falco J, Ihle JN, Paul WE 1987 B cell stimulatory factor-1/interleukin-4 mRNA is expressed by normal and transformed mast cells. Cell 50:809–818

Cherwinski HM, Schumacher JH, Brown KD, Mosmann TR 1987 Two types of mouse helper T cell clone. III. Further differences in lymphokine synthesis between Th1 and Th2 clones revealed by RNA hybridization, functionally monospecific bioassays, and monoclonal antibodies. J Exp Med 166:1229–1244

Coffman RL, Ohara J, Bond MW, Carty J, Zlotnik A, Paul WE 1986 B cell stimulatory factor 1 enhances the IgE response of lipopolysaccharide-activated B cells. J Immunol 136:4538–4541

Croft M, Carter L, Swain SL, Dutton RW 1994 Generation of polarized antigen-specific CD8 effector populations: reciprocal action of interleukin (IL)-4 and IL-12 in promoting Type 2 versus Type 1 cytokine profiles. J Exp Med 180:1715–1728

Del Prete GF, De Carli M, Mastromauro C et al 1991 Purified protein derivative of *Mycobacterium tuberculosis* and excretory–secretory antigen(s) of *Toxocara canis* expand *in vitro* human T cells with stable and opposite (type 1 T helper or type 2 T helper) profiles of cytokine production. J Clin Invest 88:346–350

Else KJ, Finkelman FD, Maliszewski CR, Grencis RK 1994 Cytokine mediated regulation of chronic intestinal helminth infection. J Exp Med 179:347–351

Erard F, Wild M-T, Garcia-Sanz JA, Le Gros G 1993 Switch of CD8 T cells to non-cytolytic CD8-CD4$^-$cells that make T_H2 cytokines and help B cells. Science 260:1802–1805

Ferrick DA, Schrenzel MD, Mulvania T, Hsieh B, Ferlin WG, Lepper H 1995 Differential production of interferon γ and interleukin 4 in response to Th1$^-$and Th2-stimulating pathogens by $\gamma\delta$ T cells *in vivo*. Nature 373:255–257

Fiorentino DF, Bond MW, Mosmann TR 1989 Two types of mouse T helper cell. IV. TH2 clones secrete a factor that inhibits cytokine production by TH1 clones. J Exp Med 170:2081–2095

Firestein GS, Roeder WD, Laxer JA et al 1989 A new murine CD4$^+$ T cell subset with an unrestricted cytokine profile. J Immunol 143:518–525

Fong TAT, Mosmann TR 1990 Alloreactive murine CD8+ T cell clones secrete the TH1 pattern of cytokines. J Immunol 144:1744–1752

Heinzel FP, Sadick MD, Mutha SS, Locksley RM 1991 Production of interferon gamma, interleukin 2, interleukin 4, and interleukin 10 by CD4+ lymphocytes *in vivo* during healing and progressive murine leishmaniasis. Proc Natl Acad Sci USA 88:7011–7015

Hsieh C-S, Macatonia SE, Tripp CS, Wolf SF, O'Garra A, Murphy KM 1993 Development of T_H1 CD4+ T cells through IL-12 produced by *Listeria*-induced macrophages. Science 260:547–549

Janeway CA Jr, Carding S, Jones B et al 1988 CD4+ T cells: specificity and function. Immunol Rev 101:39–80

Kay AB, Ying S, Varney V et al 1991 Messenger RNA expression of the cytokine gene cluster, interleukin 3 (IL-3), IL-4, IL-5, and granulocyte/macrophage colony-stimulating factor, in allergen-induced late-phase cutaneous reactions in atopic subjects. J Exp Med 173:775–778

Kelso A, Glasebrook AL 1984 Secretion of interleukin 2, macrophage-activating factor, interferon, and colony-stimulating factor by alloreactive T lymphocyte clones. J Immunol 132:2924–2931

Le Gros G, Ben Sasson SZ, Seder R, Finkelman FD, Paul WE 1990 Generation of interleukin-4 (IL-4)-producing cells *in vivo* and *in vitro*: IL-2 and IL-4 are required for *in vitro* generation of IL-4-producing cells. J Exp Med 172:921–929

Maggi E, Giudizi M-G, Biagiotti R et al 1994 Th2-like CD8+ T cells showing B cell helper function and reduced cytolytic activity in human immunodeficiency virus type 1 infection. J Exp Med 180:489–495

Mosmann TR, Coffman RL 1989 Th1 and Th2 cells: different patterns of lymphokine secretion lead to different functional properties. Annu Rev Immunol 7:145–173

Mosmann TR, Cherwinski H, Bond MW, Giedlin MA, Coffman RL 1986 Two types of murine helper T cell clone. I. Definition according to profiles of lymphokine activities and secreted proteins. J Immunol 136:2348–2357

Sad S, Mosmann TR 1994 A single IL-2-secreting precursor CD4 T cell can develop into either Th1 or Th2 cytokine secretion phenotype. J Immunol 153:3514–3522

Sad S, Marcotte R, Mosmann TR 1995 Cytokine-induced differentiation of precursor mouse CD8+ T cells into cytotoxic CD8+ cells secreting Th1 or Th2 cytokines. Immunity 2:271–279

Sanderson CJ, O'Garra A, Warren DJ, Klaus GG 1986 Eosinophil differentiation factor also has B-cell growth factor activity: proposed name interleukin 4. Proc Natl Acad Sci USA 83:437–440

Seder RA, Boulay J-L, Finkelman F et al 1992 CD8+ T cells can be primed *in vitro* to produce IL-4. J Immunol 148:1652–1656

Sher A, Coffman RL 1992 Regulation of immunity to parasites by T cells and T cell-derived cytokines. Annu Rev Immunol 10:385–409

Street NE, Schumacher JH, Fong TAT et al 1990 Heterogeneity of mouse helper T cells: evidence from bulk cultures and limiting dilution cloning for precursors of Th1 and Th2 cells. J Immunol 144:1629–1639

Swain SL, Huston G, Tonkonogy S, Weinberg A 1991 Transforming growth factor-β and IL-4 cause helper T cell precursors to develop into distinct effector helper cells that differ in lymphokine secretion pattern and cell surface phenotype. J Immunol 147:2991–3000

Taylor-Robinson AW, Phillips RS, Severn A, Moncada S, Liew FY 1993 The role of Th1 and Th2 cells in a rodent malaria infection. Science 260:1931–1934

Tsicopoulos A, Hamid Q, Varney V et al 1992 Preferential messenger RNA expression of Th1-type cells (IFN-gamma+, IL-2+) in classical delayed-type (tuberculin) hypersensitivity reactions in human skin. J Immunol 148:2058–2061

von der Weid T, Langhorne J 1993 The roles of cytokines produced in the immune response to the erythrocytic stages of mouse malarias. Immunobiology 189:397–418

Wegmann TG, Lin H, Guilbert LJ, Mosmann TR 1993 Bidirectional cytokine interactions in the maternal–fetal relationship: is successful pregnancy a Th2 phenomenon? Immunol Today 14:353–356

DISCUSSION

Fitch: Do the $CD8^+$ cells that produce interleukin 4 (IL-4) express CD40 ligand on activation?

Mosmann: We're trying to do that experiment at the moment but we don't have any results yet.

Abbas: Is the unique function of the IL-4-producing $CD8^+$ cells just a peculiarity of culturing them in high concentrations of IL-4?

Mosmann: It may be a peculiarity but they probably have a particular function, which is that they kill virus-infected cells and, at the same time, provide cytokines that do not interfere with an ongoing antibody response but indeed help it. In contrast, the Tc1 (T cytotoxic 1) phenotype makes γ-interferon (IFN-γ) and other cytokines while it's killing virus-infected targets that might inhibit a local antibody response. If one thinks of immune responses as having many different modalities, this is a particularly interesting modality.

Ramshaw: I would like to raise a point about cytotoxicity that may be somewhat controversial. You have talked about cytotoxicity and cytokines, and we've postulated elsewhere that the lytic function of $CD8^+$ cells is just an *in vitro* artefact and that it serves no purpose *in vivo* under those circumstances (Ramsay et al 1993). We have also suggested that the role of $CD8^+$ cells is not to kill target cells but to deliver cytokines that are important in controlling viral or bacterial infections (Ramsay et al 1993). We have numerous results which suggest that lysis is not important in the recovery from most viral infections (Ramshaw et al 1992). Also in the perforin knockout mice there's only one example of it being important in clearing viruses and that is infection with lymphocytic choriomeningitis virus (LCMV). It has no effect on other viruses that have been studied. LCMV does not cause lysis of the infected cell, so one may ask what is the function of the cytotoxic lymphocytes? It is possible that they require a lytic function to get out into the tissues where these virus infections occur. But the lytic function is not important for clearing viruses.

Flavell: Which viruses does it not affect?

Ramshaw: Vaccinia virus.

Liew: What about influenza?

Ramshaw: Influenza has not been tested.

Romagnani: I would like to mention some results that we have obtained using $CD8^+$ clones generated from humans (Maggi et al 1994). We found that T helper 2 (Th2)-like clones generated from HIV-infected patients have a reduced cytolytic activity in comparison with Th1-like clones. They also facilitate the production of immunoglobulins, including IgE. The system that we used was to activate with insoluble anti-CD3 antibody. We also found that they express CD40 ligand, and that this expression was up-regulated by IL-4. These cells may also exert a helper activity *in vivo*, in as much as these clones were generated from HIV-infected patients in advanced stages of infection. These patients had high levels of IgE but were virtually devoid of circulating $CD4^+$ T cells.

Mosmann: I am a little confused over the specificity of that experiment. One can imagine that B cell help is normally focused around the antigen. However, it is difficult to imagine how the antigen recognized by the B cell antibody can also be recognized by the $CD8^+$ cell because antibody capture does not normally result in antigen presentation in major histocompatibility complex (MHC) class I.

Dutton: McGee's group looked at cytokine secretion *in vitro*, and they showed that $CD8^+$ cells in the mucosal sites make IL-5 (Taguchi et al 1990).

Mitchison: This does not disagree with Tim Mosmann's results. $CD8^+$ cells evidently make cytokines, but whether they make a major contribution is another matter. After all, they are kept busy by other tasks, for example making perforin and granzymes.

Swain: $CD8^+$ cells are class I restricted, so they have the potential to recognize other cells, including other T cells. Consequently, they may have important and unique regulatory functions.

Dutton: But Tim Mosmann is implying that this is not the case, rather that they kill the B cells that they recognize directly.

Mosmann: It is possible that $CD8^+$ cells are not actually cytotoxic *in vivo*, which generates a new interpretation of the cell-mediated functions of the T cells. I'm not totally convinced of that principle yet. However, if we accept that they are cytotoxic, death of the target cell will pre-empt other things that the $CD8^+$ cells might do to the cell being recognized. This does not apply to the bystanders.

Fitch: Are you working with tumour cell targets?

Mosmann: Yes, but we've looked at normal targets as well.

Fitch: What is your classification of normal?

Mosmann: Lipopolysaccharide (LPS)-expanded blasts or concanavalin A-expanded blasts.

Fitch: I rest my case. Investigators interested in cytolytic activity generally choose sensitive target cells in order to have as big a window as possible. Just because a cell can be cytolytic doesn't necessarily mean that it is cytolytic for all target cells. There are multiple mechanisms whereby a cell can resist cytolysis. A small resting B cell doesn't necessarily have the same sensitivity for the various cytolytic mechanisms as does an LPS blast. Also, the argument that $CD8^+$ cells recognize MHC class I molecules is a partial truth: some $CD8^+$ cells may be able to recognize class II molecules in a class II-restricted way.

Certainly, some CD4$^+$ cells are able to recognize class I molecules (McKisic et al 1991).

Mitchison: There are numerous reports on T cell cytokines in the rheumatoid synovium, which is a fairly representative site of chronic inflammation. The first one mentions T cell cytokines detected only in CD4$^+$ T cells by *in situ* hybridization (Simon et al 1994), but later reports use immunostaining and detect these cytokines also in CD8$^+$ T cells (Ulfgren et al 1995, Steiner et al 1995).

Mosmann: Graham Le Gros and collaborators (Coyle et al 1995) have also described Th2-like CD8$^+$ cells in a virus infection but I haven't seen the results yet.

Fitch: There is evidence that helper-independent CD8$^+$ cells are responsible for rejection of class I disparate murine skin grafts (Rosenberg et al 1986, Rosenberg & Singer 1992).

Abbas: But if the system is forced, then that may be all that is left. In a reaction where all the components are present and intact, what is the principal source(s) of the cytokines? It seems that CD4$^+$ cells are the principal source of nearly all the T cell cytokines.

Fitch: I am not convinced that there is a generic infectious disease model that one can apply generally to all situations. There is a variety of microbial organisms, so the immune responses must be equally as varied in order for us to survive.

Swain: We need to start looking at *Leishmania* and other kinds of diseases where there is a polarization towards a Th2 response. If one looks at the kinds of inflammatory reactions that generate Th1 cells, one cannot expect CD8$^+$ cells to be the major cytokine producers.

Ramshaw: Certainly, in virus infections the major producer of cytokines will not necessarily be antigen-specific cells because non-specific recruitment of cells occurs. In a virus-induced infiltrate one finds that only a small proportion of cells can recognize antigens of the virus. It doesn't mean that they're not important, it means that they recognize the virus and recruit other cells that do the work. They may not have a direct role in terms of either killing cells or releasing their own cytokines, but they recruit other cells to do this.

Sher: In intracellular infections the IFN-γ produced by CD8$^+$ cells is the major mechanism of resistance, particularly to infections in which class I is the only restricting element.

McMichael: I know of both chronic and acute examples where there is a high frequency of CD8$^+$ viral-specific T cells. For instance, in the cerebrospinal fluid of patients with meningitis caused by mumps, 1–10% of T cells are cytotoxic T lymphocytes specific for the mumps virus (Kreth et al 1982). Similarly, some HIV-infected patients have high levels of specific cytotoxic T lymphocytes in the peripheral blood, perhaps as high as 1% (Moss et al 1995).

Ramshaw: The influenza system has been well studied. It is possible that a recruitment of specific cells and replication at the site will occur in chronic infections (Hurwitz et al 1983).

McMichael: It depends on whether samples are taken at the appropriate site. In the case of mumps/meningitis the cerebrospinal fluid is the appropriate site. On the other hand, most adults have Epstein–Barr virus-specific cytotoxic T lymphocytes in the blood at a frequency of about 1:20 000 lymphocytes. It is not known whether there are higher frequencies in the oropharynx.

Kaufmann: Kägi et al (1994) have used perforin-deficient mutant mice to demonstrate convincingly that cytotoxic T lymphocytes are important for protection against secondary infection with the intracellular bacterium, *Listeria monocytogenes*.

Romagnani: We have recently obtained some results which suggest that, at least in the human system *in vitro*, progesterone can promote Th2 responses (Piccinni et al 1995). Is it possible that the effect that you observed *in vivo* may be related to high levels of progesterone during pregnancy?

Mosmann: We haven't examined differentiation cultures, but in short-term murine activation cultures progesterone inhibits IFN-γ synthesis selectively. We also observe that the Th2-like cytokines are less inhibited by progesterone. This therefore compares with the human system. Also, if the cells are treated with the progesterone antagonist RU486, for example, some of these effects are reversed.

Abbas: Do IL-4 or IL-10 knockout mice have a higher rate of spontaneous abortions?

Mosmann: There are relatively high levels of IL-10 in wild-type mouse and human placentas. The IL-10 knockout mouse is not particularly infertile and the homozygote can become pregnant, so IL-10 is not an absolute requirement.

Coffman: We have been able to breed IL-10 homozygous knockout mice quite successfully on the 129SvEv strain background. It is possible that it depends on the strain background of the mouse.

Mosmann: Your comment is important because one might predict that infections trigger the requirement for the local protective response. Therefore, in the absence of both a strong immune response and IL-10 there may not be a problem.

References

Coyle AJ, Erard F, Bertrand C, Walti S, Pircher H, Le Gros G 1995 Virus-specific CD8$^+$ cells can switch to interleukin 5 production and induce airway eosinophilia. J Exp Med 181:1229–1233

Hurwitz JL, Korngold R, Doherty PC 1983 Specific and nonspecific T cell recruitment in viral meningitis: possible implications for autoimmunity. Cell Immunol 76:397–401

Kägi D, Ledermann B, Bürki K, Hengartner H, Zinkernagel RM 1994 CD8$^+$ T cell-mediated protection against an intracellular bacterium by perforin-dependent cytotoxicity. Eur J Immunol 24:3068–3072

Kreth HW, Kress L, Kress HG, Ott HF, Eckert G 1982 Demonstration of primary cytotoxic T cells in venous blood and cerebrospinal fluid of children with mumps meningitis. J Immunol 128:2411–2415

Maggi E, Giudizi M-G, Biagiotti R et al 1994 Th2-like $CD8^+$ cells showing B cell helper function and reduced cytolytic activity in human immunodeficiency virus type 1 infection. J Exp Med 180:489–495

McKisic MD, Sant AJ, Fitch FW 1991 Some cloned murine $CD4^+$ T cells recognize H-2Ld Class I MHC determinants directly; other cloned $CD4^+$ T cells recognize H-2Ld Class I MHC determinants in the context of Class II MHC molecules. J Immunol 147:2868–2874

Moss PAH, Rowland-Jones SL, Frodsham PM et al 1995 Persistent high frequency of human immunodeficiency virus-specific cytotoxic T cells in peripheral blood of infected donors. Proc Natl Acad Sci USA 92:5773–5777

Piccinni M-P, Giudizi M-G, Biagiotti R et al 1995 Progesterone favors the development of human T helper cells producing Th2-type cytokines and promotes both IL-4 production and membrane CD30 expression in established Th1 cell clones. J Immunol 155:128–133

Ramsay AJ, Ruby J, Ramshaw IA 1993 A case for cytokines as effector molecules in the resolution of virus infection. Immunol Today 14:155–157

Ramshaw I, Ruby J, Ramsay A, Ada G, Karupiah G 1992 Expression of cytokines by recombinant vaccinia virus: a model for studying cytokines in virus infections *in vivo*. Immunol Rev 127:157–182

Rosenberg AS, Singer A 1992 Cellular basis of skin graft rejection: an *in vivo* model of immune mediated tissue destruction. Annu Rev Immunol 10:333–358

Rosenberg AS, Mizuochi T, Singer A 1986 Analysis of T cell subsets in rejection of K^b mutant skin allografts differing at class I MHC. Nature 322:829–831

Simon AK, Seipelt E, Sieper JS 1994 Diverse cytokine patterns in inflammatory arthritis. Proc Natl Acad Sci USA 91:8562–8566

Steiner G, Akrad N, Kunaver M et al 1995 Production of cytokines by T cells in RA synovial membranes. Clin Rheumatol 14:236

Taguchi T, McGhee JR, Coffman RL et al 1990 Analysis of Th1 and Th2 cells in murine gut-associated tissues. Frequencies of $CD4^+$ and $CD8^+$ T cells that secrete IFN-γ and IL-5. J Immunol 145:68–77

Ulfgren A, Lindblad S, Rönnelid J, Klareskog L, Andersson U 1995 Application of an immunohistochemical method for cytokine detection to the study of *in vivo* and *in vitro* production of cytokines in arthroscopically obtained synovial tissue from RA patients. Clin Rheumatol 14:234

Role for CD30 antigen in human T helper 2-type responses

Sergio Romagnani

Division of Clinical Immunology and Allergy, Institute of Clinical Medicine 3, University of Florence, Viale Morgagni 85, I-50134 Florence, Italy

Abstract. Human T helper 1 (Th1) cells develop preferentially during infections by intracellular parasites and trigger phagocyte-mediated host defence. In contrast, human Th2 cells are responsible for phagocyte-independent host response, and they predominate during helminthic infestations and in atopic humans in response to common environmental antigens. Polarized human Th1 and Th2 cell responses play different roles in protection, and they can promote different immunopathological reactions. Strong and persistent Th1 responses seem to be involved in organ-specific autoimmunity, contact dermatitis and some chronic non-allergic inflammatory disorders. Polarized Th2 responses favour reduced protection against the majority of infections, including HIV, and they are responsible for triggering allergic disorders in genetically predisposed hosts. Th1 and Th2 cells probably exhibit distinct surface markers; for example, Th2 cells express preferentially membrane CD30 and release the soluble form of CD30, which is a member of the tumour necrosis factor receptor superfamily. CD30-mediated signalling promotes the *in vitro* development of Th2-like cells. The expression of CD30 in HIV-infected T cells results in enhanced HIV replication, suggesting the existence of complex links among CD30 expression, production of Th2-type cytokines and immunopathogenesis of HIV infection.

1995 T cell subsets in infectious and autoimmune diseases. Wiley, Chichester (Ciba Foundation Symposium 195) p 55–67

In recent years, two distinct $CD4^+$ T helper (Th)-cell subsets, Thl and Th2, showing distinct and mutually exclusive patterns of cytokine secretion have been identified in both mice and humans (Mosmann et al 1986, Del Prete et al 1991). Thl cells produce interleukin 2 (IL-2), γ-interferon (IFN-γ) and tumour necrosis factor β (TNF-β), whereas Th2 cells produce IL-4 and IL-5. Different cytokine patterns imply distinct effector functions. Phagocyte-dependent defence is mediated by Thl cells, which trigger both cell-mediated immunity and the production of opsonizing and complement-fixing antibodies. In contrast, Th2 cells are involved in phagocyte-independent defence mechanisms, which include IgE and IgG1 (or IgG4 in humans) antibody production, and differentiation and activation of mast cells and eosinophils. They also inhibit some functions of macrophages. Activation of Th2 cell clones occurs in IgE-mediated and eosinophil-mediated reactions against helminths, at least in atopic people, and in response to common environmental allergens (reviewed in Romagnani 1994a). In addition to clones that have either Thl or Th2 polarized phenotypes, $CD4^+$ Th cells have been identified (Th0 cells) that show a composite profile including production of both Thl-type and Th2-type cytokines. These cells mediate effects that depend on the ratio of cytokines produced (Romagnani 1994a,b). However, membrane markers allowing phenotypic characterization of these functionally distinct T cell subsets have not yet been identified.

CD30 is a member of the TNF/nerve growth factor receptor superfamily (Smith et al 1990), which includes other functionally relevant molecules, such as CD27, CD40, Fas and 4-lBB (Beutler & van Huffel 1994). CD30 exists as a 90 kDa or 120 kDa glycoprotein, or a molecularly distinguishable 57 kDa intracellular form. The extracellular portion of CD30 is cleaved proteolytically, producing a 88 kDa soluble form of the molecule that is released by CD30-expressing cells *in vitro* and *in vivo* (Smith et al 1993, Nadali et al 1994). CD30 was originally described as a surface molecule recognized by the Ki-1 monoclonal antibody (mAb) on Hodgkin's and Reed–Sternberg (H-RS) cells in patients with Hodgkin's disease, and was subsequently found in neoplastic cells of certain types of non-Hodgkin's lymphomas (Schwab et al 1982, Stein et al 1985). Under normal conditions, there are no $CD30^+$ cells in the peripheral blood, but they are present in limited numbers as large mononuclear cells with evident nucleoli mainly around the B cell follicles of lymphoid tissues and, to a minor degree, on the edge of germinal centres (Schwab et al 1982, Stein et al 1985). *In vitro* expression of CD30 is inducible on T and B cells transformed by viruses, such as the human T cell lymphotropic viruses I and II, and the Epstein–Barr virus (Stein et al 1985); and on a small proportion of anti-CD3-stimulated $CD45RO^+$ (memory) T cells (Ellis et al 1993). Although these findings suggest clearly that CD30 expression is a feature of activated lymphoid cells, the nature of the $CD30^+$ T cell subset and its possible function have remained unclear for more than a decade. Recently, it was suggested that among human peripheral blood mononuclear cells (PBMC) activated for three days with anti-CD3 antibody plus IL-2, CD30 defines a subset of human T cells that produces IFN-γ and IL-5, and exhibits enhanced B cell helper activity (Alzona et al 1994). These suggestions imply a possible Th0-like phenotype. Results from my laboratory (Manetti et al 1994, Del Prete et al 1995a), as well as additional results reported here, provide compelling evidence that expression of CD30 (and release of soluble CD30) is preferentially associated both *in vitro* and *in vivo* with T cell responses characterized by the production of Th2-type cytokines.

In vitro CD30 is preferentially expressed by T cell clones producing Th2-type cytokines

CD30 expression was first analysed on a large panel of $CD4^+$ T cell clones with established cytokine secretion profiles (20 Thl clones, 20 Th0 clones and 20 Th2 clones). T cell blasts from Thl clones expressed virtually no membrane CD30, whereas CD30 was strongly expressed by most T cell blasts from Th2 clones. Noticeable proportions of T cell blasts from the majority of Th0 clones also expressed membrane CD30 (Del Prete et al 1995a). CD30 expression was maximal in the few days following activation of Th2 and Th0 clones, and was decreased when clones were grown in medium containing IL-2 alone (Del Prete et al 1995a). Activation of Th2 and Th0 but not of Th1 clones also resulted in the release of detectable amounts of soluble CD30 in clonal supernatants. The kinetics of CD30 expression by PBMC, derived freshly from the same donor, was assessed following *in vitro* activation with three separate antigens: purified protein derivative (PPD); *Toxocara canis* excretory/secretory antigen (TES); and *Lolium perenne* group I. These are known for their ability to induce prevalent Thl (PPD) or Th2 (TES and *L. perenne* group I) cytokine profiles (Del Prete et al 1991). Production of high concentrations of IFN-γ by PPD-specific T cells was detectable on Day 5, but neither expression of CD30 nor production of IL-4 was observed at any time after activation. In contrast, both expression of CD30 and production of IL-4 and IL-5 were detectable after Day 10 in both TES-stimulated and *L. perenne* group I-stimulated cultures. This suggests a temporal relationship between the expression of CD30 antigen and the start of Th2-type cytokine production (Del Prete et al 1995a).

The preferential expression of CD30 by T cells producing Th2-type cytokines can also be observed in $CD8^+$ T cell clones. Indeed, the majority of $CD8^+$ T cell clones, which usually exhibit a Thl-like cytokine profile, did not express membrane CD30 (Manetti et al 1994). However, a few $CD8^+$ clones, which were generated from healthy donors and produced IL-4 and IL-5 in addtion to IFN-γ, showed consistent expression of membrane CD30 and they released substantial amounts of soluble CD30 in their supernatant. This was more consistently observed in $CD8^+$ clones exhibiting a Th2-like profile of cytokine secretion generated from the peripheral blood or the skin of patients with AIDS (Manetti et al 1994). Taken together, these results support strongly the concept that, at least *in vitro*, CD30 is preferentially expressed on, and its soluble form is released by, $CD4^+$ and $CD8^+$ T cells prone to the production of Th2 cytokines.

CD30-mediated signalling promotes the development of Th2-like $CD4^+$ T cells *in vitro*

To investigate whether CD30 is only associated with the Th2 phenotype or whether it also has a role in Th2-type responses, we first screened $CD4^+$

$CD30^+$ T cell clones for their responsiveness to co-stimulation with specific antigen and insolubilized anti-CD30 mAb (M44). Antigen-driven proliferation was significantly increased by insolubilized anti-CD30 mAb in both Th2-like and Th0-like T cell clones, but it was not affected in Thl-like T cell clones. Likewise, production of both IL-4 and IL-5 by Th2-like clones, as well as production of IL-4, IL-5 and IFN-γ by Th0-like clones, was significantly increased in the presence of the anti-CD30 mAb in comparison to stimulation with antigen alone (Del Prete et al 1995b). This suggests that CD30 co-stimulation promotes selective growth and cytokine production by established Th0-like and Th2-like T cell clones.

We also investigated the possibility that agonistic signals delivered through CD30 also induce the preferential development of T cells producing Th2-type cytokines. We generated tetanus toxoid-specific or Rye grass pollen Group I (Rye I)-specific T cell lines from PBMC bulk cultures in the absence or presence of insolubilized anti-CD30 mAb. Incubation of PBMC with insoluble anti-CD30 mAb at the time of antigen stimulation resulted in the expansion of antigen-specific T cell lines with higher IL-4/IFN-γ ratios compared to antigen-specific T cell lines generated in the absence of insoluble anti-CD30 mAb. Accordingly, higher proportions of tetanus toxoid-specific T cell clones producing IL-4 and IL-5, and lower proportions of tetanus toxoid-specific T cell clones producing IFN-γ were generated from anti-CD30 mAb-conditioned tetanus toxoid-specific T cell lines and tetanus toxoid-specific T cell lines generated in the absence of insoluble anti-CD30 mAb, respectively. When tetanus toxoid-specific T-cell clones, generated under the two experimental conditions, were categorized as being Th0-like, Thl-like or Th2-like, a significant increase in the proportion of Th2-like and a significant decrease in the proportion of Thl-like clones was observed (Del Prete et al 1995b).

To exclude the possibility that this effect simply reflected the selective expansion of Th2-like cells induced by CD30 co-stimulation, we blocked the interaction between CD30 and its natural ligand (CD30L) at various times. We assessed the effect of this blockade on the *in vitro* development of antigen-specific T cell clones. CD30L has recently been found to be constitutively expressed by a proportion of B lymphocytes, as well as activated T cell clones (Younes et al 1994, Maggi et al 1995). We raised Rye I-specific T cell lines in the absence or presence of a blocking anti-CD30 mAb (Ber-H2), an anti-CD30L mAb (M81) or a soluble CD30–Fc fusion protein. These were added in bulk culture at the same time as Rye I (Day 0), Day 3 or Day 6 . Early blockade of the CD30L/CD30 interaction (Day 0 or Day 3) promoted the outgrowth of Rye I-specific T cell lines that produced lower amounts of IL-4 and IL-5, and higher amounts of IFN-γ than their corresponding control T cell lines. The proportions of Th1-like and Th2-like T cell clones were significantly increased and reduced, respectively, in the series of clones generated from bulk cultures in which the CD30L/CD30 interaction was blocked early. In contrast, when the CD30L/CD30 interaction was blocked at Day 6 after antigenic stimulation, Rye-I-specific T cell lines and clones showed a cytokine production comparable to that exhibited by their control lines and clones, respectively (Del Prete et al 1995b). Taken together, these results suggest that, at least *in vitro*, an early interaction with CD30L may push $CD30^+$ memory T cells stimulated with the specific antigen to develop preferentially into T cell clones producing Th2-type cytokines.

CD30 is expressed and/or soluble CD30 is released in the course of Th2-type responses *in vivo*

The association between the expression of CD30 and the production of Th2-type cytokines *in vitro* raises the question of whether CD30 represents a marker for the detection of Th2-type responses *in vivo*. In normal lymphoid tissues, CD30 is only detectable on a small population of large mononuclear cells that have an evident nucleolus. These are mainly grouped around B cell follicles, and to a minor degree at the edge of germinal centres (Stein et al 1985). There are no extra-lymphohaemopoietic CD30-expressing cells in humans, with the exception of exocrine pancreatic cells and decidual cells (Ito et al 1994).

It has been suggested that Th2-type responses against common environmental allergens play a critical triggering role in the pathogenesis of atopic disorders (reviewed in Romagnani 1994b). Therefore, we investigated the possible presence of $CD30^+$ circulating T cells in atopic patients. Virtually no $CD4^+$ $CD30^+$ cells were detected in the blood of grass-sensitive atopic donors examined before the grass pollination season. However, four out of six grass-sensitive donors examined during the pollination season, when they were suffering from allergic symptoms, showed small numbers of $CD4^+$ $CD30^+$ cells in their circulation (from 0.08 to 0.3%). We fractionated the circulating $CD4^+$ T cells from these patients into $CD30^+$ and $CD30^-$ cells by sorting with anti-CD30 mAb, and we expanded both $CD30^+$ and $CD30^-$ fractions in IL-2. We then assessed the $CD30^+$ and $CD30^-$ populations for their ability to produce cytokines and to proliferate in response to *L. perenne* group I. We found that only $CD30^+$ cells proliferated in response to *L. perenne* group I and produced IL-4 and IL-5, whereas $CD30^-$ cells produced IFN-γ and TNF-β (Del Prete et al 1995a). These findings demonstrate that $CD4^+$ $CD30^+$ Th2-like cells can circulate in the peripheral blood of sensitive patients during *in vivo* exposure to grass pollen allergens.

The clearest demonstration that $CD30^+$ cells are associated with Th2 responses *in vivo* is provided by the study of Omenn's syndrome. This is a rare congenital immunodeficiency syndrome in which abnormal Th2-like cells may play a pathogenic role (Schandené et al 1993). We investigated the presence of $CD30^+$ cells in tissue samples immunohistochemically and we measured the serum levels of soluble C30 in three children with Omenn's syndrome and one child with maternal engraftment and Omenn's-like syndrome. Large proportions of tissue infiltrating T lymphocytes from all four patients expressed CD30, whereas in control tissues $CD30^+$ T cells were extremely few or absent. In addition, high levels of soluble CD30 were found in all patients' sera. We generated T cell clones from sorted $CD30^+$ and $CD30^-$ circulating T cells of the patient with Omenn's-like syndrome, who showed unusually elevated numbers of circulating $CD30^+$ T lymphocytes (greater than 25%), and we found that most $CD4^+$ T cell clones derived from $CD30^+$ cells showed a Th2-like cytokine profile, whereas the majority of clones generated from $CD30^-$ T cells were Thl (Chilosi et al 1995). These findings support the concept that Th2-type cells are involved in the pathogenesis of Omenn's syndrome and Omenn's-like syndrome. Moreover, they provide evidence that detection of $CD30^+$ T cells in tissues and/or increased levels of soluble CD30 in biological fluids reflect the presence of immune responses characterized by prevalent activation and/or expansion of T cells producing Th2-type cytokines.

CD30 triggering by CD30 $ligand^+$ cells promotes HIV replication

It has been suggested that during HIV infection there is a bias towards Th2-like responses, and hence Thl inhibition, which may contribute to the loss of control of the immune system over HIV infection and the resultant progression to AIDS (Clerici & Shearer 1993). We (Maggi et al 1994a) and others (Graziosi et al 1994) have been unable to support the concept of a general massive alteration to a Th2 pattern in HIV-infected individuals. However, large numbers of $CD8^+$ T cell clones showing a clear-cut Th2-like cytokine profile were generated from both peripheral blood and skin biopsies of subjects in advanced phases of HIV infection (Maggi et al 1994b). Moreover, we found that the HIV replicates preferentially in Th2 and Th0 rather than Thl T cell clones infected *in vitro* (Maggi et al 1994a). Finally, elevated levels of soluble CD30 have recently been found in the serum of HIV-infected individuals (Pizzolo et al 1994) suggesting a high turnover of CD30 in the course of HIV infection. On the basis of these findings, we have recently asked whether the expression of CD30 in HIV-infected $CD4^+$ T cells might play a role in HIV replication.

Cross-linking CD30 with the insoluble anti-CD30 mAb potentiated the HIV replication induced by stimulation with insoluble anti-CD3 mAb in T cell lines generated from HIV-infected individuals. More importantly, both B lymphocytes and paraformaldehyde-fixed $CD8^+$ T cell clones expressing CD30L had a similar or even greater potentiating effect on HIV replication, which was completely inhibited by addition of either the anti-CD30L mAb or soluble CD30–Fc fusion protein. Finally, the anti-CD30L mAb also exerted a suppressive effect on the spontaneous HIV replication in lymph node cells derived freshly from an HIV-seropositive individual (Maggi et al 1995). Taken together, these data suggest that the triggering of CD30 expression by CD30L-expressing cells may play an important role in the activation of HIV expression from latently infected CD4+ T cells particularly in lymphoid organs where cell to cell contact can easily occur.

The mechanism of CD30-mediated HIV replication is still unclear. However, more recent experiments suggest that nuclear translocation and activation of the transcription activator protein NFκB rank among the early cellular responses elicited following CD30 ligation (McDonald et al 1995). Thus, inasmuch as the HIV1 long terminal repeat sequence contains consensus NFκB binding sites (Nabel & Baltimore 1987), it is not unreasonable to suggest that the triggering of CD30 expression leads to activation of HIV replication via the activation of NFκB.

Acknowledgements

The experiments reported in this paper were supported by grants provided by Ministero della Sanità (Progetto AIDS), AIRC and CNR. I thank Immunex for providing the anti-CD30 (M44) and anti-CD30L (M81) monoclonal antibodies.

References

Alzona M, Jäck H-M, Fisher RI, Ellis TM 1994 CD30 defines a subset of activated human T cells that produce IFN-γ and IL-5 and exhibit enhanced B cell helper activity. J Immunol 153:2861–2867

Beutler B, van Huffel C 1994 Unraveling function in the TNF ligand and receptor families. Science 264:667–668

Chilosi M, Facchetti F, Notarangelo L et al 1995 CD30 cell expression and abnormal soluble CD30 serum accumulates in Omenn's syndrome. Evidence for a Th2-mediated condition. J Clin Invest, submitted

Clerici M, Shearer GM 1993 A T_H1–T_H2 switch is a critical step in the etiology of HIV infection. Immunol Today 14:107–110

Del Prete GF, De Carli M, Mastromauro C et al 1991 Purified protein derivative of *Mycobacterium tuberculosis* and excretory–secretory antigen(s) of *Toxocara canis* expand *in vitro* human T cells with stable and opposite (type 1 T helper or type 2 T helper) profile of cytokine production. J Clin Invest 88:346–350

Del Prete GF, De Carli M, Almerigogna F et al 1995a Preferential expression of CD30 by human CD4+ T cells producing Th2-type cytokines. FASEB J 9:81–86

Del Prete GF, De Carli M, D'Elios MM et al 1995b CD30-mediated signalling promotes the development of human Th2-like cells. J Exp Med, in press

Ellis TM, Simms PE, Slivnick DJ, Jack H-M, Fisher RI 1993 CD30 is a signal-transducing molecule that defines a subset of human activated CD45RO+ T cells. J Immunol 151:2380–2389

Ito K, Watanabe T, Horie R, Shiota M, Kawamura S, Mori S 1994 High expression of the CD30 molecule in human decidual cells. Am J Pathol 145:276–280

Graziosi C, Pantaleo G, Gant KR et al 1994 Lack of evidence for the dichotomy of Th1 and Th2 predominance in HIV-infected individuals. Science 265:248–252

MacDonald P, Cassetella M, Bald A et al 1995 CD30 ligation induces nuclear factor-κB activation in human T cell lines. Eur J Immunol, in press

Maggi E, Mazzetti M, Ravina A et al 1994a Ability of HIV-1 to promote a Thl to Th0 shift and to replicate preferentially in Th2 and Th0 cells. Science 265:244–247

Maggi E, Giudizi MG, Biagiotti R et al 1994b Th2-like $CD8^+$ T cells showing B cell helper function and reduced cytolytic activity in human immunodeficiency virus type 1 infection. J Exp Med 180:489–495
Maggi E, Annunziato F, Manetti R et al 1995 Activation of HIV expression by CD30 triggering in $CD4^+$ T cells from HIV-infected individuals. Immunity 3:251
Manetti R, Annunziato F, Biagiotti R et al 1994 CD30 expression by $CD8^+$ T cells producing type 2 helper cytokines. Evidence for large numbers of $CD8^+CD30^+$ T cell clones in human immunodeficiency virus infection. J Exp Med 180:2407–2411
Mosmann TR, Coffman RL 1989 Thl and Th2 cells: different patterns of lymphokine secretion lead to different functional properties. Annu Rev Immunol 7:145–173
Mosmann TR, Cherwinski H, Bond MW, Giedlin MA, Coffman RL 1986 Two types of murine helper T cell clone. I. Definition according to profiles of lymphokine activities and secreted proteins. J Immunol 136:2348–2357
Nabel G, Baltimore D 1987 An inducible transcription factor activates expression of human immunodeficiency virus in T cells. Nature 326:711–713
Nadali G, Vinante F, Ambrosetti A et al 1994 Serum levels of soluble CD30 are elevated in the majority of untreated patients with Hodgkin's disease and correlate with clinical features and prognosis. J Clin Oncol 12:793–797
Pizzolo G, Vinante F, Morosato L et al 1994 High serum level of soluble form of CD30 molecule in the early phase of HIV-1 infection as an independent predictor of progression to AIDS. AIDS 8:741–745
Romagnani S 1994a Lymphokine production by human T cells in disease states. Annu Rev Immunol 12:227–257
Romagnani S 1994b Regulation of Th2 development in allergy. Curr Opin Immunol 6:838–846
Schandené L, Ferster A, Mascart-Lemone F et al 1993 T helper type 2-like cells and therapeutic effects of interferon-γ in combined immunodeficiency with hyper-eosinophilia (Omenn's syndrome). Eur J Immunol 23:56–60
Schwab U, Stein H, Gerdes J et al 1982 Production of a monoclonal antibody specific for Hodgkin's and Sternberg–Reed cells of Hodgkin's disease and a subset of normal lymphoid cells. Nature 299:65–67
Smith CA, Davis T, Anderson D et al 1990 A receptor for tumor necrosis factor defines an unusual family of cellular and viral proteins. Science 248:1019–1023
Smith CA, Gruss H-J, Davis T et al 1993 CD30 antigen, a marker for Hodgkin's lymphoma, is a receptor whose ligand defines an emerging family of cytokines with homology to TNF. Cell 73:1349–1360
Stein H, Mason DY, Gerdes J et al 1985 The expression of Hodgkin's disease associated antigen Ki-1 in reactive and neoplastic lymphoid tissue: evidence that Reed–Sternberg cells and histiocytic malignancies are derived from activated lymphoid cells. Blood 66:848–858
Younes A, Jaing S, Gruss H, Andreeff M 1994 Expression of the CD30 ligand (CD30L) on human peripheral blood lymphocytes from normal donors and patients with lymphoid malignancies. Blood 84:A161(abstr)

DISCUSSION

Lotze: Can you explain the results of Tom Ellis, who has suggested that γ-interferon (IFN-γ)-producing cells in the circulation are $CD30^+$ T cells (Alzona et al 1995)?

Romagnani: Tom Ellis used a completely different system. He stimulated peripheral blood mononuclear cells (PBMC) with phytohaemagglutinin (PHA) or anti-CD3 antibody, then after 48 h he sorted out the $CD30^+$ cells, and he found that they produced IFN-γ and interleukin 5 (IL-5), and that they exhibited enhanced B cell helper activity (Alzona et al 1994). It is possible that he is looking at T helper 0 (Th0) cells, which are able to produce IFN-γ. These cells are probably also able to produce IL-4, although it's too early to see detectable amounts of IL-4 at this stage because the production of IL-4 occurs later in PBMC bulk cultures. Therefore, Tom Ellis' results may be consistent with, and not contradictory to, our results.

Lotze: When you sort out the $CD30^+$ cells from the atopic individuals who have allergies, can you stimulate them with other cytokines, such as IL-12?

Romagnani: We are doing these sorts of experiments at the moment. So far, we have only looked at the effect of cytokines on established clones and not on this $CD30^+$ population. We have found that IL-12 acts as a co-stimulator and it increases the proliferation of these Th2 clones. This may result in an increase in the expression of CD30. It is possible that IL-12 is a good co-stimulator for the proliferation of every type of T cell.

Liew: Is there a correlation between the expression of CD30 and the progression to AIDS?

Romagnani: We have not observed $CD30^+$ cells in the circulation or in the lymph nodes of HIV-infected patients (Maggi et al 1994). However, when we stimulated $CD4^+$ T cell clones or PBMC from HIV-infected individuals with an agonistic anti-CD30 antibody or with cells expressing the CD30 ligand (CD30L), we observed a strong increase in viral replication.

Lotze: Were these immobilized B cells?

Romagnani: No, they were paraformaldehyde-fixed B cells or $CD8^+$ T cell clones.

Lotze: This is similar to putting B7-1 or B7-2 into a cell that isn't allowed to signal, because these cells are capable of making cytokines themselves. Would you observe the same effect if you didn't fix your B cells?

Romagnani: In our experience, there are no other signals that are able to give such a strong increase in P24 antigen production in this system. Tumour necrosis factor α (TNF-α), which is a powerful inducer of viral replication, did not have such a strong effect. However, we are intrigued by the observation that there are apparently no $CD30^+$ T cells in the circulation or in the lymph nodes of these patients. This may be because $CD30^+$ T cells die rapidly after activation.

Liew: What happens to the level of soluble CD30 in the serum?

Romagnani: There is an increase in the level of soluble CD30 in the serum (Pizzolo et al 1994), which means that there is a high turnover of CD30, even if $CD30^+$ cells are not detected. We believe that activation with anti-CD30 antibody is required to induce the expression of CD30, and then either

CD30L-expressing cells or anti-CD30 antibody triggers viral replication. Therefore, *in vivo* it is possible that $CD4^+$ cells are stimulated by the specific antigen to express CD30 on their membrane. If they are infected by the virus and interact with CD30L-expressing cells, they die very quickly, so that it is impossible to detect substantial numbers of these cells in the circulation or in the lymph nodes.

Trinchieri: Do you have evidence that the $CD8^+$ cells are indeed T cells?

Romagnani: We have shown that they are α^+ β^+ and $CD3^+$.

Ramshaw: What's the purpose of CD30L being expressed on $CD8^+$ cells?

Romagnani: I don't know why the CD30L is present on $CD8^+$ cells.

Ramshaw: The reason I asked that question is that the CD30L is found on T cells and, although it's involved with B cells, it also has a potent antiviral activity (Ruby et al 1995). Consequently, I wondered whether the CD30L could also have antiviral activity.

Cosman: No one has looked at whether CD30L has antiviral activity.

Abbas: Is CD30L a co-stimulator of T cells?

Ramshaw: No. It may have direct antiviral activity that does not involve TNF or IFN-γ.

Mosmann: The anti-CD30L staining of the $CD8^+$ population seemed to be different from the $CD4^+$ population. Were the $CD8^+$ cells that expressed CD30L still α^+ β^+ T cells?

Romagnani: Yes.

Mosmann: Then they are a peculiar subpopulation of cells.

Trinchieri: Sergio Romagnani's results indicate that the $CD30^+$ cells express CD8 antigens at a low density. This is reminiscent of the low level expression of CD8 on natural killer (NK) cells, where only the CD8 α chain but not the β chain is expressed.

Dutton: Is soluble CD30 cleaved proteolytically?

Romagnani: It is not yet completely clear, but it is probably a cleavage product.

Mitchison: Could you expand on the possibility of using either an antagonistic anti-CD30 antibody or an anti-CD30L antibody for manipulations in humans?

Romagnani: We need to find out the physiological importance of these results in more general terms. In HIV, for example, it is important to devise strategies to delay the viral replication without producing any side effects.

Mitchison: Have you looked at the expression of CD30L on human B cells?

Lotze: Or on other antigen-presenting cells (APCs), such as activated macrophages and dendritic cells?

Romagnani: CD30L is expressed constitutively on B cells from the tonsils or peripheral blood of normal humans, and even on B cells from patients with chronic lymphocytic leukaemia. It is likely that there are differences between CD30L expressed on B cells and T cells because CD30L expressed on B cells is apparently more stable.

The experiments suggesting that CD30 is expressed on activated B cells have not been confirmed. They were obtained using only one anti-CD30 antibody, and have not been repeated with other anti-CD30 antibodies from different sources. However, CD30 is expressed by B cells that are infected with Epstein–Barr virus.

No one knows if CD30L is expressed by dendritic cells. There is one report suggesting that the mRNA encoding CD30L is present in lipopolysaccharide-activated macrophages (Smith et al 1993). We found that a few macrophages expressed CD30L on their membranes, but we are not sure whether this is an artefact or not. In addition, we found that T cell clones activated with either CD4 or CD8 expressed mRNA encoding CD30L but that membrane CD30L was expressed only on $CD8^+$ T cell clones. CD30 is also expressed preferentially by T cells producing Th2 cytokines. We have also found that CD30 is expressed by exocrine pancreatic cells and decidual cells (Ito et al 1994), and we have found cells that express both the progesterone receptor and CD30 on their membranes, suggesting a possible link between the two.

Mitchison: Could CD30L be used in the treatment of asthma?

Romagnani: Possibly. In my opinion, soluble CD30 may be an important regulatory molecule because we have preliminary evidence that both anti-CD30L antibody and soluble CD30L are able to inhibit some phenomena induced by the CD30L/CD30 interaction. It is possible that soluble CD30 is produced just to down-regulate the effect of this interaction.

Navikas: Are $CD30^+$ cells present in patients with systemic lupus erythematosus (SLE)?

Romagnani: $CD30^+$ cells are not present in the circulation of patients with SLE. It is possible that they are limited to a particular target organ. Frederico Caligaris Cappio and colleagues have found a substantial proportion of $CD30^+$ cells in the peritoneal exudate from one of these patients, so it is possible that they are present in target organs where the disease is active. More importantly, however, they showed that high levels of soluble CD30 were present in the serum of SLE patients. These levels were related to the activity of the disease. It seems that when the disease is in a silent phase, there is no release of soluble CD30, but the levels increase with the increasing activity of the disease (Caligaris Cappio et al 1995).

Navikas: I am interested in the relationship between soluble levels of CD30 and Th2 cell activation in the disease. Do you have any knowledge of the soluble levels of CD30 in Th1 cell-mediated diseases?

Romagnani: I am still perplexed over the meaning of soluble CD30. I am confidant that the levels increase when a Th2-type response is operating, but there are no definite conclusions with regard to the possible elevation of soluble CD30 levels during a Th1-type response. Soluble CD30 may be detected in the serum of some patients with Th1-type responses, particularly when these patients have a chronic and long-lasting pathological condition.

Mason: There is evidence that if antigen is targeted specifically to B cells, Th2 responses are produced (Finkelman et al 1987). We have demonstrated this by targeting a peptide that causes experimental autoimmune encephalomyelitis to B cells with anti-IgD antibodies (Day et al 1992, Saoudi et al 1995). The perfect experiment would be to see whether anti-CD30 antibody can block that effect. This antibody doesn't exist but the experiment is waiting to be done as soon as the antibody is generated.

Locksley: Do the small population of $CD4^+$ cells in the periphery that react to pollen and express CD30 represent the conventional $\alpha\beta$ T cell receptor $CD3^+$ lineage or are they $NK1.1^+$ $CD4^+$ cells that express Vα24 with a limited β chain repertoire? It may be important to delineate the lineage of these cells.

Romagnani: We have not addressed this.

Mitchison: In at least one disease, namely reactive arthritis, there is a clear discrepancy between what's observed in the affected area and in the T cell clones (contrast Simon et al 1994 with Simon et al 1993). Do you have an opinion on the generality of this discrepancy?

Romagnani: Schlaak et al (1992) showed that clones generated from the synovial fluid of patients with reactive arthritis induced by *Yersinia enterocolitica* have a clear-cut Th1-type profile. However, the Simon et al (1994) paper disagreed with this.

Mitchison: Is it possible that there's a bias in the cloning procedure in favour of Th1 cells that secrete cytokines such as IL-2?

Romagnani: No. Th1 or Th2 clones are generated in standard conditions, i.e. stimulating with a specific antigen and using IL-2 as a growth factor. However, endogenously produced IL-12 cannot be removed, which may favour Th1 responses. Indeed, if IL-4 or anti-IL-12 antibody are added, different results are observed. In our experience, however, there are some antigens that favour a Th1-type or a Th2-type response under standard conditions. This may also reflect the situation *in vivo*.

Fitch: All cloning involves bias, it's just that only some of the bias is known.

References

Alzona M, Jäck H-M, Fisher RI, Ellis TM 1994 CD30 defines a subset of activated human T cells that produce IFN-γ and IL-5 and exhibit enhanced B cell helper activity. J Immunol 153:2861–2867

Alzona M, Jäck H-M, Fisher RI, Ellis TM 1995 IL-12 activates IFN-γ production through the preferential activation of $CD30^+$ T cells. J Immunol 154:9–16

Caligaris Cappio F, Bertero T, Converso M et al 1995 Circulating levels of CD30, a marker of cells producing Th2-type cytokines, are increased in patients with Systemic Lupus Erythematosus and correlate with disease activity. Clin Exp Rheumatol 13:339–343

Day MJ, Tse AGD, Puklavec M, Simmonds SJ, Mason DW 1992 Targeting autoantigen to B cells prevents the induction of a cell-mediated autoimmune disease in rats. J Exp Med 175:655–659

Finkelman FD, Snapper CM, Mountz JD, Katona IM 1987 Polyclonal activation of the murine immune system by a goat antibody to mouse IgD. 9. Induction of a polyclonal IgE response. J Immunol 138:2826–2830

Ito K, Watanabe T, Horie R, Shiota M, Kawamura S, Mori S 1994 High expression of the CD30 molecule in human decidual cells. Am J Pathol 145:276–280

Maggi E, Mazzetti M, Ravina A et al 1994 Ability of HIV to promote a Th1 to a Th0 shift and to replicate preferentially in Th2 and Th0 cells. Science 265:244–248

Pizzolo G, Vinante F, Morosato L et al 1994 High serum level of soluble form of CD30 molecule in the early phase of HIV-1 infection as an independent predictor of progression to AIDS. AIDS 8:741–745

Ruby J, Bluethmann H, Aguet M, Ramshaw IA 1995 CD40 ligand has potent antiviral activity. Nature Med 1:437–441

Saoudi A, Simmonds S, Huitinga I, Mason D 1995 Prevention of experimental allergic encephalomyelitis in rats by targeting autoantigen to cells: evidence that the protective mechanism depends on changes in the cytokine response and migratory properties of the autoantigen-specific T cells. J Exp Med 182:335–344

Schlaak J, Hermann E, Ringhoffer M et al 1992 Predominance of Th1-type T cells in synovial fluid of patients with *Yersinia*-induced reactive arthritis. Eur J Immunol 22:2771–2776

Simon AK, Seipelt E, Sieper JS 1994 Diverse cytokine patterns in inflammatory arthritis. Proc Natl Acad Sci USA 91:8562–8566

Simon AK, Seipelt E, Wu P, Wenzel B, Brann J, Sieper J 1993 Analysis of cytokine profiles in synovial T cell clones from chlamydial reactive arthritis patients—predominance of Th1 subset. Clin Exp Immunol 94:122–126

Simon AK, Seipelt E, Sieper J 1994 Divergent T-cell cytokine patterns in inflammatory arthritis. Proc Natl Acad Sci USA 91:8562–8566

Smith CA, Gruss H-J, Davis T et al 1993 CD30 antigen, a marker for Hodgkin's lymphoma, is a receptor whose ligand defines an emerging family of cytokines with homology to TNF. Cell 73:1349–1360

Regulation of T lymphocyte subsets

F. W. Fitch, R. Stack, P. Fields, D. W. Lancki and D. C. Cronin

Committee on Immunology, Department of Pathology and the Ben May Institute, University of Chicago, 5841 South Maryland Ave, Chicago, IL 60637, USA

Abstract. Patterns of cytokine secretion and functional differences distinguish T lymphocyte subsets. T lymphocyte subsets are also regulated differentially. Most established $CD8^+$ lymphocyte clones secrete γ-interferon (IFN-γ) but not interleukin 2 (IL-2) or IL-4. Using murine T cells which express a transgenic, antigen-specific α/β T cell receptor (TCR) specific for L^d class I major histocompatibility complex antigen, we have found that $CD8^+$ lymphocytes can be divided into functional subsets. Freshly isolated $CD8^+$ T cells are not cytolytic, do not proliferate and do not secrete cytokines. Stimulation of TCR alone does not induce cytokine secretion, but cells become responsive to exogenous IL-2 or IL-4. Stimulation of CD28 together with TCR induces secretion of IL-2 and IFN-γ, and cells proliferate without exogenous cytokines. Proliferation is necessary for the development of cytolytic activity. If IL-4 is present during initial stimulation, IL-4 is secreted following restimulation. Upon stimulation, some IL-4-producing murine $CD8^+$ T cell clones express CD40 ligand (CD40L), and they potentiate proliferation and immunoglobulin secretion by small resting B cells. Thus, the $CD8^+$ T cell subsets T cytotoxic 1 (Tc1) and Tc2 are analogous to $CD4^+$ T helper 1 (Th1) and Th2. IL-2 production by naive $CD8^+$ cells requires co-stimulation. IL-4 production by $CD8^+$ T cells requires the presence of IL-4 during initial stimulation. Some IL-4-producing $CD8^+$ T cells express CD40L following TCR stimulation and provide help for B cells.

1995 T cell subsets in infectious and autoimmune diseases. Wiley, Chichester (Ciba Foundation Symposium 195) p 68–85

T lymphocytes carry out their central role in immune responses through direct intercellular contact and through production of secreted cytokines. In recent years, multiple subsets of T lymphocytes have been distinguished by the expression of cell surface molecules and by the array of secreted cytokines. The qualitative nature of a specific immune response is determined by the relative numbers of T lymphocytes belonging to various subsets. Most peripheral T lymphocytes can be divided into two mutually exclusive sets, based on expression of CD4 or CD8. The expression of these molecules correlates generally (but imperfectly) not only with which class of major histocompatibility complex (MHC) molecules restricts the T cell response, but also with functional differences. $CD4^+$ T lymphocytes, or T helper (Th) cells, are restricted by class II MHC molecules and provide efficient help for B lymphocytes, whereas $CD8^+$

T lymphocytes, or cytotoxic T lymphocytes (CTLs), are restricted by class I MHC molecules and lyse target cells. Subsets of both $CD4^+$ and $CD8^+$ cells have been defined by the arrays of secreted cytokines.

Two polar extremes of $CD4^+$ T cells have been described (Mosmann & Coffman 1989). Th1 cells produce interleukin 2 (IL-2), lymphotoxin and γ-interferon (IFN-γ), but not IL-4, IL-5, IL-6 or IL-10. They mediate delayed-type hypersensitivity reactions and provide help to CTLs through the paracrine effects of secreted IL-2. Th2 cells produce IL-4, IL-5, IL-6 and IL-10, but not IL-2 or IFN-γ. They provide efficient help for antibody production and mediate allergic reactions through their effects on B cell antibody production and Ig isotype switching. Thus, the dominance of particular $CD4^+$ subsets can profoundly influence the characteristics of immune responses *in vivo*. These subsets appear to develop from a common precursor cell that secretes only IL-2 (Torbett et al 1990). $CD4^+$ T cell clones which secrete other arrays of cytokines are not infrequent (Gajewski et al 1989, Street et al 1990, Firestein et al 1989). These cells may not have differentiated fully or they may constitute additional subsets.

$CD8^+$ T lymphocytes were originally considered to have homogeneous characteristics because they are cytolytic and they secrete a limited array of cytokines, including IFN-γ, but not T cell growth factors. However, it is now clear that $CD8^+$ T lymphocyte subsets exist, although these subsets have not yet been fully characterized. Freshly isolated $CD8^+$ clones produce multiple cytokines (Fong & Mosmann 1990, Kelso et al 1991). Helper-independent $CD8^+$ T cells produce IL-2, and they can be maintained in short-term culture without added growth factors (Widmer & Bach 1981, Klarnet et al 1989, Otten & Germain 1991). These $CD8^+$ T cell clones usually cease secreting IL-2 with repeated passage *in vitro*. IL-4-producing $CD8^+$ cells have also been described (Seder et al 1992, Bloom et al 1992, Seder & Le Gros 1995). $CD8^+$ cells activated in the presence of IL-4 may lose cytotoxic functions, stop producing IFN-γ and develop the capacity to produce IL-4, IL-5 and IL-10 (Le Gros & Erard 1994). Some human $CD8^+$ MHC class II-restricted T cell clones have suppressive activity (Bloom et al 1992). A detailed characterization of the differentiation of $CD8^+$ subsets and their regulatory and effector functions has been hampered because of the difficulty in deriving stable long-term $CD8^+$ T cell clones that secrete different arrays of cytokines. Evidence to date suggests that at least some of the mechanisms which regulate $CD4^+$ T cell subsets differentially also appear to operate in $CD8^+$ T cell subsets. These regulatory processes shape the characteristics of immune responses following immunization and they are involved in maintaining particular patterns of established immune responses.

Differential regulation of $CD4^+$ T cell subsets

Many activities of murine Th0, Th1 and Th2 subsets are regulated differentially. These differences have been discussed in detail elsewhere and are reviewed only briefly here. Differentiation of $CD4^+$ T cell subsets is influenced profoundly by the cytokine environment in which naive T cells develop. Cytokines secreted by one subset may inhibit the development of other subsets. IFN-γ inhibits proliferation of Th2 cells, both in response to stimulation of the T cell receptor (TCR) and in response to exogenous IL-2 or IL-4 (Fitch et al 1993). However, secretion of cytokines, including IL-4, is not affected. IFN-γ also inhibits many of the agonist effects of cytokines secreted by Th2 cells. IL-2 is the dominant growth factor for T lymphocytes. Even Th2 lymphocytes, which do not produce IL-2 and which utilize IL-4 as an autocrine growth factor, proliferate in response to stimulation with IL-2. IL-2 also plays a critical role in priming naive $CD4^+$ T cells to become producers of IL-4 or IFN-γ (Seder et al 1994). IL-4 is required for the development of IL-4-producing $CD4^+$ T cells (Seder & Paul 1994). Also, IL-4 inhibits strongly the effects of IL-12 on naive murine T cells (Schmitt et al 1994). IL-10, secreted by Th2 cells but not by Th1 cells, inhibits macrophage-induced cytokine production by Th1 cells but not by Th2 cells (Mosmann 1994). IL-12 plays a major positive role in the development of Th1 cells by enhancing the level of IFN-γ produced (Trinchieri 1993). IL-1 is required, in addition to IL-2 or IL-4, for proliferation of some Th2 clones, and some Th2 clones secrete IL-1 (R. Stack & F. Fitch, unpublished observations, Greenbaum et al 1988). Thus, IL-2, IL-4, IL-10, IL-12, IFN-γ and possibly IL-1 are the dominant cytokines that regulate the effector functions of T cells elicited during an immune response.

In addition to being an autocrine growth factor for Th1 cells, IL-2 temporarily impairs the antigen-responsiveness of these cells, as well as IL-2-producing $CD8^+$ CTLs. This effect is not observed in Th2 cells or in $CD8^+$ cells that do not secrete IL-2 (Fitch et al 1993). However, IL-2-treated Th1 cells remain responsive to exogenous IL-2 and to the combination of phorbol ester and Ca^{2+} ionophore. Thus, although the interaction of IL-2 with its receptor leads to proliferation of all T cell subsets, this interaction has additional consequences in cells that secrete IL-2. The basis for this difference is not known, but communication between the TCR and the IL-2 receptor is clearly different in T cells that secrete IL-2 and in T cells that do not.

There are several other differences in the activation requirements of Th1 and Th2 T cells. Adherent spleen cells stimulate optimal proliferation of Th1 cells, whereas purified splenic B cells stimulate optimal proliferation of Th2 clones (Fitch et al 1993). Production of cytokines by each subset occurs in response to either stimulating cell, indicating that at least partial activation signals are induced in each lymphocyte subset by either stimulus. This differential effect does not correlate with the MHC restriction element, with susceptibility to inhibition by anti-CD4 monoclonal antibody (mAb) or anti-CD11 (lymphocyte function-associated antigen 1) mAb, or with differences in antigen processing by macrophages and B cells (Fitch et al 1993). Stimulation of Th1 clones with concanavalin A or anti-TCR mAb leads to an elevated intracellular Ca^{2+} concentration ($[Ca^{2+}]_i$) and to the generation of inositol phosphates. However, these second messengers are not detected following stimulation of Th2 clones, even though Th2 clones can be stimulated to secrete cytokines by treatment with active phorbol esters and Ca^{2+} ionophores, agents which mimic the effects of these second messengers (Fitch et al 1993). Also, treatment with Ca^{2+} ionophore alone induces IL-4 production in Th2 clones, whereas such treatment induces anergy in Th1 clones (Fitch et al 1993). Cytokine production by Th1 clones is substantially more sensitive to the inhibitory effects of cholera toxin, cyclosporin A, and 8-Br-cAMP than is cytokine production by Th2 clones (Fitch et al 1993). High concentrations of antigen (or anti-TCR mAb) inhibit proliferation but not cytokine production by Th1 and CTL clones. This failure to proliferate is due to an inability to respond to IL-2 because increasing levels of cytokines, including IL-2, are secreted at high antigen concentration (Fitch et al 1993).

Stimulation by co-stimulatory molecules in addition to TCR engagement is required for Th1 cells to secrete IL-2 and proliferate. In the absence of this co-stimulatory signalling, a state of unresponsiveness, termed anergy, is induced (Mueller et al 1989). Anergy has also been induced in a $CD8^+$ CTL clone that produces IL-2 by using conditions that generate anergy in Th1 cells (Otten & Germain 1991). Th2 cells do not develop unresponsiveness analogous to anergy. The essential metabolic events responsible for the inability to produce IL-2 have not been defined. The majority of Th1 and Th2 T cells lyse target cells with an efficiency similar to that of conventional $CD8^+$ CTLs (Fitch et al 1993). However, not all T cells use the same lytic mechanisms. Most $CD8^+$ and Th2 T cells lyse sheep erythrocytes coated with anti-CD3 mAb while Th1 clones do not. This lytic ability correlates with expression of perforin and CTLA-4 mRNA (Fitch et al 1993). Th1 cells can lyse target cells through the Fas-mediated pathway, and this mechanism is also utilized by $CD8^+$ CTLs and Th2 cells (Henkart 1994, D. Lancki, unpublished observations).

Conditions for activation of $CD8^+$ T cell subsets

In this section, emphasis will be placed on observations that we have made using $CD8^+$ T cells that express an antigen-specific, transgenic, α/β TCR which reacts with H-2L^d alloantigen (Sha et al 1988). T cells that respond in this model express a single TCR and are present at a high frequency. This can be utilized to define more precisely the relationships among TCR stimulation, cytokine production, proliferation and the acquisition of cytolytic capability. Following depletion of $CD4^+$ cells by administration of anti-CD4 mAb *in vivo*, all $CD8^+$ cells express the transgenic TCR. These cells constitute 16–20% of the spleen cells (results not shown). We have used tumour cells as a homogeneous source of stimulating alloantigen, and we have used immobilized anti-CD3ε mAb to stimulate the TCR in the absence of other

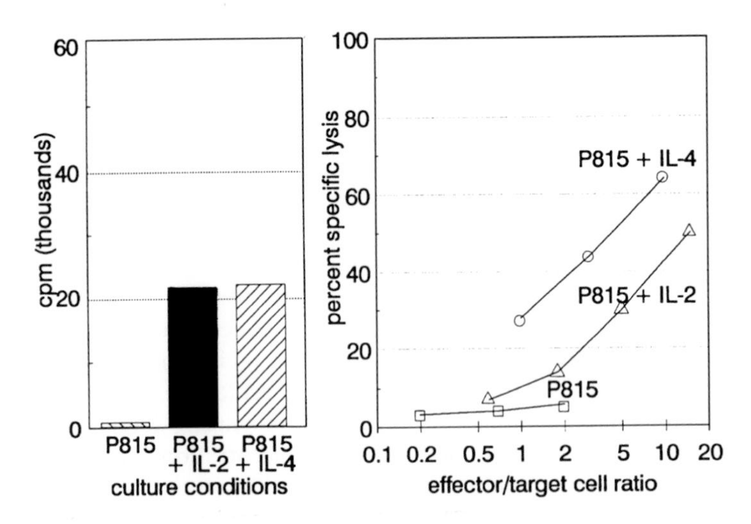

FIG. 1. Proliferative and cytolytic responses of naive $CD8^+$ T cell receptor (TCR) transgenic T cells to stimulation with allogeneic P815 mouse mastocytoma cells, alone or in the presence of exogenous interleukin 2 (IL-2) or IL-4. Naive $CD8^+$ TCR transgenic T cells stimulated for three days with mitomycin C-treated P815 cells, which express the class I major histocompatibility complex H-$2L^d$ alloantigen (with which the TCR reacts), did not incorporate [^{3}H]thymidine or express cytolytic activity in the absence of exogenous cytokines. However, naive $CD8^+$ TCR transgenic T cells stimulated with mitomycin C-treated P815 cells for three days in the presence of either 15 U/ml of human recombinant IL-2 or 400 U/ml of murine recombinant IL-4 showed appreciable incorporation of [^{3}H]thymidine and developed moderate cytolytic activity (methods used to determine the values are described in Gajewski et al 1989). Neither IL-2 nor IL-4 alone stimulated proliferation of naive $CD8^+$ TCR transgenic T cells (results not shown).

interactions. Other cell populations were used in some studies, but heterogeneous populations of stimulating cells make it difficult to characterize the molecules involved in cellular interactions.

Freshly isolated, naive $CD8^+$ TCR transgenic T cells do not display cytolytic activity, do not contain serine esterases implicated in CTL-mediated cytolysis, do not proliferate and do not secrete cytokines (results not shown). Similar findings have been obtained using another model system (Croft et al 1994). $CD8^+$ TCR transgenic T cells stimulated through the TCR by H2-L^d-positive P815 mouse mastocytoma cells in the absence of co-stimulation do not proliferate, secrete cytokines or acquire cytolytic activity. However, these cells proliferate and become cytolytic in the presence of exogenous IL-2 or IL-4 (Fig. 1). Under these conditions, these T cells do not secrete cytokines (Table 1); however, upon restimulation they secrete IFN-γ but not IL-2 (results not

TABLE 1 Stimulation of cytokine secretion under various conditions

Stimulating conditions	*IL-2*	*IL-4*	*IFN-γ*
Culture medium alone	0	0	0
P815 cells	0	0	0
P815 cells transfected with B7-1	7	0	150
P815 cells plus anti-CD28 monoclonal antibodies	1	0	250
Anti-CD3ε mAb	0	0	0
T cell depleted DBA/2 splenocytes	2–10	0–1	130–250
Small splenic B cells	5	0	160
Splenic dendritic cells	53	0	160
Cultured epidermal cells (Langerhans cells)	3	0	150

Naive $CD8^+$ T cell receptor transgenic T cells were cultured for 24 h with the indicated stimuli. Culture supernatants were collected and assayed for the presence of interleukin-2, interleukin-4 and γ-interferon.

shown). Similar results were observed with $CD8^+$ TCR transgenic T cells stimulated by anti-CD3ε mAb (results not shown). Thus, stimulation of $CD8^+$ TCR transgenic T cells with alloantigen or anti-CD3ε mAb alone does not induce proliferation or cytolytic activity. However, such stimulation does induce responsiveness to IL-2 and to IL-4.

Proliferation seems to be required for the acquisition of cytolytic activity by naive $CD8^+$ T cells. $CD8^+$ TCR transgenic T cells stimulated with alloantigen and IL-2 in the presence of pharmacological agents that inhibit progression through the cell cycle did not incorporate [^{3}H]thymidine or develop cytolytic activity (Cronin et al 1994). The failure of naive $CD8^+$ TCR transgenic T cells to generate cytolytic activity did not reflect a general failure of T cell activation because these cells produced IFN-γ (Cronin et al 1994). Although the ability of a conventional $CD8^+$ CTL clone to incorporate [^{3}H]thymidine was impaired by these agents, the cytolytic activity of established $CD8^+$ CTLs was not affected by any of the cell cycle inhibitors (Cronin et al 1994). Progression through the G1/S phase of the cell cycle was inhibited by either aphidicolin, hydroxyurea or mimosine, and progression through M phase was inhibited by either colcemid or nocodazole. Similar results were observed with these multiple agents acting at different phases of the cell cycle, so it is unlikely that the observed effects were uniquely attributable to a particular agent.

If co-stimulation through CD28 is provided in addition to stimulation of the TCR, either by B7-1 expressed on the stimulating cell or by anti-CD28 mAb bound to the stimulating cell through Fc receptors, $CD8^+$ TCR transgenic T cells are activated to secrete IL-2 and IFN-γ (Table 1). These cells also

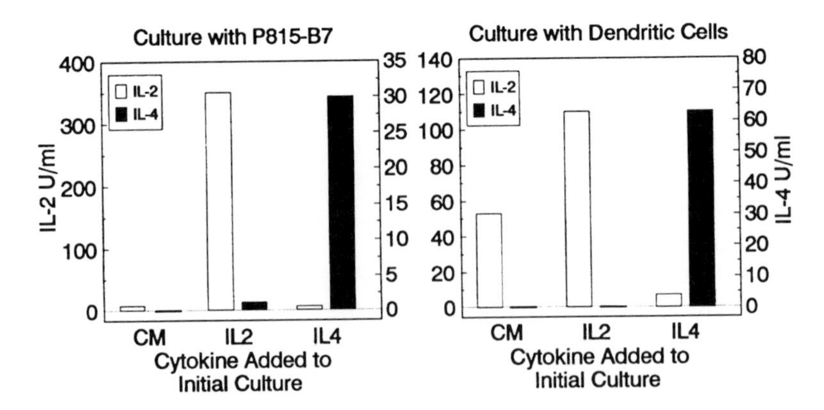

FIG. 2. Production of interleukin 2 (IL-2) and IL-4 in response to immobilized anti-CD3ε monoclonal antibody (mAb) seven days after initial stimulation with splenic dendritic cells or P815 mouse mastocytoma cells transfected with B7-1 (P815-B7) in the presence of either IL-2 or IL-4. Naive $CD8^+$ T cell receptor transgenic T cells were stimulated initially with either splenic dendritic cells or P815 cells transfected with B7-1 in the presence of either 15 U/ml human recombinant IL-2 or 1000 U/ml murine recombinant IL-4. After culture for seven days, cells were collected, washed thoroughly and stimulated with immobilized anti-CD3ε mAb. Culture supernatants were collected 24 h later and assayed for the presence of IL-2 and IL-4. CM, culture medium.

proliferate and become cytolytic under these conditions (results not shown). Different stimulating populations (allogeneic spleen cells, splenic B cells, dendritic cells and cultured Langerhans cells) vary in their capacity to stimulate proliferation and IL-2 production by $CD8^+$ TCR transgenic T cells (Table 1). Blocking studies suggest that there may be other co-stimulatory molecules in addition to B7-1, B7-2/CD28 and CTLA-4.

We have explored the influence of IL-2 and IL-4 present during the initial culture on the response to various stimulating cell populations. After seven days, cells were stimulated with immobilized anti-CD3ε mAb, and supernatants were assayed for secreted IL-2 and IL-4. Representative results obtained with dendritic cells and B7-1-transfected P815 cells are shown in Figure 2. Two conclusions can be drawn: (1) the initial presence of IL-2 enhances the ability of $CD8^+$ T cells to secrete IL-2 on restimulation; and (2) the initial presence of IL-4 both enhances the ability of cells to secrete IL-4 on restimulation and inhibits their ability to secrete IL-2. Similar effects have been observed by Croft et al (1994).

$CD4^+$ T cells that secrete IL-4 provide efficient help for antibody production by B cells. Consequently, we studied IL-4-producing $CD8^+$ T cell clones generated by stimulating $CD8^+$ T cell clones from the mouse strain B6.CH-2^{bm-1} with C57BL/6 mouse splenocytes, for their helper activity (Cronin et al

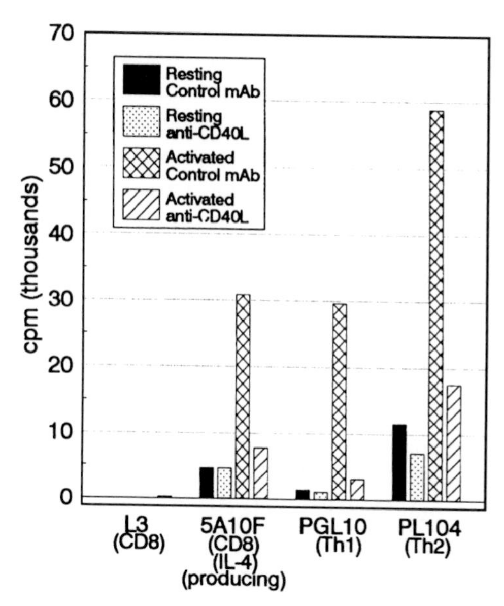

FIG. 3. Incorporation of [^{3}H]thymidine by small resting B cells to interleukin 4 (IL-4) induced by both activated IL-4-producing CD8$^+$ T cell clones and activated T helper 1 (Th1) and Th2 clones is inhibited by anti-CD40 ligand (anti-CD40L) monoclonal antibody (mAb). IL-4-producing CD8$^+$ T cytotoxic 2 (Tc2) cell clone (5A10F) cells, CD4$^+$ Th1 clone (PGL10) cells, Th2 clone (PL104) cells and conventional CD8$^+$ cytotoxic T lymphocyte clone (L3) cells were obtained from maintenance cultures. Either resting or activated (stimulated with immobilized anti-CD3ε mAb for 12–16 h) cloned T cells were incubated with purified small resting splenic B cells. Either anti-CD40L mAb or a control mAb was added at the beginning of co-culture. Cultures were incubated for a total of 72 h and pulsed for the final 9 h with [^{3}H]thymidine. Control cultures containing small resting B cells and IL-4 only did not show significant [^{3}H]thymidine incorporation (378–1655 cpm and 188–1896 cpm, respectively). These results are representative of four separate experiments. Reprinted from Cronin et al 1995, with permission.

1995). The response in this system is limited to the wild type K^b class I MHC molecule. After depletion of CD4$^+$ T cells, CD8$^+$ T cells from B6.CH-2^{bm-1} mice were repeatedly stimulated with C57BL/6 splenocytes in the absence of exogenous cytokines to favour the growth of T cells that utilize autocrine T cell growth factors. After five successive passages, it was possible to derive T cell clones that secreted IL-2 or IL-4 following TCR stimulation. Clones that secreted IL-4 in response to stimulation with alloantigen or with anti-CD3 mAb also expressed CD40 ligand (CD40L) after stimulation (results not shown). After stimulation, these CD8$^+$ clones were also able to activate small resting B cells as measured by the augmentation of proliferation to suboptimal amounts of IL-4 (Fig. 3) and increased secretion of Ig (results not shown). This proliferation required expression of CD40L on the T cell because anti-CD40L mAb inhibited this proliferative response (Fig. 3). Neither a conventional murine $CD8^+$ CTL clone nor an IL-2-producing $CD8^+$ CTL clone expressed CD40L upon activation. Also, these clones did not potentiate proliferation of small resting B cells exposed to IL-4. Thus, it appears that a subset of $CD8^+$ T cells, characterized by activation-induced expression of CD40L, can provide help for B cells. Such cells may play a role in antibody responses to antigenic challenges that are restricted only by class I MHC antigens.

Implications of differential regulation of T lymphocyte subsets

Subsets of murine $CD4^+$ and $CD8^+$ T lymphocyte clones respond differently to various regulatory influences. The potential effects of these immunoregulatory processes on $CD4^+$ and $CD8^+$ T cells are summarized diagrammatically in Table 2. Naive $CD4^+$ and $CD8^+$ T cells respond in similar ways to several of these regulatory influences, but it is likely that differential expression of co-stimulatory and accessory molecules by various antigen-presenting cells will affect subset development significantly. Conditions during initial stimulation influence the subsets of effector cells generated and, therefore, the kinds of cytokines produced on restimulation (see above, Fitch et al 1993, Swain et al 1991, Croft et al 1994). For example, the presence of IL-2 leads to secretion of predominantly IL-2 and IFN-γ. In contrast, the development of cells secreting IL-4 and IL-5 requires IL-4, and this cytokine suppresses the development of effector cells secreting IL-2 and IFN-γ. IFN-γ enhances the development of effector cells that secrete substantially more IFN-γ, somewhat more IL-2, and considerably less IL-4 than cells stimulated without the addition of IFN-γ. IL-12 is a dominant factor in directing the development of $CD4^+$ T cells that produce high levels of IFN-γ (Hsieh et al 1993). It probably has a similar effect on $CD8^+$ cells (Croft et al 1994). IL-4 influences the effects of IL-12 on naive murine $CD4^+$ T cells. In the presence of high levels of IL-4, IL-12 promotes the development of T cells that produce minimal IFN-γ and maximal IL-4 and IL-10. However, low levels of IL-4 and high levels of IL-12 favour the development of Th0 cells that secrete medium levels of IFN-γ and IL-4 (Schmitt et al 1994). IFN-γ produced by both Th1 and many $CD8^+$ T cells acts directly on Th2 cells to inhibit proliferation. In contrast, IL-10 acts indirectly by affecting macrophages, which inhibit cytokine production and consequent IL-2-mediated proliferation of Th1 cells. Through these mechanisms, T cell subsets cross-regulate each other. Several immunological phenomena may be explained by these regulatory processes. The inhibitory effect of high concentrations of antigen on the proliferation of Th1 clones but not of Th2 clones may account for the observation that delayed hypersensitivity reactions mediated primarily by Th1 cells are favoured by low daily doses of antigen,

TABLE 2 Factors predicted to favour or inhibit the activation of CD4$^+$ (T helper 1 [Th1] cells and Th2 cells) and CD8$^+$ (T cytotoxic 1 [Tc1] cells and Tc2 cells) T cell subsets. (Modified from figure published in Fitch et al 1993.)

Influences on the development of CD4$^+$ and CD8$^+$ T cells that secrete IL-2 and IFN-γ (Th1 cells and Tc1 cells)	*Influences on the development of CD4$^+$ and CD8$^+$ T cells that secrete IL-4 (Th2 cells and Tc2 cells)*
Positive	Positive
Presence of IL-2 and IFN-γ	Presence of IL-4, IL-10 (and IL-1?)
Presence of IL-12	Absence of IL-12
Presence of activated Th1 cells or Tc1 cells	Absence of activated Th1 cells or Tc1 cells
Adherent cells as antigen-presenting cells	B cells as stimulating cells
Antigen requiring phagocytosis	Antigen uptake by pinocytosis
Intracellular pathogens	Low dose of cyclosporin A
Low antigen dose	High antigen dose
Co-stimulation through CD28	Interference with co-stimulation through CD28
Negative	Negative
Presence of IL-10 and IL-4	Presence of IFN-γ
Presence of activated Th2 cells or Tc2 cells	Presence of activated Th1 cells or Tc1 cells

whereas high antibody responses mediated by Th2 cells are favoured by high daily doses of antigen (Parish 1972). Phenomena attributed to CD8$^+$ suppressor cells probably also involve some of the immunoregulatory mechanisms described above.

The interaction between CD28 on T cells and B7-1 or B7-2 on stimulating cells co-stimulates the secretion of multiple cytokines significantly following TCR engagement (June et al 1994). CTLA-4 is structurally related to CD28, and it also binds to both B7-1 and B7-2. It is expressed only on activated T cells and it may negatively regulate immune responses (June et al 1994). An attractive model for the functions of the B7 and CD28 families is that interactions involving CD28 lead to co-stimulation, whereas interactions involving CTLA-4 lead to termination of ongoing responses (June et al 1994). TCR stimulation together with CD28/B7 interaction is sufficient for generation of class I MHC-specific, IL-2-secreting CD8$^+$ CTLs in the absence of exogenous help. In the absence of CD28/B7 interactions, TCR stimulation generates CD8$^+$ CTLs only if exogenous T cell growth factors are provided.

Two explanations may account for the observation that most $CD8^+$ CTLs do not secrete IL-2. Naive $CD8^+$ T cells may have interacted initially with stimulating cells that express class I MHC antigens but not members of the B7 family. Growth factors for such cells may be provided either by $CD4^+$ Th1 cells or by a few $CD8^+$ cells that have received co-stimulation through CD28. Alternatively, conventional $CD8^+$ CTLs may be anergic. In the absence of co-stimulation, anergy is only induced in Th1 (Mueller et al 1989), Th0 (Gajewski et al 1994) and $CD8^+$ CTLs that produce IL-2 (Otten & Germain 1991). Anergy involves the selective impairment of IL-2 production. In contrast, the secretion of IFN-γ is much less inhibited (Mueller et al 1989). However, anergic T cells remain responsive to exogenous IL-2 (Mueller et al 1989). Thus, conventional $CD8^+$ CTLs may represent anergic cells that have been maintained by exogenous growth factors. We are currently evaluating these two possibilities experimentally.

Although it is evident that multiple subsets exist among both $CD4^+$ and $CD8^+$ T cells, the events leading to their differential development have not been completely characterized. Understanding the mechanisms that regulate the various subsets of T cells differentially should: (1) provide insights into the molecular events associated with activation of those subsets; (2) facilitate modulation of the activities of these subsets; and (3) make it possible to influence the favourable outcome of disease processes.

Acknowledgements

This research was supported by USPHS Research Grants CA-44372, AI-29531, AI-35294 and by the Cancer Center Support Grant CA-14599 from the US Public Health Service. D. C. was supported by USPHS Training Grant HL-07665. The authors wish to thank Dennis Loh for providing the α/β TCR transgenic mice and the anti-clonotypic mAb which reacts with this receptor, Randolph Noelle for providing the anti-CD40L mAb, Lewis Lanier for providing P815 mouse mastocytoma cells transfected with murine B7-1 and Robert Schreiber for providing the reagents for the IFN-γ ELISA assay. The authors wish to acknowledge the technical assistance of Yukio Hamada. Also, we wish to thank Frances Mills for help in preparing this manuscript.

References

Bloom BR, Modlin RL, Salgame P 1992 Stigma variations: observations on suppressor T cells and leprosy. Annu Rev Immunol 10:453–488

Croft M, Carter L, Swain SL, Dutton RW 1994 Generation of polarized antigen specific CD8 effector populations: reciprocal action of interleukin (IL)-4 and IL-12 in promoting type 2 versus type 1 cytokine profiles. J Exp Med 180:1715–1728

Cronin DC, Lancki DW, Fitch FW 1994 Requirements for activation of $CD8^+$ murine T cells. I. Development of cytolytic activity. Immunol Res 13:215–233

Cronin DC, Stack R, Fitch FW 1995 IL-4-producing, $CD8^+$ T cell clones can provide B cell help. J Immunol 154:3118–3127

Firestein GS, Roeder WD, Laxer JA et al 1989 A new murine $CD4^+$ T cell subset with an unrestricted cytokine profile. J Immunol 143:518–525
Fitch FW, McKisic MD, Lancki DW, Gajewski TF 1993 Differential regulation of murine T lymphocyte subsets. Annu Rev Immunol 11:29–48
Fong TA, Mosmann TR 1990 Alloreactive murine $CD8^+$ T cell clones secrete the Th1 pattern of cytokines. J Immunol 144:1744–1752
Gajewski TF, Joyce J, Fitch FW 1989 Anti-proliferative effect of IFN-γ in immune regulation. III. Differential selection of Th1 and Th2 murine helper T lymphocyte clones using recombinant IL-2 and recombinant IFN-γ. J Immunol 143:15–22
Gajewski TF, Lancki DW, Stack R, Fitch FW 1994 'Anergy' of Th0 helper T lymphocytes induces downregulation of Th1 characteristics and a transition to a Th2-like phenotype. J Exp Med 179:481–491
Greenbaum LA, Horowitz JB, Woods A, Pasqualini T, Reich E-P, Bottomly K 1988 Autocrine growth of $CD4^+$ T cells. Differential effects of IL-1 on helper and inflammatory T cells. J Immunol 140:1555–1560
Henkart PA 1994 Lymphocyte-mediated cytotoxicity: two pathways and multiple effector molecules. Immunity 1:343–346
Hsieh C-S, Macatonia SE, Tripp CS, Wolf SF, O'Garra A, Murphy KM 1993 Development of T_H1 $CD4^+$ T cells through IL-12 produced by *Listeria*-induced macrophages. Science 260:547–549
June CH, Bluestone JA, Nadler LM, Thompson CB 1994 The B7 and CD28 receptor families. Immunol Today 15:321–331
Kelso A, Troutt AB, Maraskovsky E et al 1991 Heterogeneity in lymphokine profiles of $CD4^+$ and $CD8^+$ T cells and clones. Immunol Rev 123:85–114
Klarnet JP, Kern DE, Dower SK, Matis LA, Cheever MA, Greenberg PD 1989 Helper-independent $CD8^+$ cytotoxic T lymphocytes express IL-1 receptors and require IL-1 for secretion of IL-2. J Immunol 142:2187–2191
Le Gros G, Erard F 1994 Noncytotoxic, IL-4, IL-5, IL-10 producing $CD8^+$ T cells: their activation and effector function. Curr Opin Immunol 6:453–457
Mosmann TR 1994 Properties and functions of interleukin-10. Adv Immunol 56:1–26
Mosmann TR, Coffman RL 1989 Th1 and Th2 cells: different patterns of lymphokine secretion lead to different functional properties. Annu Rev Immunol 7:145–173
Mueller DL, Jenkins MK, Schwartz RH 1989 Clonal expansion versus functional clonal inactivation: a costimulatory signalling pathway determines the outcome of T cell antigen receptor occupancy. Annu Rev Immunol 7:445–480
Otten GR, Germain RN 1991 Split anergy in a $CD8^+$ T cell: receptor-dependent cytolysis in the absence of interleukin-2 production. Science 251:1228–1231
Parish CR 1972 The relationship between humoral and cell-mediated immunity. Transplant Rev 13:35–66
Schmitt E, Hoehn P, Germann T, Rüde E 1994 Differential effects of interleukin-12 on the development of naive mouse $CD4^+$ T cells. Eur J Immunol 24:343–347
Seder RA, Le Gros GG 1995 The functional role of $CD8^+$ T helper type 2 cells. J Exp Med 181:5–7
Seder RA, Paul WE 1994 Acquisition of lymphokine-producing phenotype by $CD4^+$ T cells. Annu Rev Immunol 12:635–673
Seder RA, Boulay J-L, Finkelman F et al 1992 $CD8^+$ T cells can be primed *in vitro* to produce IL-4. J Immunol 148:1652–1656
Seder RA, Germain RN, Linsley PS, Paul WE 1994 CD28-mediated costimulation of interleukin 2 (IL-2) production plays a critical role in T cell priming for IL-4 and interferon gamma production. J Exp Med 179:299–304

Sha WC, Nelson CA, Newberry RD, Kranz DM, Russell JH, Loh DY 1988 Selective expression of an antigen receptor on CD8-bearing T lymphocytes in transgenic mice. Nature 335:271–274

Street NE, Schumacher JH, Fong TAT et al 1990 Heterogeneity of mouse helper T cells: evidence from bulk cultures and limiting dilution cloning for precursors of Th1 and Th2 cells. J Immunol 144:1629–1639

Swain SL, Bradley LM, Croft M et al 1991 Helper T-cell subsets: phenotype, function and the role of lymphokines in regulating their development. Immunol Rev 123:115–144

Torbett BE, Laxer JA, Glasebrook AL 1990 Frequencies of T cells secreting IL-2 and/or IL-4 among unprimed $CD4^+$ populations. Evidence that clones secreting IL-2 and IL-4 give rise to clones which secrete only IL-4. Immunol Lett 23:227–234

Trinchieri G 1993 Interleukin-12 and its role in the generation of T_H1 cells. Immunol Today 14:335–338

Widmer MB, Bach FH 1981 Antigen driven helper cell-independent cloned cytolytic T lymphocytes. Nature 294:750–752

DISCUSSION

Dutton: We've done some similar experiments with T cell receptor (TCR) transgenic mice (Croft et al 1994). We've published some results on T cells from TCR anti-H-Y transgenic mice, and we have also looked at T cells from transgenic mice with anti-influenza HA2 TCR constructs that were made and kindly provided by Roland Lieblau and Linda Sherman. We found that it is possible to generate polarized populations of $CD8^+$ cells. We cultured $CD8^+$ cells from these transgenic mice with antigen-presenting cells (APCs) and the influenza peptide in the presence of interleukin 2 (IL-2) and IL-12, and we observed cells that produce γ-interferon (IFN-γ) but not IL-5. In contrast, if we cultured the same cells with APCs and the influenza peptide in the presence of both IL-2 and IL-4, we observed cells that produce IL-5 but only a small amount of IFN-γ. In our original experiments, where we used the TCR anti-H-Y transgenic mice, the polarization was clearer in the sense that they either made IFN-γ or both IL-4 and IL-5. The stimulation in the influenza model was more potent in these experiments because we loaded the APCs with 1 μg/ml of the influenza peptide. We tended to observe variable amounts of IFN-γ in the T cytotoxic 2 (Tc2) population, so we tried to reduce the production of IFN-γ. We found that one way to do this was to add IL-4 to the cells at Day 2 instead of IL-2; however, this also resulted in an increased production of IL-5. This suggests that there are Tc2 cells that make IL-4 and IL-5, and almost no IFN-γ.

We examined the cytotoxic activities of these two populations. Originally, we did this in an allogeneic situation because we found that anti-H-Y mice did not generate any cytotoxicity. However, the anti-influenza transgenics did generate cytotoxic activity, both in the Tc1 and the Tc2 cell population. Generally, the Tc2 cell populations were slightly more active than Tc1 cell populations. A decrease in cytotoxic activity was observed when IL-2 was present without IL-4 or IL-12. Therefore, in our hands, both IL-4 and IL-12 boost the generation of cytotoxic activity.

In some of these experiments, but not all of them, the Tc2 cell populations killed non-specifically, i.e. in the allogeneic situation they killed syngeneic cells, and in the influenza situation they killed the targets in the absence of the antigenic peptide. This may suggest that the mechanism of cytotoxicity is different in the two cell populations, but so far we haven't been able to demonstrate this conclusively. We looked at the generation of Tc1 and Tc2 cells in perforin knockout mice (Walsh et al 1994a). We could generate polarized Tc1 and Tc2 cell populations using C57BL/6 (H-2^b) perforin knockout mice responding against B10.D2 (H-2^d). We found that both the Tc1 and the Tc2 cell populations killed the Fas^+ targets in the perforin knockout mice (the targets were L12-10 lymphocytic leukaemia cells transfected either with sense or antisense DNA encoding Fas) (Walsh et al 1994b). Therefore, both cell populations can kill via the Fas pathway; although we had hoped to observe Fas-mediated killing in one of the cell populations but not the other.

Sher: Therefore, the involvement of the Fas pathway is only by inference. Have you examined Fas ligand knockout mice?

Dutton: No, but the effectors from perforin knockout mice score zero on the Fas^- targets.

We expected that only the Tc2 cell population which mediated non-specific cytotoxicity would be Fas ligand-mediated. However, it turned out that cytotoxicity in the $CD8^+$ cells parallels the expression of Fas ligand on Th1 cells and Th2 cells, i.e. in both cases the Th1 cells expressed higher levels of ligand than the Th2 cells.

Mosmann: We also observed non-specific killing of Tc cells, particularly in the Tc2 populations in our earlier experiments (S. Sad & T. R. Mosmann, unpublished observations). This killing was variable, and as we have refined the conditions it's disappeared for some unknown reason.

Dutton: We observed non-specific killing in one complete series of experiments, and then when we decided to analyse it further, it faded away. However, it comes back again once in a while.

Mosmann: A Tc2 cell population at its happiest and cleanest can be a specific population, but something else can occur in culture that induces non-specific killing.

Dutton: We have also worried about natural killer cell-mediated killing but the cytotoxic populations are at least 98% $CD8^+$ at the time of the assay.

Mosmann: I would like to raise another point in relation to the degree of killing by Tc1 and Tc2 cells. I agree that Tc2 cells kill more efficiently in the absence of added cytokines, but if we grow Tc1 cells in IL-12, they are just as cytotoxic as Tc2 cells grown in IL-4.

Dutton: The experiment that I described where Tc2 was more active than Tc1 was the most extreme case. I agree that they kill almost equally, but in seven out of eight experiments the Tc2 cells were more efficient than the Tc1 cells.

Mosmann: This is interesting because Sergio Romagnani's results suggest that human Tc1 cells are more efficient than Tc2 cells (Maggi et al 1994).

Romagnani: Yes. $CD8^+$ clones generated from HIV-infected patients were also cytotoxic. Cytotoxicity indices of Tc1 clones are usually greater than 50%, whereas those of Tc2 clones are 30–50%, so there was a decrease in cytotoxicity in the Tc2 clones but they were still cytotoxic.

Trinchieri: We have preliminary results which suggest that in $CD8^+$ clones from normal individuals, Tc2 cells are less efficient at killing than Tc1 cells (D. Perrit & G. Trinchieri, unpublished results).

Dutton: Erard et al (1993) showed that CD8 and cytotoxic activity could be lost by stimulating the cells with 12-*O*-tetradecanoylphorbol 13-acetate (TPA) or ionomycin. This suggests that *in vivo* there are different ways of stimulating Tc2 activity that may or may not down-regulate cytotoxic activity. Therefore, the down-regulation of CD8 may be physiological, and it may be due to different co-stimulatory activators.

Mosmann: But losing CD8 would be a good reason to lose cytotoxicity because they would also lose recognition of the target.

Flavell: Sylvie Gnerder in my lab has some unpublished results that support Frank Fitch's results. These results are related not to the Tc1 and Tc2 cells themselves, but to the differentiated effector cells and the requirements for this process. She has used an SV40-specific TCR transgenic mouse in which all $CD8^+$ cells react with a peptide of SV40. If those $CD8^+$ cells are cultured *in vitro*, they make a substantial amount of IL-2 and become cytotoxic T lymphocytes (CTLs). She used IL-2 and IFN-γ to assay for cytokine production, and either CTL1 mRNA or cytotoxicity as a marker of end-stage differentiation. She also measured the levels of IFN-γ. She was trying to address the role of co-stimulation in this process. She found that if she blocked with CTLA4–Ig, CTL function was completely wiped out but the levels of IFN-γ were unaffected. It is possible that this effect occurred because clonal expansion was blocked. Therefore, she repeated the experiment in the presence of IL-2, and she observed that CTL function was still blocked even though the cells had proliferated. This suggests that the development of the CTL function from naive cells requires co-stimulation.

Fitch: In our hands it apparently does not.

Flavell: We all have different TCR transgenic mice. However, only about 60% of CTL function is blocked, so there is some residual CTL function, despite the fact that all the B7 molecules are blocked. The experiment has been performed many times and it is a clear result. We don't know what this residual CTL function represents. It is possible that it is the activity that is normally obtained when T cell clones are derived from a normal target.

Fitch: I suspect that this residual activity may relate, at least in part, to the affinity of the TCR. Cai & Sprent (1994) have studied the same transgenic mice as ours. They found that H2-K^{bm11}-stimulating cells, which express a H2-K^b mutant MHC molecule, appeared to interact with TCR with a lower avidity than did stimulating cells which express H2-L^d alloantigen. Also, Eisen and colleagues (Sykulev et al 1994a,b) have found that H2-L^d molecules which have bound different peptides appeared to interact with the same transgenic TCR with different affinities. Therefore, this model system gives us the opportunity to look at the consequences of stimulating the same TCR with MHC/peptide complexes that react with different affinities.

Mitchison: During selection in the thymus, do thymocytes have different TCRs or do they all have the same TCR at different densities?

Fitch: In transgenic mice all thymocytes are presumed to have the same TCR. The question is how many TCRs have to interact with how many MHC/peptide complexes to account for either positive or negative selection. It is possible that all T cells which express a transgenic TCR may not have experienced the same signalling events during positive and negative selection in the thymus. There must be a large array of different peptides presented on the selecting MHC molecules, which in this case we know is H2-K^b.

Mitchison: What is the source of variation in the responding T cells?

Fitch: There is no variation in the TCR of the responding T cells. There is a variation in the types of peptides expressed in combination with the MHC molecules on the selecting cells, i.e. on the cell that is expressing H2-K^b.

Mitchison: But the selecting cell can't select anything unless there's some variation in the responding cells, which you say are TCR transgenics.

Locksley: There must be selection on natural ligands, presumably serving as altered T cell ligands, in the TCR transgenic mice.

Allen: Jameson et al (1994) showed that the CD8 levels on T cells from transgenic mice depend on the ligand for which they have been positively selected.

Mitchison: Do they stay tuned in the periphery?

Allen: Yes. Jameson and colleagues would argue that this is how they are prevented from being reactivated by the positive selecting ligand.

Fitch: These T cells have a range of reactivities but I don't know how that's going to be reflected.

Flavell: There will definitely be a variation in the numbers of TCR molecules in each cell. This will be influenced somewhat by the other receptors that the cells contain.

Mason: It is necessary to knock out the endogenous α chain because the β chain transgene can associate with this endogenous α chain, and the proper $\alpha\beta$ TCR that the transgene encodes will not be fully represented. The level of expression of the endogenous α chain paired with the β chain is variable from one cell to another, which generates an apparently homogeneous population that is not strictly homogeneous with regard to expression of the α chain transgene.

Abbas: One simple way of distinguishing between anergy and differentiation may be to take the T cells that have seen P815 cells without B7 and culture them in the presence of IL-2. They should recover if our 'rules' about anergy are correct.

Fitch: The rules for anergy have been established for $CD4^+$ T cells in short-term cultures. Conventional $CD8^+$ clones are maintained by weekly antigen stimulation in the presence of IL-2. Co-stimulatory molecules under these conditions are absent or expressed on relatively few cells because most of the spleen cells that are generally used for stimulation don't express co-stimulatory molecules. No one, to my knowledge, has cultured $CD4^+$ cells with weekly antigen stimulation in the absence of co-stimulation but with added exogenous IL-2 to determine whether IL-2 production can be rescued. I suspect that after four weeks in culture, they cannot be rescued. We do know that some established conventional $CD8^+$ clones that don't make IL-2 cannot be persuaded to make IL-2 if co-stimulation is provided. In contrast, naive $CD8^+$ cells that express the transgenic TCR continue to make IL-2 if co-stimulation is provided, but the IL-2 production gradually goes away if the co-stimulation is removed.

Abbas: You could just culture them in the presence of IL-2 for a week or two and see if they recover.

Fitch: It may take more than that to get an answer. Repeated passage may be necessary to get to a stage at which the cytokine secretion pattern is fully differentiated.

Swain: Is there something else on P815 cells that can potentially provide co-stimulation? Because we have found that CD54 (intracellular adhesion molecule 1) can be as good a co-stimulator of naive cells as B7 (Dubey et al 1995).

Fitch: We don't know, and we also don't know if the P815 stimulating cells are producing cytokines. The interaction between the stimulating cell and the T cell is like a continuing conversation. We don't know how the stimulating cell responds to the messages given by the T cell. We're looking at the effect of neutralizing anti-IL-12 monoclonal antibody. We only know that P815 cells do not express significant levels of B7-1 or B7-2 in the absence of transfection, and that they do express CD11 (lymphocyte function-associated antigen 1). It is possible that they express CD54 and other molecules that may have a role in co-stimulation. Clearly, we should examine the role of these and other molecules using limiting dilution precursor frequency analysis. Initially, I was not convinced that we would be able to get every cell to react at limiting dilution, but we have. Therefore, we now have a benchmark against which some of these possible influences can be tested.

Flavell: Are there any B7-positive cells in the P815 culture?

Fitch: We have obtained comparable results with purified $CD8^+$ T cells, and we have observed positive wells where the likelihood of contaminating B7-positive cells is very low, so it is unlikely that B7-positive cells in the responding cell populations are contributing to the results.

References

Cai Z, Sprent J 1994 Resting and activated T cells display different requirements for CD8 molecules. J Exp Med 179:2005–2015

Croft M, Carter L, Swain SL, Dutton RW 1994 Generation of polarized antigen-specific CD8 effector populations: reciprocal action of interleukin (IL)-4 and IL-12 in promoting type 2 versus type 1 cytokine profiles. J Exp Med 180:1715–1728

Dubey C, Croft M, Swain SL 1995 Costimulatory requirements of naive $CD4^+$ T cells: ICAM-1 or B7-1 can costimulate naive CD4 T cell activation but both are required for optimum response. J Immunol 155:45–57

Erard F, Wild M-T, Garcia-Sanz JA, Le Gros G 1993 Switch of CD8 T cells to noncytolytic $CD8^-CD4^+$ cells that make T_H2 cytokines and help B cells. Science 260:1802–1805

Jameson SC, Hogquist KA, Bevan MJ 1994 Specificity and flexibility in thymic selection. Nature 369:750–752

Maggie E, Guidizi MG, Biagiotti R et al 1994 Th2-like $CD8^+$ cells showing B cell helper function and reduced cytolytic activity in human immunodeficiency virus type I infection. J Exp Med 180:489–495

Sykulev Y, Brunmark A, Jackson M, Cohen RJ, Peterson PA, Eisen HN 1994a Kinetics and affinity of reactions between an antigen-specific T cell receptor and peptide-MHC complexes. Immunity 1:15–22

Sykulev Y, Brunmark A, Tsomides TJ et al 1994b High-affinity reactions between antigen-specific T-cell receptors and peptides associated with allogeneic and syngeneic major histocompatibility complex class I proteins. Proc Natl Acad Sci USA 91:11487–11491

Walsh CM, Matloubian M, Liu C-C et al 1994a Immune function in mice lacking the perforin gene. Proc Natl Acad Sci USA 91:10854–10858

Walsh CM, Glass AA, Chiu V, Clark WR 1994b The role of the *fas* lytic pathway in a perforin-less CTL hybridoma. J Immunol 153:2506–2514

General discussion II

Anergy and peripheral tolerance

Mitchison: Some of us are sceptical about the importance of anergy in normal self-tolerance (Mitchison 1993, Miller 1993), whereas others are more enthusiastic. Much of the recent work has been done in double transgenic systems, in which a T cell receptor (TCR) transgene is combined with a transgene encoding the peptide recognized by the TCR. An excellent recent study from Zinkernagel's laboratory describes such a system, and reviews similar previous work (Bachmann et al 1994). Another useful review is from Hämmerling and colleagues (1993). A general finding is that 'anergy' is generated quickly when non-anergic TCR transgenic cells are transplanted into an environment containing the peptide that they recognize, and it is lost quickly when they are transplanted out of that environment. The question is whether these changes represent an altered state induced in all the TCR transgenic cells, rather than a selective loss of activated T cells. The loss of anergy after transplantation into the peptide-free environment would result, according to the selectionist view, from environmental priming. There are good reasons for suspecting that negative selection operates in the periphery in the same manner, but with slower kinetics, as it does in the thymus (Mitchison 1993). There is no direct evidence of selective loss, although its consequences could probably be detected only after a long while. Another reason for being suspicious about anergy being a change of state is that nobody has yet found a marker for it.

Mosmann: We need to try to define exactly what we are talking about. For instance, I'm a little confused about the relationship between the loss of the ability to make interleukin 2 (IL-2) and the state of anergy. Numerous results, over many years, suggest that both $CD4^+$ and $CD8^+$ T cell clones can lose the ability to make IL-2. This is a relatively permanent loss, although they continue to proliferate and make other cytokines.

Mitchison: Let's confine ourselves to what happens in mice, rather than getting bogged down in what happens in clones, because it's unclear as to whether any of the anergy phenomena that are seen in clones are applicable *in vivo*.

Mosmann: What do you mean by the term 'anergy'?

Mitchison: The anergy hypothesis is that a subset of potentially self-reactive T cells exists in the body. These cells are unreactive because they have been induced to go into a state of anergy.

Mosmann: What do you mean by unreactivity? Do they make cytokines? Do they proliferate?

Mitchison: Is it necessary to make these distinctions?

Fitch: In my opinion it is essential to make these distinctions because anergy is an unresponsive state that has been defined recently as a specific defect in IL-2 production.

Mitchison: That definition was made by the clonalists, but I was talking more about the double transgenic model.

Allen: Anergy is basically the inability to be restimulated. Is there a long-lasting phenotype *in vivo* that approximates the anergy that we observe *in vitro*? It is possible that it is related to the time-frame of the experiment. A cell could exist in an anergic state for two or three weeks *in vivo* but its fate is basically to die. Therefore, this state is not long lived, but it is what we're recapitulating *in vitro* in our anergic systems. If that cell is committed to die, is it dead, alive or anergic? We could compare this with ourselves because we're all committed to die but I don't think that any of us would call ourselves dead or anergic!

Mitchison: How can you explain Jenkins' mice, in which the whole cell population is in that state (Kearney et al 1994)?

Allen: The situation is complicated by the constant renewal of T cells.

Abbas: Jenkins and his colleagues have performed a key experiment that is most difficult to explain if you argue against anergy. The experiment involves the adoptive transfer of ovalbumin and I-A^d-specific transgenic T cells into a normal BALB/c mouse, which was primed with ovalbumin peptide, either aqueous or in incomplete Freund's adjuvant. They found that the transferred cells don't expand *in vivo*, and that if they take them out of the mouse and re-stimulate them, the cells respond less well than naive cells that do not see the antigen.

Mitchison: But the selection argument can also be applied. This may be a cell population which has had its activated cells deleted.

Flavell: The cells do expand *in vivo*, but they expand then disappear (presumably because they die).

Abbas: Yes, but it is the residual cells that, on a per cell basis, respond less well than naive cells.

Mitchison: That may be because those naive cells have been stimulated by environmental antigens.

Mason: Consider the physiology of the immune response. The stimulated lymph node contains all the necessary antigen-reactive cells and co-stimulatory molecules. When an effector cell is generated, it leaves the lymph node and performs its function in the periphery. It may encounter the antigen on a cell that presents the peptide but not necessarily on a cell that presents the co-stimulatory molecule. In this case, the cell should not necessarily proliferate in the lesion because this could be potentially dangerous. It makes more sense to prevent it from proliferating, even though it continues to secrete the cytokines that it was instructed to produce in the lymph node.

Swain: But that cell is not anergic because it produces cytokines when it is stimulated.

Mitchison: The principle reason for postulating anergy is to account for self-antigens that will be encountered outside the thymus. Your suggestion is a perfectly reasonable addition to that. The immune system certainly needs a mechanism for preventing responses to self-antigens that are first encountered in the periphery. The question is whether that mechanism is inductive or selective.

Abul Abbas, you've tried to demonstrate that Fas-mediated death occurs at the periphery but not in the thymus. If I accepted that conclusion in full I would be more willing to accept the notion of anergy.

Abbas: We showed that in LPR mice, deletion with antigen or superantigen can occur in the periphery but not in the thymus (Singer & Abbas 1994). We also injected an LPR mouse with antigen in incomplete Freund's adjuvant and we looked at the induction of tolerance to subsequent priming with antigen in complete Freund's adjuvant. We found that the LPR mouse becomes tolerant just as well as a normal mouse. Therefore, it is possible to induce tolerance in the absence of at least one pathway of deletion. We believe that this model of tolerance induced by antigen in incomplete Freund's adjuvant represents anergy.

Locksley: This may also be important for oral tolerance because many peptides are ingested through the gut, but the T cell responses to these peptides are minimal.

Abbas: But isn't suppression involved in oral tolerance?

Locksley: No. It's equivalent to Kearney et al (1994) injecting peptides without adjuvant in the T cell transfer model.

Allen: Av Mitchison, how do you imagine that tolerance can be induced to an antigen that does not reach the thymus?

Mitchison: We first have to understand the differences between the thymus and the periphery. One view is that there is a programme of T cell development in the thymus that is not repeated in the periphery, although not everyone accepts this idea (Mitchison 1993). Others believe that everything in the thymus can be performed in the periphery, but more slowly. This argument is more difficult to explain. It is possible that the periphery represents a smaller compartment, not because it is anatomically smaller, but because the right conjunction of adhesion molecules and auxiliary stimulators occurs much less frequently in the periphery than it does in the thymus. The main evidence supporting the second view are the experiments which show that T cell tolerance can be obtained in the periphery of a thymectomized mouse T cell, and that this tolerance is indistinguishable from the tolerance induced in the thymus (Mitchison 1993).

Abbas: By what criteria?

Mitchison: Dose-response criteria and the duration of tolerance.

Allen: But one could argue that tolerance in the thymus has different kinetics. A mature $CD4^+$ cell is deleted more slowly, and that deletion event may be what we are calling anergy *in vitro*.

Flavell: We have some evidence for peripheral tolerance in a thymectomized mouse to islet antigens. We don't know what the mechanism is, but programmed cell death is one possibility. We have also found that if the mice are not thymectomized, the opposite result is obtained. There is an autoimmune attack on the tissue that carries B-7. However, if immunogenic islets containing a B-7 molecule are grafted into a normal mouse that has been thymectomized, it will not reject it. If the mouse is not thymectomized, it's gone in three weeks.

Mitchison: You have to do that experiment in either MRL or GLD mice. The prediction is that it won't work in these mice.

Flavell: Yes, this would be an interesting experiment to do.

Dutton: With respect to suppression, it is difficult to distinguish between an inherently unreactive cell and a cell that's being switched off by another cell.

Abbas: If one assumes that suppressor cells are T helper 2 (Th2) cells, the way to rule out suppression within that context would be to induce tolerance in the presence of anti-IL-4 antibody, which will prevent the development of Th2 cells, and then see whether you get tolerance. To rule out deletion by the Fas pathway, one would induce tolerance in an LPR mouse.

Mitchison: The subjects of bystander suppression and oral tolerance are currently of great interest. A subset of T cells that has not yet been discussed is the one associated with mucosal tissue. This is the population that is derived not from gut epithelial cells but from Peyer's patches, and it has been studied by the Weiner group (Khoury et al 1992, Trentham et al 1993). The results are highly controversial and many people have not been able to reproduce them. However, these cells are supposed to mediate suppression by transforming growth factor β and by IL-4.

Liew: We found that oral tolerance can be induced in IL-4 knockout mice (Garside et al 1995). Therefore, the IL-4 mechanism is not the only mechanism that is involved in oral tolerance.

Romagnani: What is the general opinion on the source of the cell that provides IL-4 at the beginning of Th2 cell differentiation?

Abbas: Radbruch's experiments suggest that they are T cells (Schmitz et al 1994).

Kaufmann: I believe that the $CD4^+$ $NK1.1^+$ liver lymphocytes are a major source of early IL-4. These cells disappear soon after *Listeria monocytogenes* infection, so this pathogen seems to improve Th1 development not only via stimulation of early IL-12 production but also via curtailment of early IL-4 production.

Romagnani: How is this $CD4^+$ $NK1.1^+$ cell stimulated by the specific antigen during the primary infection?

Kaufmann: I don't know.

Locksley: Most NK1.1$^+$ CD4$^+$ T cells are thought to be restricted by non-polymorphic major histocompatibility complex class I molecules.

Abbas: Have CD4$^+$ NK1.1$^+$, IL-4-producing cells ever been found in humans?

Romagnani: As far as I know, there isn't an equivalent marker for NK1.1 in humans. The closest is CD56.

Kaufmann: The NK1.1 marker is not an important feature of CD4$^+$ NK1.1$^+$ cells, because similar cells, which lack NK1.1, exist in BALB/c mice.

Locksley: Such cells have been described in humans, where they also express invariant TCR α chains (Porcelli et al 1993, Lantz & Bendelac 1994).

Mitchison: We have the general picture that Th0 cells differentiate initially into Th1 or Th2 cells. The simplest view, although it may not be true, is that this commitment is totally irreversible. A rat that grows up in an antigen-free environment has no peripheral T cells in its lymph node. Therefore, it is clear that T cells do not develop in the absence of antigenic stimulation, which creates a paradox because every T cell that you claim is a virgin must have been stimulated sometime in the past by environmental antigens. Therefore, how can Th0 cells be produced?

Mosmann: You can't say that antigen stimulation is necessary for those T cells to go into lymph nodes in that experiment. The statement that they've all been stimulated by antigen is an extrapolation.

Mitchison: It is an extrapolation, but it suggests that T cells are not produced in the absence of antigen.

Mosmann: Something has to be stimulated, but not necessarily the majority of naive cells. The T cells could still be naive, but their neighbours may have been stimulated and, therefore, that's the stimulus for their maintenance or proliferation.

Dutton: Could it be bacterial stimulation of IL-1?

Mitchison: That's possible, but it's a bit far fetched.

Swain: In the TCR transgenic mice all the α/β T cells are naive and they have not seen a pigeon cytochrome c analogue, so it's unlikely that T cell production requires T cell stimulation through the TCR. It is possible that a different kind of stimulation is required.

Mosmann: Perhaps this could be resolved by repeating the TCR transgenic experiments in germ-free mice.

Signal transduction in T helper 1 and T helper 2 cells

Liew: We have mentioned the differential requirements for generating a Th1 or Th2 response but we haven't talked about possible differences in the signalling mechanisms. Does anyone have any results which suggest that these two responses have different signalling pathways?

Fitch: Yes. The inositol phosphate pathway is not stimulated when Th2 cells are stimulated. There is a higher level of basal Ca^{2+} following TCR signalling but not a significant increase in intracellular Ca^{2+} (Gajewski et al 1990). In addition, the tyrosine phosphorylation patterns are different following stimulation of Th1 and Th2 clones (Gajewski et al 1995). However, these signalling pathways are not understood well enough to determine the exact mechanism that accounts for these differences.

Abbas: We have made similar observations in Th2 clones (M. E. Williams & A. K. Abbas, unpublished results, Abbas et al 1991). However, when we tried to repeat these results, we found that the level of inositol phosphate and the influx of intracellular Ca^{2+} varied not only in different Th2 clones, but also within the same clone, depending on how it was maintained. We also did a few phosphotyrosine Western blots which convinced us that we were not looking at reproducible differences. It is possible that we had these problems simply because we were studying clones.

Cantrell: I have never seen anyone do a detailed analysis of signal transduction in Th1 or Th2 cells. A comprehensive study would require both a qualitative analysis and a quantitative analysis of kinetics, such as those done by Goldsmith & Weiss (1988), who showed that the length of Ca^{2+} flux in a T cell is crucial for IL-2 production, i.e. if the flux is only for a couple of minutes the cell does not make IL-2. One relevant question is what is known about the IL-4 promoter versus an IL-2 promoter? For example, is it possible to predict, based on the analysis of transcription factors and AP-1 sites, that Ca^{2+} is more important for transcription of the gene encoding IL-2 rather than IL-4?

Flavell: There are many factors involved with both promoters. For example, the IL-4 nuclear factor of activated T cells (NFAT) is responsive to Ca^{2+} alone, and it is present in Th2 cells. This is contrary to what you were just talking about, and it doesn't require the TPA (12-*O*-tetradecanoylphorbol 13-acetate)-type signal, whereas the equivalent molecule in the case of IL-2 requires both Ca^{2+} and TPA. We've made transgenic reporter mice with IL-4 NFAT and IL-2 NFAT and they seem to have the same transcriptional activity. There's not a unique answer because the activation requirements depend on whether it's a naive cell, an effector cell or a cell that's rested and is re-stimulated.

Cantrell: But that's clearly where the differences are going to lie. Different transcription factors will be involved in the regulation of Th2-type cytokines and Th1-type cytokines. The pathways could be the same, but different transcription factors may be expressed in the different cells. This is certainly a possibility for the NFAT family, where different members may be expressed differentially.

Locksley: Bluestone and Matis have described a kinase called RLK that's a member of the Tek kinase family that is preferentially expressed in Th1 clones (Hu et al 1995). There is also some interest in the inducible T cell kinase (ITK). These kinases have Src homology domains and pleckstrin homology domains. They're related and some are T cell specific. ITK has been immunoprecipitated with CD28 and it is tyrosine phosphorylated following CD28 cross-linking with TCR. These kinases may all be candidates for differential signalling.

Lotze: The NFκB knockout mice also have some unusual immunological defects (Sha et al 1995).

Abbas: The p50 knockout, which Bill Sha and David Baltimore made (Sha et al 1995), generates normal Th2 cells and Th1 cells, depending on how it is stimulated (A. H. Lichtman, unpublished results).

Liew: We need to obtain T cell clones that recognize the same antigen epitope in order to compare the qualitative differences in signalling by Th1 and Th2 clones.

Allen: We have to do quantitative studies, using transgenic T cells with the same antigen and antigen-presenting cells, to differentiate Th precursor cells into Th1 or Th2 cells because it seems as though I could pick any two clones and come up with any story that I wanted.

Abbas: But we only have some of the technology to do this. We don't have antibodies against most of the transcription factors, for example, so we can't do immunoprecipitations.

Fitch: I would like to make an obvious point. A clone is a clone only once in its lifetime, i.e. when it was generated. Cells in a clone divide at least every 24 h, so that in three months time it's possible to have 90 generations. 2^{90} is a large number. Variant cells do emerge. Fortunately for us, many of them die but some may also persist, and this probably explains some of the observed differences in the laboratories of different investigators.

Mitchison: Andrew McMichael, can you give us an update on the cytokines and transcription factors involved in multifactorial diseases? John Bell (University of Oxford), in connection with his analysis of the multifactorial genetics of rheumatoid arthritis, hopes to find microsatellite markers associated with groups of cytokine genes. Has that work made any progress yet?

McMichael: Not that I know of for rheumatoid arthritis. The strongest association is with HLA class II polymorphisms.

Mitchison: Is there a polymorphism in the gene that encodes IL-2?

Flavell: Yes. There's a polymorphism in non-obese diabetic (NOD) mice but it is not known if this also exists in humans.

Mitchison: Is this in the cytokine gene itself or in its promoter?

Flavell: It is in the gene that encodes IL-2 and not in the promoter. There are variations in the number of glutamine repeats in each NOD mouse.

Abbas: Is the polymorphism associated with a difference in the activity of IL-2?

Flavell: Not that I know of. There are only a few strains that have this polymorphism. It doesn't prove that this is the gene causing the phenotype, but we've made mice with the normal version to see if we can protect against diabetes. The effect is quite small, so we would have to look at a large number of mice to obtain statistical significance.

Mitchison: But wouldn't differences in the levels of IL-2 expression be related to differences in the promoter sequence and not the structural gene?

Flavell: Are you suggesting that polymorphisms should be related to the ability to make NFAT, for example?

Mitchison: That is what Doreen Cantrell suggested.

Cantrell: Not exactly. NFAT is not just one protein, it is a multigene family. Each member is clearly not expressed in all the same cells. What controls this differential expression is not known, and this expression has not been fully mapped. Members of the NFAT family could be expressed differentially during development because this is true for many other transcription factors. If they were, this could have an effect on the Th1/Th2 switch and the cytokines that the cells would produce. We should look at members of the NFAT family because most of the genes encoding cytokines have NFAT-binding sites.

References

Abbas AK, Williams ME, Burstein HJ, Chang T-L, Bossu P, Lichtman AH 1991 Activation and functions of $CD4^+$ T cell subsets. Immunol Rev 123:5–22

Bachmann MF, Rohrer UH, Steinhoff U et al 1994 T helper cell unresponsiveness: rapid induction in antigen-transfected and reversion in non-transgenic mice. Eur J Immunol 24:2966–2973

Gajewski TF, Schell SR, Fitch FW 1990 Evidence implicating utilization of different T cell receptor-associated signalling pathways by T_H1 and T_H2 clones. J Immunol 144:4110–4120

Gajewski TF, Alegre M-L, Fitch FW 1995 Th2 cells can be induced to revert to an IL-2-producing, Th0 phenotype, submitted

Garside P, Steel M, Worthey EA et al 1995 T helper 2 cells are subject to high dose oral tolerance and are not essential for its induction. J Immunol 154:5649–5655

Goldsmith MA, Weiss A 1988 Early signal transduction by the antigen receptor without commitment to T cell activation. Science 240:1029–1031

Hämmerling GJ, Schonrich G, Ferber I, Arnold B 1993 Peripheral tolerance as a multi-step mechanism. Immunol Rev 133:93–104

Hu Q, Davidson D, Schwartzberg PL et al 1995 Identification of rlk, a novel protein tyrosine kinase with predominant expression in the T cell lineage. J Biol Chem 270:1928–1934

Kearney ER, Pape KA, Loh DY, Jenkins MK 1994 Visualization of peptide-specific T cell immunity and peripheral tolerance induction *in vivo*. Immunity 1:327–339

Khoury SJ, Hancock WW, Weiner HL 1992 Oral tolerance to myelin basic protein and natural recovery from experimental autoimmune encephalomyelitis are associated with downregulation of inflammatory cytokine and differential upregulation of transforming growth factor β, interleukin 4 and prostaglandin E expression in the brain. J Exp Med 176:1355–1364

Lantz O, Bendelac A 1994 An invariant T cell receptor alpha chain is used by a unique subset of major histocompatibility complex class I-specific $CD4^+$ and $CD4^-8^-$ T cells in mice and humans. J Exp Med 180:1097–1106

Lichtman AH, Chin J, Schmidt JA, Abbas AK 1988 The role of interleukin-1 in the activation of T lymphocytes. Proc Natl Acad Sci USA 85:9699–9703

Miller JF 1993 Self–nonself discrimination and tolerance in T and B lymphocytes. Immunol Res 12:115–130

Mitchison NA 1993 T-cell activation states: the next breakthrough in signaling? Immunologist 1:78–80

Porcelli S, Yockey CE, Brenner MB, Balk SP 1993 Analysis of T cell antigen receptor (TCR) expression by human peripheral blood $CD4^-8^-$ alpha/beta T cells demonstrates preferential use of several v-beta genes and an invariant TCR alpha chain. J Exp Med 178:1–16

Schmitz J, Thiel A, Kuhn R et al 1994 Induction of interleukin 4 (IL-4) expression in T helper (Th) cells is not dependent on IL-4 from non-Th cells. J Exp Med 179:1349–1353

Sha WC, Liou H-C, Tuomanen EI, Baltimore D 1995 Targeted disruption of the p50 subunit of NF-κB leads to multifocal defects in immune responses. Cell 80:321–330

Singer GG, Abbas AK 1994 The Fas antigen is involved in peripheral but not thymic deletion of T lymphocytes in T cell receptor transgenic mice. Immunity 1:365–371

Trentham DE, Synesius-Trentham RA, Orav EJ et al 1993 Effects of oral administration of type II collagen on rheumatoid arthritis. Science 261:1727–1730

Induction and regulation of host cell-mediated immunity by *Toxoplasma gondii*

Alan Sher, Eric Y. Denkers and Ricardo T. Gazzinelli

Immunobiology Section, Laboratory of Parasitic Diseases, National Institute of Allergy and Infectious Diseases, Building 4, Room 126, National Institutes of Health, Bethesda, MD 20892, USA

Abstract. *Toxoplasma gondii* is a highly infectious intracellular parasite which, if left unchecked by the immune system, rapidly overwhelms its intermediate hosts, as illustrated by the pathogenesis of toxoplasmic encephalitis in patients with AIDS. In order to insure both its host's and consequently its own survival simultaneously, *T. gondii* induces a potent γ-interferon (IFN-γ)-dependent cell-mediated immunity early in infection that controls the replication of the protozoan and facilitates transformation into the dormant cyst stage. The protective IFN-γ is derived from three sources: natural killer cells; and $CD4^+$ and $CD8^+$ T lymphocytes, which can partially compensate for each other in knockout mice lacking the appropriate major histocompatibility complex-restricting elements. At least two properties of the parasite appear to be responsible for the early induction of these effector cells. The first is a hydrophobic molecule (or group of related molecules) that triggers interleukin 12 (IL-12), tumour necrosis factor α and IL-1β synthesis in macrophages. This response can also promote HIV replication in the same cells. The second is a superantigen activity that drives IFN-γ-producing $V\beta5^+$ $CD8^+$ T cells. These potentially lethal responses are later regulated through the triggering of IL-10 and by the induction of anergy in the superantigen-stimulated $V\beta5^+$ T cell population.

1995 T cell subsets in infectious and autoimmune diseases. Wiley, Chichester (Ciba Foundation Symposium 195) p 95–109

Cell-mediated immunity (CMI), a response driven by lymphocytic cells possessing a T helper 1 (Th1)-type cytokine secretion pattern, serves as a major defence against a variety of important pathogens. Consequently, its selective induction is often a major goal in vaccine design. Several microbial agents are themselves good stimuli of CMI, and they can be used to study the specific mechanisms which both trigger and regulate the response. *Toxoplasma gondii*, which is a zoonotic apicomplexan protozoan parasite and the causative agent of toxoplasmosis in man, has in recent years emerged as an important model for these studies (Gazzinelli et al 1993a).

T. gondii in its replicative tachyzoite stage invades a wide variety of nucleated cells, and it is therefore capable of inducing the rapid destruction of tissues and death of both its murine and human intermediate hosts (Frenkel 1988). The parasite avoids this consequence, detrimental to its own successful transmission, by inducing a strong cell-mediated immune response that controls tachyzoite replication while promoting transformation into a dormant tissue cyst stage (Gazzinelli et al 1993a). The importance of CMI in maintaining *T. gondii* in a benign chronic form is evident in HIV infection where, after the onset of AIDS, the parasite reverts to the tachyzoite stage causing an acute, and if untreated lethal, disease, namely toxoplasmic encephalitis (Luft et al 1984).

γ-Interferon (IFN-γ) plays a crucial role in host resistance against acute *T. gondii* infection as well as in the maintenance of the chronic state (Suzuki et al 1988, 1989). A dramatic demonstration of this requirement comes from recent experiments (Fig. 1) in which IFN-γ knockout mice were shown to die rapidly from a normally avirulent infection with the ME-49 *T. gondii* strain. The presumed mode of action of IFN-γ (Fig. 2) is to activate macrophages for intracellular killing of tachyzoites (Adams et al 1990) or induce nutrient starvation in fibroblasts and other host cells invaded by the parasite (Pfefferkorn 1984). The focus of our recent work has been to determine the cellular requirements for this protective cytokine response and to analyse the properties of the parasite responsible for its induction early in infection.

Multiple effector cells mediate IFN-γ-dependent immunity

$CD4^+$ lymphocytes, $CD8^+$ lymphocytes and natural killer (NK) cells have each been shown to produce IFN-γ in response to *in vitro* stimulation with tachyzoites or tachyzoite extracts (Gazzinelli et al 1993a). *In vivo* studies indicate that acquired immunity to *T. gondii* is influenced strongly by class I major histocompatibility complex (MHC) genes, and it is dependent on $CD8^+$ T cells (Brown & McLeod 1990). $CD4^+$ lymphocytes are required as helper cells for the generation and maintenance of the $CD8^+$ effectors (Gazzinelli et al 1991, Hakim et al 1991). Nevertheless, experiments in class I and class II MHC gene knockout mice have revealed considerable redundancy in this response (Fig. 2). Thus, in *T. gondii*-vaccinated, class I-deficient, β_2 microglobulin knockout mice, NK cells are expanded and they replace $CD8^+$ T lymphocytes as the primary effectors of IFN-γ-dependent immunity (Denkers et al 1993). Parallel experiments (E. Y. Denkers, unpublished results) in class II MHC knockout mice reveal that the latter restricting element required for the generation of conventional $CD4^+$ helper T cells is not essential for the

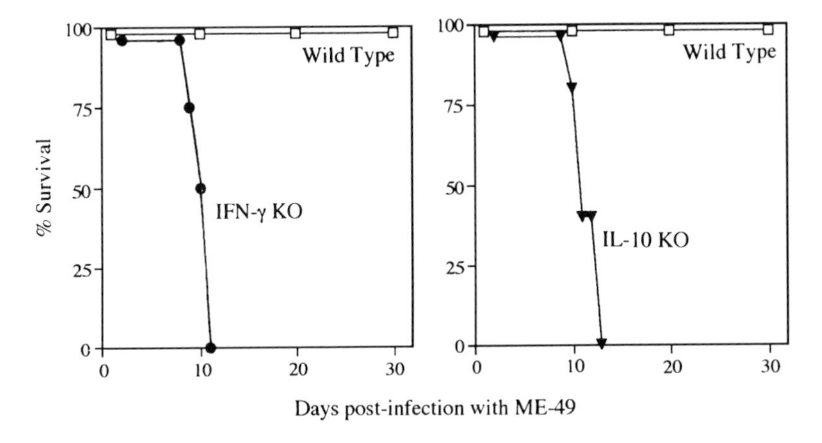

FIG. 1. Enhanced susceptibility of γ-interferon (IFN-γ) and interleukin 10 (IL-10) knockout (KO) mice to infection with the avirulent strain of *Toxoplasma gondii*, ME-49. IFN-γ and IL-10 knockout mice and wild-type littermate controls were infected with cysts of the ME-49 strain and the survival of the mice was monitored for the following month. Both cytokine defects cause rapid attrition of the mice. However, mortality of the IFN-γ knockout mice is associated with enhanced parasite replication, whereas death of the IL-10 knockout mice appears to result from overproduction of T helper 1-type cytokines (R. T. Gazzinelli, M. Wysoka & W. Muller, unpublished results).

production of vaccine-induced $CD8^+$ effector cells. Helper function in these mice is provided by a population of class II non-restricted interleukin 2 (IL-2)-producing $CD4^+$ T cells with a phenotype closely resembling the $NK\text{-}1^+$ thymic precursor population described by Bendelac et al (1994).

Resistance is initiated through both T cell-independent and T cell-dependent pathways

The ability of NK cells to substitute for $CD8^+$ cells as effectors of IFN-γ-dependent resistance suggested the existence of T cell-independent mechanisms by which *T. gondii* stimulates CMI. Consistent with this hypothesis was the demonstration that tachyzoites or tachyzoite extracts stimulate IFN-γ synthesis by NK cells in cultures of spleen cells from T lymphocyte-deficient, severe combined immunodeficiency mice (Sher et al 1993). The response was shown to require a macrophage accessory cell population and its production of three monokines, IL-12, tumour necrosis factor α (TNF-α) (Gazzinelli et al 1993b) and IL-1β (Hunter & Remington 1994). These act in synergy to trigger NK cell function (Fig. 2). Evidence for a similar pathway *in vivo* was obtained in experiments in which peritoneal cells recovered from mice as early as two days

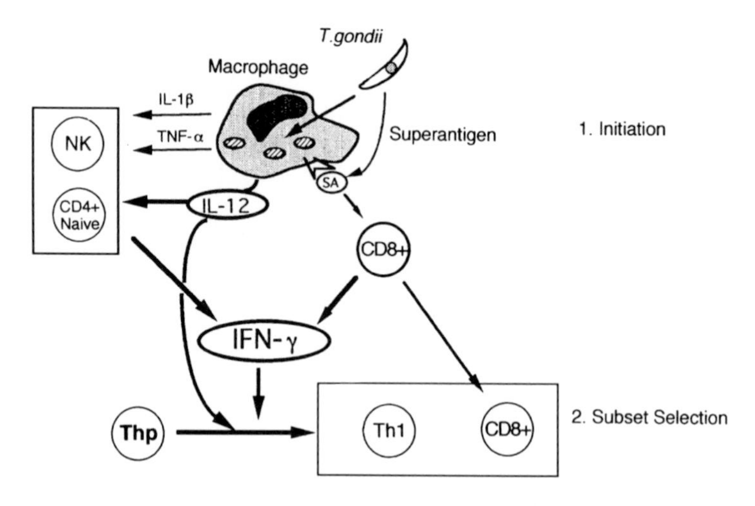

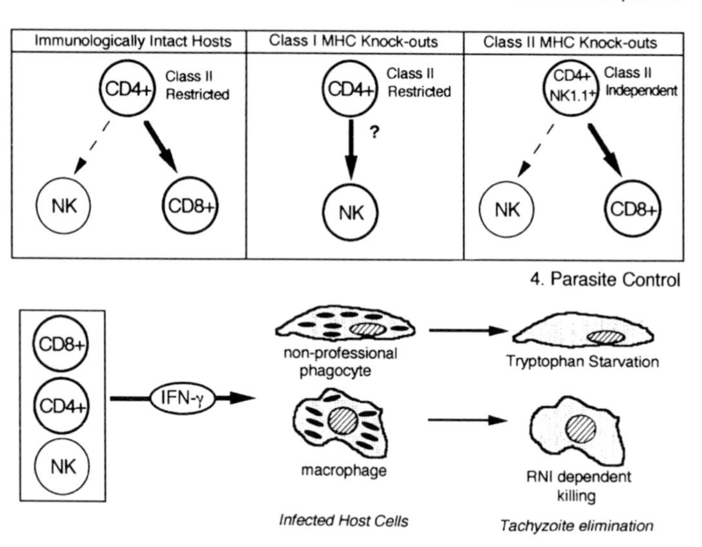

FIG. 2. T cell responses to *Toxoplasma gondii* infection. (1) Initiation of resistance to *T. gondii*. Upon initial encounter with host macrophages, tachyzoites trigger interleukin 12 (IL-12), IL-1β and tumour necrosis factor α (TNF-α) which in turn stimulate γ-interferon (IFN-γ) synthesis by natural killer (NK) cells and naive CD4$^+$ T cells. At the same time a parasite superantigen (SA) binds to macrophage class II molecules and induces the expansion of an IFN-γ-producing CD8$^+$ lymphocyte population. (2) Subset selection. The IFN-γ from both sources, together with IL-12, promotes the selective differentiation of T helper 1 (Th1) CD4$^+$ cells also producing IFN-γ. Thp, T helper precursor. (3) Effector cell expansion. Parasite-reactive Th1 cells provide help for the expansion of CD8$^+$ and NK effector cells. This pathway has several redundancies, as revealed by studies in major histocompatibility complex (MHC) knockout mice. In class I

after infection with ME-49 were shown to synthesize IFN-γ concurrently with IL-12 p40 and TNF-α. This is also supported by the ability of neutralizing anti-IL-12 antibody to both inhibit IFN-γ production *in vivo* and cause a dramatic increase in susceptibility of mice to the parasite (Gazzinelli et al 1994, Khan et al 1994). In addition to NK cells, $CD4^+$ cells were also shown to secrete IFN-γ early in infection, a response which may result from the direct effects of IL-12 during priming of immature $CD4^+$ cells and/or stimulation by a parasite superantigen (see below). The early initiation of IFN-γ production is clearly essential for control of acute infection, and it is probably a key host factor in determining the virulence of different *T. gondii* strains. It may also play an important role in driving T cell responses toward a Th1-type phenotype (Fig. 2) (Gazzinelli et al 1994).

Induction of monokine synthesis by *Toxoplasma gondii*

As discussed above, the induction of monokines following exposure of macrophages to *T. gondii* is a crucial determinant in the subsequent initiation of CMI. Our *in vitro* studies indicate that the response triggered by the parasite in macrophages is selective, involving a subset of the genes normally transcribed after exposure to classical activation stimuli (e.g. endotoxin). Genes so far characterized as belonging to this subset are those that encode TNF-α, IL-1β, IL-12 p40, IL-10, TNF receptor 2 (Gazzinelli et al 1993b, Li et al 1994) and most recently, IL-15 (M. Doherty, unpublished results 1995). In every case examined, *T. gondii*-induced gene expression is amplified in the presence of IFN-γ, the lymphokine end-product of the macrophage-triggered response. As discussed below, this positive amplification loop must be tightly regulated to avoid host tissue damage.

We are currently characterizing both the host cell signalling pathways and the parasite molecule(s) responsible for monokine induction. The parasite stimulus is clearly distinct from that provided by bacterial lipopolysaccharide (LPS) in that a different subset of macrophage genes are induced, and that the LPS receptor is not required. Nevertheless, *T. gondii* induces a similar membrane tyrosine phosphorylation/dephosphorylation pattern as LPS and may, therefore, use common signalling pathways downstream from receptor engagement (Li et al 1994). The relevant parasite molecules that trigger this

---

knockout mice, NK cells emerge as the major effector cell of resistance, whereas in class II knockout mice, class II-independent $CD4^+$ cells replace the normal class II restricted population as the source of help. (4) Parasite control. The IFN-γ produced by the $CD8^+$ or NK cell effectors controls tachyzoite growth by activating infected macrophages for parasite killing by reactive nitrogen intermediates (RNI) or by depleting infected non-professional phagocytes of tryptophan, an essential amino acid for the parasite.

response exist in both soluble and membrane-bound forms, they are heat stable at 100 °C and they appear to be hydrophobic lipoproteins or glycolipids.

A parasite superantigen activity also contributes to the induction of cell-mediated immunity

In addition to the parasite molecules that trigger monokine production, *T. gondii* possesses an immunological activity that appears to drive the early induction of cell-mediated responses against the parasite (Fig. 2). Thus, live tachyzoites as well as soluble parasite extracts induce significant proliferative responses in unprimed lymphocytes from uninfected mice (Denkers et al 1994). This non-specific stimulation appears to be due to a parasite superantigen. This is based on several criteria. (1) The T cell receptor population expressing the Vβ5 chain is preferentially expanded. (2) In common with classical superantigens (e.g. staphylocccal enterotoxins), the *T. gondii* activity is dependent on class II (either syngeneic or allogeneic) presentation but does not require cellular antigen processing. (3) Mice of different inbred strains differ in their parasite-driven proliferative responses based on the endogenous expression of Vβ5 (Denkers et al 1994).

The cell populations stimulated *in vitro* by the *T. gondii* superantigen have two additional properties consistent with their possible function in the early cell-mediated immune response to the parasite. First, within one week, the majority of the Vβ5 T cells expanded are CD8$^+$. As already discussed, CD8$^+$ lymphocytes play a major role as effector cells in resistant mice. More importantly, the expanded CD8$^+$ population produces substantial levels of IFN-γ but none of the lymphokines (e.g. IL-4 and IL-5) associated with Th2-type responses (Denkers et al 1994). Thus, the *T. gondii* superantigen activity may stimulate CD8$^+$ development whilst providing an additional source of protective IFN-γ.

Control of *Toxoplasma gondii*-induced cellular responses

The early cellular and cytokine responses to *T. gondii* are clearly both potent and dramatic. Nevertheless, if left unregulated they could become detrimental to the host. In particular, both IFN-γ and TNF-α are toxic in high doses, and activated macrophages can themselves be autoreactive against neighbouring host cells. Recent work suggests the existence of at least three mechanisms for controlling the initial and potentially lethal end-product-driven cellular response to *T. gondii*. Firstly, as demonstrated *in vitro* (Candolfi et al 1994), reactive nitrogen intermediates produced as a consequence of macrophage activation could dampen lymphocyte proliferation, thereby controlling the expansion of IFN-γ-producing T cells. Secondly, IL-10, which is also a product of *T. gondii*-activated macrophages, may suppress the overproduction of

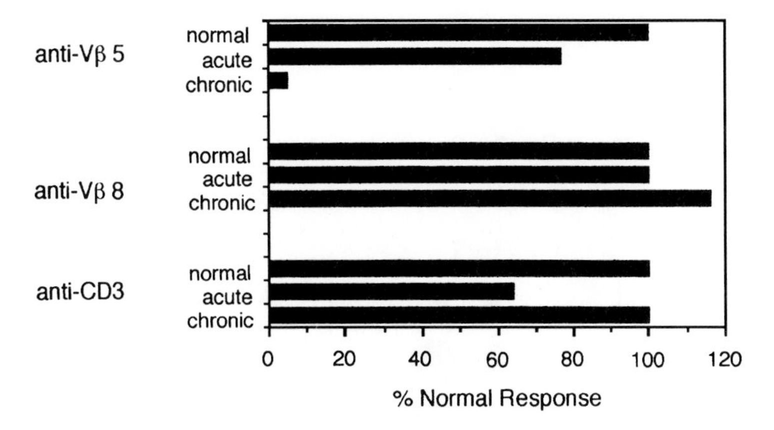

FIG. 3. Vβ5-bearing T cells are anergized during chronic infection with *Toxoplasma gondii*. Spleen cells from C57BL/6 mice at seven (acute) or 30 days (chronic) of infection with the ME-49 strain of *T. gondii* were stimulated *in vitro* with anti-Vβ5, Vβ8 or CD3 antibodies, and the resulting proliferative response was compared to that obtained with uninfected control mice. Spleen cells become non-responsive to anti-Vβ5 stimulation during the chronic stage of infection, although anti-Vβ8 and anti-CD3 responses are unaffected.

TNF-α and IL-12 in an autocrine fashion. Experiments with IL-10 knockout mice support an *in vivo* role for this mechanism (R. T. Gazzinelli, M. Wysoka & W. Muller, unpublished observations). Thus, when infected with the ME-49 strain, IL-10 knockout mice show enhanced mortality (Fig. 1) and increased systemic levels of both IFN-γ and IL-12. They also show overproduction of both IL-12 and TNF-α at the macrophage level.

A third control mechanism occurs at the level of the superantigen-driven response. Recent studies reveal that a loss in responsiveness to stimulation with anti-Vβ5 antibody occurs during the transition from acute to chronic infection. This suggests that T cells expressing Vβ5 are specifically anergized (Fig. 3). Overproduction of cytokines as a consequence of chronic superantigen exposure is, therefore, averted.

HIV induction: a potential detrimental consequence of the macrophage response to *Toxoplasma gondii*

Macrophages can serve as a major reservoir of HIV1 in asymptomatic individuals. Previous studies have indicated that the growth of monotropic strains of the virus in macrophages is greatly enhanced by cytokines such as TNF-α and IFN-γ, which activate these cells *in vitro* (Koyangi et al 1988, Fauci 1993). Because of the potency of *T. gondii* as a macrophage-activating stimulus, we asked whether exposure to parasite extracts or live tachyzoites would

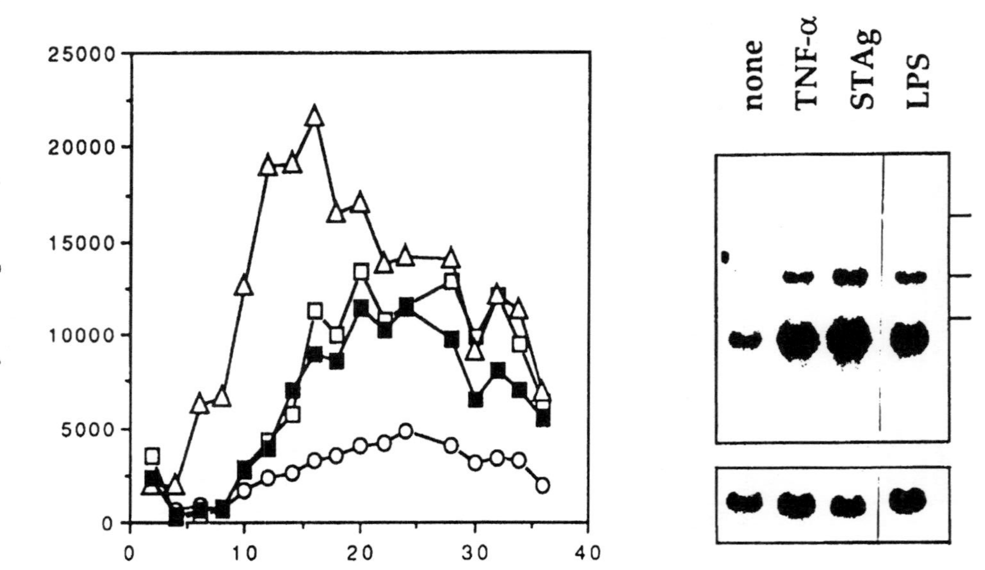

FIG. 4. Induction of HIV expression in macrophages by *Toxoplasma gondii* extracts or live parasites. (A) Human monocyte-derived macrophages from peripheral blood of HIV$^-$ donors were infected with a monotropic strain of HIV1 in the absence (○–○) or the presence of tachyzoite extract (STAg) (△–△) or live tachyzoites at a parasite to macrophage ratio of 1:1 (■–■) or 10:1 (□–□). Enhancement of viral replication is evident from the increase in reverse transcriptase (RT) activity. rnx, reaction. (B) Splenic macrophages from a transgenic mouse line (166) incorporating the HIV1 proviral genome were exposed to STAg *in vitro* and proviral gene expression measured by Northern hybridization. The parasite extract induced HIV transcripts at levels comparable to those triggered by tumour necrosis α (TNF-α) and lipopolysaccharide (LPS). Proviral transcription was also observed in tissues recovered

stimulate HIV expression in either human monocytes infected with the virus or uin transgenic mice containing integrated HIV provirus. *In vitro* stimulation by parasite extracts or live infection caused a marked enhancement of HIV replication in monocyte cultures (Fig. 4A, also R. T. Gazzinelli, G. Englund, S. Bala, M. Martin & A. Sher, unpublished results). Perhaps more revealing was the induction of proviral transcription in tissues of transgenic mice during acute *T. gondii* infection as well as in macrophages from these mice exposed to parasite extract *in vitro* (Fig. 4B, also R.T. Gazzinelli, A. Sher, A. Cheever, S. Gerstberger, M. Martin & P. Dickie, unpublished results).

Conclusions

It is clear that the induction of CMI by *T. gondii*, although largely beneficial to both host and parasite, represents a delicate balancing act with possible detrimental consequences to both partners in the relationship. Therefore, the regulation of this response is crucial and is probably a key factor in determining parasite virulence. The elucidation of the *T. gondii* molecules and intracellular signalling pathways that control the response should offer important basic information relevant to the selective induction of cellular immunity and Th1-type cytokine responses in vaccines and other forms of immunological intervention. In both *in vitro* and *in vivo* experimental models, *T. gondii* promotes growth of HIV in macrophages. Consequently, the hypothesis that HIV progression is influenced by non-retroviral microbial co-stimulation warrants further investigation in the specific context of pathogens which trigger monokine synthesis.

Acknowledgements

We thank Giorgio Trinchieri, Malcolm Martin, Stephanie Vogel, Peter Dickie, George Englund, Carl Manthey and Maria Wysocka for their major collaborative contributions.

References

Adams LB, Hibbs JB Jr, Taintor RR, Krahenbuhl JL 1990 Microbiostatic effect of murine-activated macrophages for *Toxoplasma gondii*. Role for synthesis of inorganic nitrogen oxides from L-arginine. J Immunol 144:2725–2729

Bendelac A, Killeen N, Littman DR, Schwartz RH 1994 A subset of $CD4^+$ thymocytes selected by MHC class I molecules. Science 263:1774–1778

Brown CR, McLeod R 1990 Class I MHC genes and $CD8^+$ T cells determine cyst number in *Toxoplasma gondii* infection. J Immunol 145:3438–3441

Candolfi E, Hunter CA, Remington JS 1994 Mitogen- and antigen-specific proliferation of T cells in murine toxoplasmosis is inhibited by reactive nitrogen intermediates. Infect Immun 62:1995–2001

Denkers EY, Gazzinelli RT, Martin D, Sher A 1993 Emergence of $NK1.1^+$ cells as effectors of immunity to *Toxoplasma gondii* in MHC class I-deficient mice. J Exp Med 178:1465–1472

Denkers EY, Caspar P, Sher A 1994 *Toxoplasma gondii* possesses a superantigen activity that selectively expands murine T cell receptor Vβ5-bearing $CD8^+$ lymphocytes. J Exp Med 180:985–994
Fauci AS 1993 Multifactorial nature of human immunodeficiency virus disease: implications for therapy. Science 262:1011–1018
Frenkel JK 1988 Pathophysiology of toxoplasmosis. Parasitol Today 4:273–278
Gazzinelli RT, Hakim FT, Hieny S, Shearer GM, Sher A 1991 Synergistic role of $CD4^+$ and $CD8^+$ T lymphocytes in IFN-γ production and protective immunity induced by an attenuated *Toxoplasma gondii* vaccine. J Immunol 146:286–292
Gazzinelli RT, Denkers E, Sher A 1993a Host resistance to *Toxoplasma gondii*: a model for studying the selective induction of cell-mediated immunity by intracellular parasites. Infect Agents Dis 2:139–149
Gazzinelli RT, Hieny S, Wynn TA, Wolf S, Sher A 1993b Interleukin 12 is required for the T-lymphocyte-independent induction of interferon-γ by an intracellular parasite and induces resistance in T-cell-deficient hosts. Proc Natl Acad Sci USA 90:6115–6119
Gazzinelli RT, Wysocka M, Hayashi S et al 1994 Parasite-induced IL-12 stimulates early IFN-γ synthesis and resistance during acute infection with *Toxoplasma gondii*. J Immunol 153:2533–2543
Hakim FT, Gazzinelli RT, Denkers E, Hieny S, Shearer GM, Sher A 1991 $CD8^+$ T cells from mice vaccinated against *Toxoplasma gondii* are cytotoxic for parasite-infected or antigen-pulsed host cells. J Immunol 147:2310–2316
Hunter CA, Remington JS 1994 Regulation of IL-12 and TNF-α induced NK cell production of IFN-γ by IL-1β and TGF-β. Eur Cytokine Network 5:145 (abstr A198)
Khan IA, Matsuura T, Kasper LH 1994 Interleukin-12 enhances murine survival against acute toxoplasmosis. Infect Immun 62:1639–1642
Koyangi YW, O'Brian WA, Zhao JQ, Golde DW, Gasson JC, Chen ISY 1988 Cytokines alter production of HIV-1 from primary mononuclear phagocytes. Science 241:1673–1675
Li ZY, Manthey CL, Perera PY, Sher A, Vogel SN 1994 *Toxoplasma gondii* soluble antigen induces a subset of lipopolysaccharide-inducible genes and tyrosine phosphoproteins in peritoneal macrophages. Infect Immun 62:3434–3440
Luft BJ, Brooks RG, Conley FK, McCabe RE, Remington JS 1984 Toxoplasmic encephalitis in patients with acquired immune response deficiency syndrome. JAMA 252:913–917
Pfefferkorn ER 1984 Interferon γ blocks the growth of *Toxoplasma gondii* in human fibroblasts by inducing the host cells to degrade tryptophan. Proc Natl Acad Sci USA 81:908–912
Sher A, Oswald IP, Hieny S, Gazzinelli RT 1993 *Toxoplasma gondii* induces a T-independent IFN-γ response in natural killer cells that requires both adherent accessory cells and tumor necrosis factor-α. J Immunol 150:3982–3989
Suzuki Y, Orellana MA, Schreiber RD, Remington JS 1988 Interferon-γ: the major mediator of resistance against *Toxoplasma gondii*. Science 240:516–518
Suzuki Y, Conley FK, Remington JS 1989 Importance of endogenous IFN-γ for prevention of toxoplasmic encephalitis in mice. J Immunol 143:2045–2050

DISCUSSION

Swain: You have observed a preferential outgrowth of $CD8^+$ cells. Are you sure that the $CD4^+$ cells don't proliferate rapidly then die?

Sher: No, I am not completely sure. The observed CD8 response is indeed CD4 dependent. $CD4^+$ cells are also stimulated, but cytokine-triggered $CD8^+$ cells eventually predominate. Eric Denkers has also taken purified $CD8^+$ cells and stimulated them directly with the superantigen in the presence of antigen-presenting cells and interleukin 2 (IL-2) (Denkers et al 1994).

Swain: Could viral replication be the signal for CD4 activation?

Sher: In the HIV transgenic model, HIV provirus expression after *Toxoplasma gondii* infection occurs in the adherent cell fraction of the spleen, i.e. in the macrophages. We would like to argue that this is a useful model for looking at the effects of various stimuli on the induction of HIV expression in this dormant macrophage reservoir population. It's obviously a poor model for the induction of HIV expression in T cells because HIV doesn't infect mouse T cells.

Kaufmann: Are $CD4^+$ T cells activated by the superantigen in the absence of $CD8^+$ T cells?

Sher: We have not tested purified $CD4^+$ cells. All I can say is that the final proliferative response is inhibited by anti-CD4 antibody.

Kaufmann: The study by Bendelac et al (1994) focused on thymocytes. Have you shown that, in major histocompatibility (MHC) class II-deficient mice, the $NK1.1^+$ $CD4^+$ cells produce γ-interferon (IFN-γ)?

Sher: No. We suggested that the $NK1.1^+$ $CD4^+$ cells in MHC class II knockout mice provide help for the generation of $CD8^+$ cells which then protect through the production of IFN-γ.

Kaufmann: In our experience, the peripheral $NK1.1^+$ $CD4^+$ cells only produce IL-4, so they are not like T helper 1 (Th1) cells (M. Emoto & S. H. E. Kaufmann, unpublished results).

Sher: The cells we are studying have a similar but distinct phenotype. They are functionally depleted with anti-NK1.1 or anti-IL-2 antibody treatment, which suggests that they are functioning as helpers. These cells are also $CD4^+$ even though they arise in an MHC class II knockout mouse.

Mitchison: Can you clarify the role of $CD8^+$ cells in chronic infection? The mouse models that you have described involved short-term HIV infections. What is the situation in chronic control in humans? If reactivation occurs in HIV-infected patients, is that compatible with it being mediated by $CD8^+$ cells?

Sher: Initially, our group characterized the role of $CD4^+$ and $CD8^+$ subsets in established rather than acute infections (Gazzinelli et al 1991, 1992). We found that depletion of $CD8^+$ cells had the largest effect on resistance. The $CD8^+$ cells probably act through the production of IFN-γ. Suzuki & Remington (1990) did an adoptive transfer experiment with purified $CD8^+$ cells from immune mice. They were able to block the transfer of resistance by treating the recipients with anti-IFN-γ antibody. Therefore, they argued that the effect of the $CD8^+$ cells was mediated through IFN-γ. However, this experiment needs to be repeated in severe combined immunodeficiency mice to eliminate possible interactions with recipient lymphocytes. Also, these cells may be functioning through an IFN-γ-dependent cytolytic pathway, resulting in an apparent synergy between IFN-γ and cytolytic function. Stefan Kaufmann proposed a model for *Listeria* immunity in which $CD8^+$ cells lyse the infected cell resulting in the release of intracellular organisms, and at the same time, the $CD8^+$ cells release IFN-γ which activates macrophages that phagocytose and kill the free parasites (Kaufmann 1988). The available evidence does not rule out this hypothesis.

Your last point concerned the loss of CD8 function in HIV infection. Reactivation in HIV-infected patients occurs primarily in the late stages of AIDS, when $CD4^+$ cell counts have dropped and loss of CD8 function begins (Luft et al 1984). This is consistent with the hypothesis that $CD8^+$ cells are the major effectors of resistance and that reactivation depends primarily on their loss.

Mitchison: If you look in the lesions where the parasite is located in chronic disease, do you see $CD8^+$ cells and detectable levels of IFN-γ?

Sher: Yes. There is also a mouse model of this neuropathology. Chronically infected mice develop focal inflammatory lesions in the brain around tissue cysts. These lesions contain substantial numbers of $CD8^+$ cells and the brains express elevated levels of IFN-γ mRNA (Hunter et al 1994).

Mitchison: Have you also shown this in humans?

Sher: No. There's a shortage of suitable clinical material.

Locksley: Experiments with β_2 microglobulin knockout mice show that the $NK1.1^+$ lineage of T cells is not required for control, because these mice do not have this MHC class I-restricted T cell lineage.

Kaufmann: But this is not a problem because the deficiency may be compensated for.

Flavell: You would need to make a double knockout mouse for that.

Sher: We are planning to do that experiment.

Mosmann: Which subsets were unresponsive in the Vβ5 experiments?

Sher: It is probably a mixture of $CD4^+$ and $CD8^+$ T cells but we haven't really looked at this in detail. Eric Denkers has tried to find out whether plate-bound anti-Vβ5, anti-Vβ8 or anti-CD3 antibodies can stimulate lymphocyte proliferation or IFN-γ production. He has found that the $V\beta5^+$ cells become non-responsive as infection proceeds to the chronic stage. However, they're not deleted because he stained the spleens for Vβ5 and he found that there is only a small reduction in the numbers of $V\beta5^+$ cells (E. Denkers, unpublished observations).

Abbas: Your results indicate that the source of IL-10 is from macrophages, and that tumour necrosis factor (TNF) and IL-1 stimulate natural killer (NK) cells directly. However, the interpretation of the *Listeria* experiments performed by Rogers et al (1994) was that IL-1 and probably TNF are responsible for the early neutrophil reaction in *Listeria* infection, and that IL-1, at least, is not an NK cell stimulator.

Sher: This is the only point where we disagree with the results of Rogers et al (1994). They do not believe that IL-1 plays a role in the induction of IFN-γ expression in NK cells following infection with *Listeria*.

Trinchieri: Human lymphocytes require IL-1β for IFN-γ production (D'Andrea et al 1993). We have also shown the same requirement for murine Th1 clones; although, because Th1 clones are not considered to be responsive to IL-1, the requirement for IL-1 may represent an indirect effect on accessory cells or on the antigen-presenting cells (Murphy et al 1994).

Abbas: Most Th1 clones don't have high affinity receptors for IL-1, so it is difficult to understand how IL-1 can have an effect on Th1 cells (Lichtman et al 1988).

Lotze: Alan Sher, you believe that IL-10 is relatively unimportant. However, it plays an anti-inflammatory role in terms of suppressing adverse consequences that are presumably due to nitric oxide (NO) production (Alleva et al 1994). Have you tried to block these adverse effects in an IL-10 knockout mouse by adding *N*-methylaspartate or aminoguanidine, which would block the production of NO? Also, have you added exogenous IL-10, either in IL-10 knockout mice or normal mice, to see what it does in the establishment of active infection in normal mice? It may promote specific immunity.

Sher: We haven't done any of those experiments. It has been difficult to prove that the overproduction of these cytokines in IL-10 knockout mice is important for pathology. This is because IL-12, TNF and IFN-γ are all important for resistance. If they are depleted, the mouse becomes susceptible and dies of infection. We have to play with the concentrations of these cytokines to get them down to a threshold level where they control the infection but don't induce pathology.

Lotze: It is possible that IL-10 plays a more interesting role in your model because IL-10 knockout mice die rapidly on Day 10. IL-10 is a $CD8^+$ cell chemoattractant, so do you see decreased $CD8^+$ cells at the sites of active infection? Also, IL-10 induces NK cell activity and lymphokine-activated killer (LAK)-like activity. Have you examined the role of NK and LAK cells in your model?

Sher: No.

Coffman: Alan Sher is interpreting a dysregulated Th1 inflammatory response as being the main source of pathology. This is consistent with studies in Donna Rennick's lab (D. Berg & D. Rennick, personal communication). They looked at various cutaneous inflammatory reactions in both normal and IL-10 knockout mice. These included chemically induced inflammation and contact sensitivity. In all cases, a more severe and/or more prolonged response was produced in IL-10 knockout mice. The primary site of action of IL-10 in these responses appears to be modulation of the production of TNF and other monokines.

Liew: Do you know what antigen induces the production of IL-10?

Sher: Not yet. We are currently purifying the major monokine-inducing molecules from the parasite. The IL-10 molecule will probably be one of them.

Liew: Is there a relationship between the level of the IL-10-inducing factor and the virulence of the parasite?

Sher: That's a lovely question. We are hoping to use transfection technology as an approach for examining the relationships between cytokines, superantigen responses and virulence in *T. gondii.*

Kaufmann: In the discussion after Tim Mosmann's presentation, a correlation between pregnancy and Th2 responses was suggested. We know that *T. gondii* and *Listeria monocytogenes* are frequent causes of abortion. On the other hand, they are also potent inducers of Th1 responses. Is there any relationship between these two phenomena?

Sher: I know of no direct evidence that supports such a relationship.

Mosmann: There is evidence that both *Toxoplasma* and *Listeria* cause more serious infections in pregnant mice, which is consistent with the idea that pregnancy weakens the required Th1 response against these parasites (Luft & Remington 1982).

Kaufmann: L. monocytogenes and *T. gondii* both suppress Th2 responses. Is there any influence on Th2 responses during pregnancy?

Sher: Spontaneous abortion is one consequence of congenital toxoplasmosis. Tim Mosmann might argue that a weakening of the Th2 response induced by infection contributes to this phenomenon.

References

Alleva DG, Burger CJ, Elgert KD 1994 Tumor-induced regulation of suppressor macrophage nitric oxide and TNF-α production—role of tumor-derived IL-10, TGF-β, and prostaglandin E2. J Immunol 153:1674–1686

Bendelac A, Killeen N, Littman DR, Schwartz RH 1994 A subset of $CD4^+$ thymocytes selected by MHC class I molecules. Science 263:1774–1778

D'Andrea A, Aste-Amezaga M, Valiante NM, Ma XJ, Kubin M, Trinchieri G 1993 Interleukin-10 (IL-10) inhibits human lymphocyte interferon-γ production by suppressing natural killer cell stimulatory factor/IL-12 synthesis in accessory cells. J Exp Med 178:1041–1048

Denkers EY, Caspar P, Sher A 1994 *Toxoplasma gondii* possesses a superantigen activity that selectively expands murine T cell receptor Vβ5-bearing $CD8^+$ lymphocytes. J Exp Med 180:985–994

Gazzinelli RT, Hakim FT, Hieny S, Shearer GM, Sher A 1991 Synergistic role of $CD4^+$ and $CD8^+$ T lymphocytes in IFN-γ production and protective immunity induced by an attenuated *Toxoplasma gondii* vaccine. J Immunol 146:286–292

Gazzinelli R, Xu Y, Hieny S, Cheever A, Sher A 1992 Simultaneous depletion of $CD4^+$ and $CD8^+$ T lymphocytes is required to reactivate chronic infection with *Toxoplasma gondii.* J Immunol 149:175–180

Hunter CA, Litton MJ, Remmington JS, Abrams JS 1994 Immunocytochemical detection of cytokines in the lymph nodes and brains of mice resistant or susceptible to toxoplasmic encephalitis. J Infect Dis 170:939–945

Kaufmann SHE 1988 CD8$^+$ T lymphocytes in intracellular microbial infections. Immunol Today 9:168–174

Lichtman AH, Chin J, Schmidt JA, Abbas AK 1988 The role of interleukin-1 in the activation of T lymphocytes. Proc Nat Acad Sci USA 85:9699–9703

Luft BJ, Brooks RG, Conley FK, McCabe RE, Remington JS 1984 Toxoplasmic encephalitis in patients with acquired immune response deficiency syndrome. JAMA 252:913–917

Luft BJ, Remington JS 1982 Effect of pregnancy on resistance to *Listeria monocytogenes* and *Toxoplasma gondii* infections in mice. Infect Immun 38:1164–1171

Murphy EE, Terres G, Macatonia SE et al 1994 B7 and IL-12 cooperate for proliferation and IFN-γ production by mouse T helper clones that are unresponsive to B7 costimulation. J Exp Med 180:223–231

Rogers HW, Tripp CS, Schreiber RD, Unanue ER 1994 Endogenous IL-1 is required for neutrophil recruitment and macrophage activation during murine listeriosis. J Immunol 153:2093–2101

Suzuki Y, Remmington JS 1990 The effect of anti-IFN-gamma on the protective effect of LYT-2$^+$ immune T cells against toxoplasmosis in mice. J Immunol 144:1954–1956

The development of effector T cell subsets in murine *Leishmania major* infection

Richard M. Locksley, Adil E. Wakil, David B. Corry, Sabine Pingel, Mark Bix and Debbie J. Fowell

Departments of Medicine and Microbiology/Immunology, University of California, Box 0654, C-443, San Francisco, CA 94143, USA

Abstract. Leishmania major infection has proven an exceptional model for $CD4^+$ subset development in inbred mice. Most strains contain infection coincident with the appearance of T helper 1 (Th1) cells that produce γ-interferon (IFN-γ) required for macrophage activation. In contrast, mice on the BALB background are unable to control infection due to the development of Th2 cells that produce counter-regulatory cytokines, particularly interleukin 4 (IL-4), capable of abrogating the effects of IFN-γ. Selective gene disruption studies in mice have illustrated critical components of the host response to *L. major*. Mice deficient in β_2 microglobulin, which have no major histocompatibility complex (MHC) class I or $CD8^+$ T cells, control infection as well as wild-type mice, whereas mice deficient in MHC class II (and $CD4^+$ T cells) suffer fatal infection. Mice with disruption of the gene encoding IFN-γ are also incapable of containing infection, reflecting absolute requirements for this cytokine. A number of interventions have been demonstrated to abrogate Th2 cell development in BALB mice, enabling these mice to control infection. Each of these—IL-12, anti-IL-4, anti-IL-2, anti-CD4 and CTLA4–Ig—has in common the capacity to make IL-4 rate limiting at the time of $CD4^+$ cell priming.

1995 T cell subsets in infectious and autoimmune diseases. Wiley, Chichester (Ciba Foundation Symposium 195) p 110–122

Leishmania are flagellated protozoa deposited into the blood pool on the skin that is created by the probing of the vector sandfly. The organisms infect only macrophages in the host, primarily by an interaction between the major promastigote surface molecule, lipophosphoglycan, and complement receptors CR1 and CR3 (Turco & Descoteaux 1992). The organisms undergo a marked reduction in total body mass, transforming to the intracellular amastigotes (the form in which they remain for their life in the vertebrate host). Most studies suggest that organisms persist for years, possibly for the life of the individual, even after spontaneous cure (Aebischer et al 1993). This is demonstrated by their recrudescence at the time of immunosuppression through drugs or illness, such as infection with HIV (Berenguer et al 1989).

Studies with mouse macrophages have delineated characteristics of the intracellular compartment inhabited by amastigotes that are consistent with the major histocompatibility complex (MHC) class II peptide loading compartment (Tulp et al 1994, Amigorena et al 1994). Organisms persist in an acidic, protease-rich environment that contains MHC class II molecules devoid of mature invariant chains (Lang et al 1994, Russell et al 1992), suggesting that *Leishmania* may flood the peptide loading compartment for newly synthesized class II molecules. This interpretation is consistent with observations that parasitized macrophages present non-parasite-derived, class II-dependent antigens poorly, despite normal levels of surface class II molecules (Prina et al 1993, Fruth et al 1993). MHC class I antigens are not detected within the compartment inhabited by *Leishmania*, and class I-dependent epitopes expressed as transgenes within the parasite are not presented by infected macrophages, suggesting that the class I processing pathway is not part of this compartment (Lopez et al 1993). The MHC class II dependence of *Leishmania* host defence was emphasized in studies in mice with disrupted MHC class II or class I molecules. MHC class II-deficient mice were unable to control parasite replication and suffered fatal infection (Locksley et al 1993), whereas MHC class I-deficient mice (through disruption of β_2 microglobulin) controlled infection as well as normal littermates (Wang et al 1993).

Invasion of bone marrow (Reiner et al 1994) or splenic (Hsieh et al 1995) macrophages by *Leishmania major* promastigotes was not associated with the induction of cytokines, including interleukin 12 (IL-12), in contrast to similar experiments using other intracellular organisms such as *Listeria* (Hsieh et al 1993) or *Toxoplasma* (Gazzinelli et al 1993). The identification of IL-12 as a primary determinant for T helper 1 (Th1) cell differentiation, together with the adverse effects of anti-IL-12 antiserum on the course of *L. major* infection in resistant strains of mice (Sypek et al 1993), raised questions regarding the signals for Th1 development during the host response to this organism. Studies *in vivo* demonstrated the appearance of transcripts for IL-12 p40 late in the first week after infection, suggesting that the intracellular amastigotes might be inducing IL-12 from macrophages. Indeed, incubation of purified amastigotes with bone marrow macrophages caused IL-12 p40 mRNA and protein production that was comparable to that induced by *Listeria* or endotoxin (Reiner et al 1994). These results suggest that promastigotes evade induction of cytokines from host macrophages until developmental maturation to the mammalian form, the amastigotes, which are more capable of withstanding the macrophage microbicidal armamentarium.

Although macrophages invaded by promastigotes generated little inflammatory cytokines, the commandeering of MHC class II molecules by

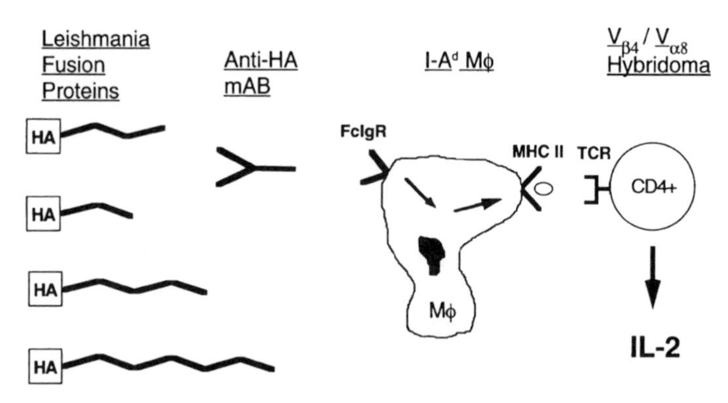

FIG. 1. Strategy for identification of major histocompatibility complex (MHC) class II *Leishmania major* antigens. Pools of expressed *L. major* cDNA sequences are tagged with an epitope from haemagglutinin (HA) and incubated with macrophages in the presence of anti-HA monoclonal antibody (mAB) prior to the addition of a T cell hybridoma that expressed the immunodominant Vα8/Vβ4 heterodimeric T cell receptor (TCR). Supernatants are screened for interleukin 2 (IL-2) production to identify the antigen capable of activating the hybridoma in the context of MHC class II I-A^d molecules (I-A^d Mϕ). FcIgR, FcIg receptor.

the organism led to marked activation of CD4$^+$ T cells in the host (Reiner et al 1994). In all inbred strains of mice investigated, CD4$^+$ T cells in the lymph nodes draining the site of inoculation were induced to produce IL-2 and IL-4, as well as γ-interferon (IFN-γ), at high levels that peaked at Day 4 of infection. Expansion of reactive CD4$^+$ T cells was not random; indeed, the selective appearance of CD4$^+$ T cells expressing the Vα8/Vβ4 heterodimeric T cell receptor (TCR) could be demonstrated by fluorescence cytometry (Reiner et al 1993). Similarly, analysis of T cell hybridomas made from *Leishmania*-activated CD4$^+$ T cells revealed the relative overusage of the Vα8/Vβ4 TCR with substantial charge conservation within the complementarity-determining region 3 deduced amino acid sequence.

These results suggested that an immunodominant antigen, perhaps by virtue of avidity for MHC class II or abundance following invasion, was the focus of the initial host CD4$^+$ response. To identify this immunodominant antigen, we expressed an *L. major* cDNA library in *Escherichia coli* as fusion proteins that were epitope tagged with a region from influenza haemagglutinin recognized by the monoclonal antibody 12CA5. We incubated bone marrow macrophages from BALB/c mice with pools of expressed proteins in the presence of 12CA5 to facilitate processing of antigens to the MHC class II pathway via the FcIg receptor. We incubated macrophages with IFN-γ to induce the MHC class II pathway, and we screened the macrophages using a T cell hybridoma with the Vα8/Vβ4 TCR to assay IL-2 production (Fig. 1). We subdivided the active pools until an individual active clone was identified. This was used subsequently to obtain the full sequence by rescreening the *L. major* cDNA library.

Using this strategy, we identified the immunodominant *Leishmania* antigen as a 36 kD homologue of RACK1, the receptor for activated protein kinase C (PKC), that had been cloned previously from rat brain (Mougneau et al 1995). RACK1 is a member of the ancient family of Trp-Asp repeat proteins that consist of a class of highly conserved proteins expressed in all eukaryotes but not prokaryotes (Neer et al 1994). In general, Trp-Asp repeat proteins serve as scaffolds for the assembly of multicomponent complexes, but they are neither substrates nor enzymes for their binding partners. RACK1 binds phospholipase C $\gamma 1$ and activated PKC, and it is thought to be the receptor on the inner side of the membrane to which PKC binds during its translocation from the cytosol to the membrane upon activation (Ron et al 1994). The *Leishmania* homologue, designated LACK, is highly homologous to RACK1, and was present in all *Leishmania* species studied, including *L. major*, *L. chagasi*, *L. donovani* and *L. amazonensis*. Although LACK was the target of the early restricted $V\alpha 8/V\beta 4$ response, recombinant RACK1 had no reactivity with the T cell hybridoma (Mougneau et al 1995).

Identification of the target of the early $CD4^+$ response suggested that LACK might be an effective vaccine candidate for leishmaniasis. Prior studies had established the capacity of recombinant IL-12, given as two doses with soluble *Leishmania* antigens, to protect susceptible BALB/c mice subsequently challenged with a fatal inoculum (Afonso et al 1994). Similarly, BALB/c mice immunized twice with LACK plus IL-12 are protected from *L. major* challenge, and this protection correlates with a redirection of the underlying $CD4^+$ response from the non-healer Th2 to the healer Th1 phenotype (Mougneau et al 1995). The abundant *Leishmania* membrane protease, gp63, is not protective when used as an immunogen in the same experiments. This suggests that successful vaccine antigens must be immunodominant so that sufficient numbers of committed T cells can be targeted during subsequent infection to allow the redirection of the otherwise underlying immune responses.

Studies that identified the peak of $CD4^+$ cell cytokine production at Day 4 revealed the correlation between failure of BALB/c mice to down-regulate IL-4 mRNA and a fatal outcome (Reiner et al 1994). Increasing evidence suggests that the BALB genetic defect is expressed at the level of the $CD4^+$ T cell and is manifest as the production of augmented amounts of IL-4 during priming. Two recent reports using TCR transgenic mice bred onto the BALB background support this hypothesis. In the first, mice expressing influenza haemagglutinin in pancreatic β cells were crossed with mice containing a MHC class II-restricted TCR transgene specific for haemagglutinin (Scott et al 1994). These doubly transgenic mice were backcrossed onto the BALB/c and B10.D2 strains. The B10.D2 mice suffered islet cell infiltration and diabetes in association with Th1-like lymphokine generation by the infiltrating $CD4^+$ T cells. In contrast, the BALB/c mice suffered infiltration but not diabetes, and their lymphokine profile contained substantially more IL-4, reflecting a Th0-like profile. In the second report a $CD4^+$, MHC class II-restricted, ovalbumin-specific TCR transgene was backcrossed onto BALB/c and B10.D2 mice (Hsieh et al 1995). Priming of these cells with antigen-presenting cells (APCs) and antigen *in vitro* caused the development of T cells that produced substantially more IL-4 upon restimulation when cells from BALB/c, rather than B10.D2, were used. Crossing the respective APCs revealed that the defect segregated clearly with the $CD4^+$ T cell, and not with the background of the APC.

These results suggest that $CD4^+$ T cells from BALB/c mice have an inherent genetic defect causing them to produce too much IL-4 during priming. The requirement for IFN-γ in curing *L. major* infection was demonstrated using mice with disruption of the genes encoding IFN-γ or its receptor (Wang et al 1994a, Swihart et al 1995). However, BALB/c mice frequently have levels of IFN-γ comparable to healer strains, but the capacity of IL-4 to interfere with the macrophage-activating capacity of IFN-γ presumably accounts for the susceptibility of these mice and the success of anti-IL-4 antibodies in mediating protection (Sadick et al 1990). *Leishmania* causes marked $CD4^+$ T cell priming in the absence of APC-derived IL-12. This might compensate for aberrant IL-4 production (Wang et al 1994b), and it might account for the normal host response of BALB/c mice to pathogens such as *Listeria*. Analysis of all interventions that abrogate susceptibility of BALB/c mice to *L. major* (including sublethal irradiation, anti-CD4, anti-IL-2, anti-IL-2 receptor [R], anti-IL-4, anti-IL4R, IL-12 and CTLA4-Ig) reveals that each reduces the amount of IL-4 that would be available during T cell priming (Reiner & Locksley 1995). Once $CD4^+$ T cell priming to the Th1 phenotype has occurred, BALB mice are able to control *L. major* infection, demonstrating that the defect is subtle and remedial. This effect is observed only if interventions are used during the critical period of initial priming with antigen. Later interventions in the *L. major* system are ineffective.

Consideration of these various interventions in experimental leishmaniasis in the context of signals generated through the IL-2R and IL-4R suggests a possible focus for efforts at localizing the genetic defect in BALB mice (Fig. 2). The IL-2R has three components—IL-2Rα, IL-2Rβ and IL-2Rγ—the latter of which is shared with components of the IL-4R, IL-7R, IL-9R and IL-15R. The IL-4R has a unique chain, the IL-4Rα, that is paired with IL-2Rγ and transduces signals through a unique transcription factor, termed IL-4 STAT (signal transducer and activator of transcription) (Hou et al 1994). Several of the Janus tyrosine kinase (JAK) proteins, including JAK1 and JAK3, are shared components of the signalling cascade generated through the IL-2R and

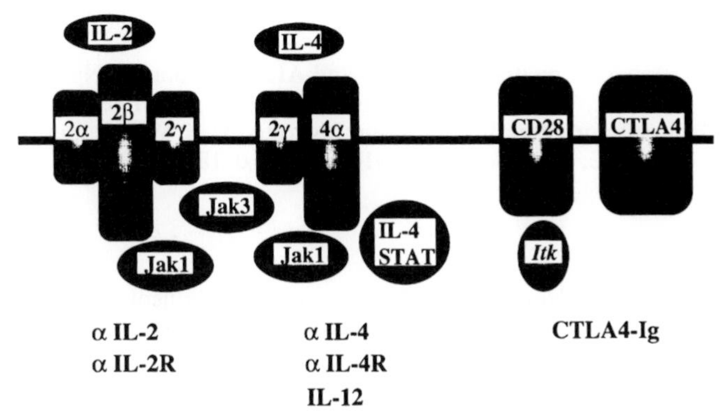

FIG. 2. Signalling through the interleukin 2 receptor (IL-2R) and IL-4 receptor (IL-4R). Members of the IL-2R and IL-4R complexes are shown, together with some of the identified intracellular signal transducing molecules. Interventions capable of abrogating the susceptibility of BALB/c mice given *Leishmania major* are shown at the bottom of the figure. α, antibody against; IL-4 STAT, IL-4 signal transducer and activator of transcription; Itk, inducible T cell kinase, JAK, Janus tyrosine kinase.

IL-4R (Witthuhn et al 1994). In the IL-2 knockout mouse IL-4 signalling becomes prominent, as shown by the skewed production of high levels of IgG1 and IgE, which are IL-4-driven immunoglobulin isotypes that disappear when these mice are crossed with IL-4 knockout mice (Sadlack et al 1994). The IL-2Rβ chain knockout mice have an even more severe phenotype and they die with autoimmune haemolytic anaemia, other autoantibodies, elevated IgG1 and IgE levels and colitis (T. Mak, personal communications). Interestingly, crossing the IL-2 knockout mice onto the BALB/c background causes a phenotype identical to the IL-2Rβ chain knockout mice, demonstrating that the BALB gene complemented the milder phenotype of the IL-2 knockout mice (W. Muller, personal communication). These experiments suggest that the BALB defect may lie within shared components of the IL-2R/IL-4R signalling systems, such that disruption of the IL-2 pathway leaves unopposed components of the signal transduction system that are shared, such as the JAKs, for association with components of the IL-4 pathway.

The *Leishmania major* system has proven fertile soil for immunologists seeking to understand the host response to intracellular pathogens. The organisms' unique residence within MHC class II loading compartments provides an exceptionally robust model for examining CD4$^+$ T cell effector development *in vivo*. The next few years should see the elucidation of the BALB genetic defect itself, which should provide even further insights towards the pathways that determine how effector cells make critical decisions in response to discrete classes of organisms.

Acknowlegements

This work was supported by grants AI26918 and AI30663 from the National Institutes of Health to R. M. L., who is a Burroughs Wellcome Fund Molecular Parasitology Scholar. M. B. is supported by the Irvington Institute. A. E. W. is supported by NIH T32 DK07007. D. B. C. is supported by NIH T32 HL07185.

References

Aebischer T, Moody SF, Handman E 1993 Persistence of virulent *Leishmania major* in murine cutaneous leishmaniasis: a possible hazard for the host. Infect Immun 61:220–226

Afonso LCC, Scharton TM, Vieira LQ, Wysocka M, Trinchieri G, Scott P 1994 The adjuvant effect of interleukin-12 in a vaccine against *Leishmania major*. Science 263:235–237

Amigorena S, Drake JR, Webster P, Mellman I 1994 Transient accumulation of new class II MHC molecules in a novel endocytic compartment in B lymphocytes. Nature 369:113–120

Berenguer J, Moreno S, Cercenado E, Bernaldo e Quiros JL, Garcia del al Fuente A, Bouza E 1989 Visceral leishmaniasis in patients infected with human immunodeficiency virus (HIV). Ann Intern Med III:129–132

Fruth U, Solioz N, Louis JA 1993 *Leishmania major* interferes with antigen presentation by infected macrophages. J Immunol 150:1857–1864

Gazzinelli RT, Hieny S, Wynn TA, Wolf S, Sher A 1993 Interleukin 12 is required for the T-lymphocyte-independent induction of interferon-γ by an intracellular parasite and induces resistance in T-cell-deficient hosts. Proc Natl Acad Sci USA 90:6115–6119

Hou J, Schindler U, Henzel WJ, Ho TC, Brasseur M, McKnight SL 1994 An interleukin-4-induced transcription factor: IL-4 Stat. Science 265:1701–1706

Hsieh C-S, Macatonia SE, Tripp CS, Wolf SF, O'Garra A, Murphy KM 1993 Development of T_H1 $CD4^+$ T cells through IL-12 produced by *Listeria*-induced macrophages. Science 260:547–549

Hsieh CS, Macatonia SE, O'Garra A, Murphy KM 1995 T cell genetic background determines default T helper phenotype development *in vitro*. J Exp Med 181:713–721

Lang T, Hellio R, Kaye PM, Antoine JC 1994 *Leishmania donovani*-infected macrophages: characterization of the parasitophorous vacuole and potential role of this organelle in antigen presentation. J Cell Sci 107:2137–2150

Locksley RM, Reiner SL, Hatam F, Littman DR, Killeen N 1993 Helper T cells without CD4: control of leishmaniasis in CD4-deficient mice. Science 261:1448–1451

Lopez JA, LeBowitz JH, Beverley SM, Rammensee HG, Overpath P 1993 *Leishmania mexicana* promastigotes induce cytotoxic T lymphocytes *in vivo* that do not recognize infected macrophages. Eur J Immunol 23:217–223

Mougneau E, Altare F, Wakil AE et al 1995 Expression cloning of a protective *Leishmania* antigen. Science 268:563–566

Neer EJ, Schmidt CJ, Nambudripad R, Smith TF 1994 The ancient regulatory-protein family of WD-repeat proteins. Nature 371:297–300

Prina E, Jouanne C, de Souze Lao S, Szabo A, Guillet JG, Antoine JC 1993 Antigen presentation capacity of murine macrophages infected with *Leishmania amazonensis* amastigotes. J Immunol 151:2050–2061

Reiner SL, Locksley RM 1995 The regulation of immunity to *Leishmania major*. Annu Rev Immunol 13:151–177

Reiner SL, Wang ZE, Hatam F, Scott P, Locksley RM 1993 Th1 and Th2 cell antigen receptors in experimental leishmaniasis. Science 259:1457–1460

Reiner SL, Zheng S, Wang ZE, Stowring L, Locksley RM 1994 *Leishmania* promastigotes evade interleukin 12 (IL-12) induction by macrophages and stimulate a broad range of cytokines from CD4$^+$ T cells during initiation of infection. J Exp Med 179:447–456

Ron D, Chen CH, Caldwell J, Jamieson L, Orr E, Mochly-Rosen D 1994 Cloning of an intracellular receptor for protein kinase C: a homolog of the β subunit of G proteins. Proc Natl Acad Sci USA 91:839–843

Russell DG, Xu S, Chakroborty P 1992 Intracellular trafficking and the parasitophorous vacuole of *Leishmania mexicana*-infected macrophages. J Cell Sci 103:1193–1210

Sadick MD, Heinzel FP, Holaday BJ, Pu RT, Dawkins RS, Locksley RM 1990 Cure of murine leishmaniasis with anti-interleukin-4 monoclonal antibody. J Exp Med 171:115–127

Sadlack B, Kühn R, Schorle H, Rajewsky K, Müller W, Horak I 1994 Development and proliferation of lymphocytes in mice deficient for both interleukins-2 and -4. Eur J Immunol 24:281–284

Scott B, Liblau R, Degermann S et al 1994 A role for non-MHC genetic polymorphism in susceptibility to spontaneous autoimmunity. Immunity 1:73–82

Swihart K, Fruth U, Messmer N et al 1995 Mice from a genetically resistant background lacking the interferon-γ receptor are susceptible to infection with *Leishmania major* but mount a polarized T helper cell 1-type CD4$^+$ T cell response. J Exp Med 181:961–971

Sypek JP, Chung CL, Mayor SEH et al 1993 Resolution of cutaneous leishmaniasis: interleukin 12 initiates a protective T helper type 1 immune response. J Exp Med 177:1797–1802

Tulp A, Verwoerd D, Dobberstein B, Ploegh H, Pieters J 1994 Isolation and characterization of the intracellular MHC class II compartment. Nature 369:120–126

Turco SJ, Descoteaux A 1992 The lipophosphoglycan of *Leishmania* parasites. Annu Rev Microbiol 46:65–94

Wang ZE, Reiner SL, Hatam F et al 1993 Targeted activation of CD8 cells and infection of β2-microglobulin-deficient mice fail to confirm a primary protective role for CD8 cells in experimental leishmaniasis. J Immunol 151:2077–2086

Wang ZE, Reiner SL, Zheng S, Dalton DK, Locksley RM 1994a CD4$^+$ effector cells default to the Th2 pathway in interferon γ-deficient mice infected with *Leishmania major*. J Exp Med 179:1367–1371

Wang ZE, Zheng S, Corry DB et al 1994b Interferon γ-independent effects of interleukin 12 administered during acute or established infection due to *Leishmania major*. Proc Natl Acad Sci USA 91:12932–12936

Witthuhn BA, Silvennolnen O, Miura O et al 1994 Involvement of the Jak-3 Janus kinase in signalling by interleukins 2 and 4 in lymphoid and myeloid cells. Nature 370:153–157

DISCUSSION

Mitchison: Could you clarify what the transgenic T cell receptor (TCR) recognizes? Does it recognize the self-protein or the *Leishmania* protein?

Locksley: The transgenic TCR recognizes an 18 amino acid motif in the fourth Gly-His-Trp-Asp repeat of the immunodominant antigen (Mougneau et al 1995).

Mitchison: Do the TCR transgenic cells respond to the self-protein?

Locksley: No, there is no response to the self-protein.

Mitchison: Does the self-protein drive negative selection in the thymus?

Locksley: It was not negatively selected, it was positively selected.

Mitchison: Were the mice that expressed CD4 and the transgene stimulated by the self-protein, and do they respond?

Locksley: The transgenic T cells express the TCR but not CD4, and the question is, does this represent a partial deletion phenotype? We generated mice that were forced to express CD4 under a CD2 promoter, so that all the transgenic cells made CD4 artificially (R. Locksley, S. Reiner & N. Killeen, unpublished results). However, these cells also populated the periphery and interacted equivalently with the antigen.

Flavell: Is there an interaction between the self-RACK1 (receptor for activated protein kinase C) and the *Leishmania* homologue, LACK?

Locksley: No. There is no interaction between the self-RACK1 and LACK. They differ at six out of 13 amino acids. The complementarity determining region 3 of reactive T cell hybridomas has a positive charge in the middle of the sequence, and there is a corresponding negative charge in the middle of the T cell epitope. This charge is missing in the self-antigen.

Flavell: Have you looked at human serum albumin staining in the thymus?

Locksley: Yes. Neither human serum albumin nor CD69 staining have been particularly informative.

Lachmann: If you gave anti-*Leishmania* IgE passively to the transgenic BALB/c mice that do heal, would you then expect them not to heal? Interleukin 4 (IL-4) promotes IgE formation, but IgE (bound to mast cells and triggered by antigen) in its turn gives rise to IL-4 release from mast cells. If IL-4 is required early after infection, its lack may be because these mice cannot make IgE and, therefore, cannot recruit mast cell IL-4.

Locksley: Passive antibody transfers in *Leishmania* have not produced any effects. We have also tried to add IL-4 directly by giving large amounts of IL-4 and anti-IL-4 antibody in a complex to see if this shifts the phenotype. Our first attempt at this did not shift the phenotype.

Sher: A normal IL-4 response is initiated in schistosome-infected Fcε receptor knockout mice (A. Sher, unpublished observations).

Coffman: The BALB/c transgenic mice heal but this may not mean that there is a defect in the early source of IL-4. The phenotype suggests that T helper 2 (Th2) cells are being made for the first four weeks. Are Th2 cytokines produced during this early phase of the response?

Locksley: IL-4 and γ-interferon (IFN-γ) are produced in mice on the BALB/c background.

Coffman: The anti-IL-4 antibody experiment suggests that the Th2 phenotype is due to the early presence of IL-4. If the source of the early IL-4 was depleted in the transgenic mice, it seems as though they should make a more Th1-like response from the beginning.

Locksley: One hypothesis is that BALB/c mice produce too much IL-4 in two CD4$^+$ cell populations: a traditional lineage and the NK1.1$^+$ lineage. Hsieh et al (1995) have worked on the ovalbumin-specific TCR transgene. They have backcrossed these transgenic mice to BALB/c and B10.D2, and they have observed the same phenotype as ours. Scott et al (1994) have observed similar findings in the insulitis diabetes model. IL-4 is produced without diabetes on the BALB/c background, and diabetes and a strong Th1 response are produced on the B10.D2 background. There may be a polymorphism in the amount of IL-4 that CD4$^+$ cells make, and BALB/c mice may have this aberrant polymorphism. We suggest that the polymorphism has to be present in both the major histocompatibility complex (MHC) class II lineage and the NK1.1$^+$ lineage of CD4$^+$ T cells in order to generate mice susceptible to *Leishmania*. Shankar & Titus (1995) showed recently that C57BL/6 T cells are cured when they are placed into BALB/c mice, and that BALB/c T cells are cured when they are placed into C57BL/6 mice. They concluded that the gene defect was in both the T cell compartment and the non-T cell compartment because the C57BL/6 mice were reconstituted with their own bone marrow cells. However, the NK1.1$^+$ CD4$^+$ cells are generated from the bone marrow lineage, whereas the MHC class II-restricted CD4$^+$ T cells are generated in the thymus. Therefore, they did not reconstitute the two T cell lineages in concert. The hypothesis that this phenotype must be expressed in all of the T cell lineages for BALB/c mice to become susceptible to *Leishmania* is testable. We're crossing BALB/c mice with β_2 microglobulin knockout mouse because that will eliminate the population of NK1.1$^+$ CD4$^+$ cells, which are MHC Class I restricted.

Shearer: If one studies murine lupus induced either by a graft-versus-host reaction or with the 16.6 idiotype, C57BL/6 mice are genetically resistant and both BALB/c and DBA/2 mice are susceptible (Via & Shearer 1988). Have you looked at any BXD recombinant inbred strains that could permit genetic mapping?

Locksley: Genetic crosses have mapped *Leishmania* susceptibility to chromosome 11. This is interesting because many other relevant genes are located on chromosome 11, i.e. the genes that encode IL-4, IL-12 p40, IL-13, the C–C chemokine family, inducible T cell kinase, IL-2 receptor β chain and inducible nitric oxide synthase. This mapping has to be done more precisely if we want to determine what this gene defect is. I predict that it will be a defect in the IL-2 or IL-4 signalling pathway.

Allen: Your transgenic mice are reminiscent of the AND transgenic mice that Kaye et al (1989) generated, where the two TCR chains did not exist in an actual T cell clone.

Locksley: That's not quite correct. Although other T cells had the same α chains that are in the starting clone 9.12, we used the paired α and β chains.

Allen: There was a problem with the T cells in H-2^k homozygous mice because there was too much ligand for the developing T cells. Therefore, Kaye et al (1985) made H-$2^{k/b}$ heterozygotes, which had half the level of I-E^k, and they found that T cells developed normally. Have you done a similar experiment to see if the MHC density is lowered in the thymus?

Locksley: That's a good experiment. We haven't done it.

Abbas: You presented some circumstantial evidence that either the levels of IL-2 or the responsiveness to IL-2 is the key determinant for a Th1 response, as opposed to a Th2 response. Hsieh et al (1995) have also observed correlations between the amount of IL-2 and Th1/Th2 cell development. It may be possible to change the response pattern by reintroducing IL-2 into your cultures. If the amount of IL-2 produced is the defect, the susceptibility will map close to the gene encoding IL-2.

Locksley: This circumstantial evidence relates to studies of IL-2 and IL-2 receptor β chain knockout mice. These mice have high levels of IgE, which suggests that IL-4 is dysregulated when IL-2 is deficient. The IL-2 receptor β chain knock-out is more greatly affected (Schorle et al 1991, Suzuki et al 1995) but the IL-2 knockout behaves like the IL-2 receptor β chain knockout when crossed onto the BALB/c background (W. Müller, personal communication). However, I'm not sure that a defect in the level of IL-2 is responsible. There may be a defect in the level of IL-4, although Hsieh et al (1995) suggest that it is IL-2. They have neutralized IL-2 and then added exogenous human IL-2 to control the amount of IL-2 in each culture during priming. They claim that they can reduce the amount of IL-4 that a BALB/c T cell makes down to the levels made by B10.D2 cells. I should point out to those who are not so familiar with the system that if antigen-presenting cells are crossed with T cells during priming, the BALB/c defect segregates with the $CD4^+$ cells. The obvious experiments that we must do are to cross the T cells in our mouse strains and experiments similar to those of Kearney et al (1994), i.e. put a known number of transgenic T cells into normal BALB/c mice to determine whether they become Th2 cells.

Mitchison: Is someone doing microsatellite genetics on this particular type of disease susceptibility?

Locksley: We are doing these experiments, and I'm sure that other people are too.

Mosmann: Do $CD4^-$ $CD8^-$ cells appear in normal mice, i.e. is this a normal T cell population?

Locksley: All of the original clones in these experiments expressed CD4. The original analysis of the Vα8/Vβ4 heterodimer was in a typical $CD4^+$ T cell. We don't observe double negatives expanding in wild-type BALB/c mice infected with *Leishmania*.

Mosmann: So how do the CD4$^+$ cells survive negative selection in the thymus in the normal situation?

Locksley: This could be explained by transgenesis. We have introduced a TCR transgene that is expressed at a slightly higher level and slightly earlier. It is possible that this causes them to modulate the CD4$^+$ co-receptor.

Flavell: Is it a standard TCR construct?

Locksley: Yes. It has the appropriate enhancer and promoter, and it is 8–12 kb long. It is essentially genomic.

Flavell: They may skip the double-positive stage altogether.

Locksley: No. They go through a double-positive stage.

Mitchison: Another polymorphism has been described by Asherson et al (1990) in which there are variations in IFN-γ and also IL-5. This polymorphism is also MHC linked, which distinguishes BALB/c (H-2^d) mice from BALB/k (H-2^k) mice. Have you looked at BALB/b, BALB/d and BALB/k?

Locksley: BALB/b, BALB/c and BALB/k mice all die of leishmaniasis but there are slight differences in sensitivity—BALB/c mice are more sensitive than BALB/b, and both are more sensitive than BALB/k mice.

Mitchison: In your hands gp63 did not seem to be involved. This molecule has been advocated as a vaccine antigen, especially when encoded in mycobacteria.

Locksley: It depends on the strain. Our strain of *Leishmania* expresses relatively small amounts of gp63 on the surface, but other strains express a much higher level.

Mitchison: So from the vaccine angle there are now two candidate antigens: your protein, and gp63?

Locksley: There isn't any evidence that gp63 is a major target of the early immune response to natural infections. I agree that there are various systems in which people have demonstrated protection by gp63. It is probable that these complex parasites will require multiple antigens. At the moment there are two candidates but there will be many others, including some stage-specific ones.

Liew: We are testing this in the *Salmonella* system by delivering gp63, p24 and a new antigen, gp55.

Lamb: What happens when you reinfect the BALB/c transgenics?

Locksley: They're resistant.

Sher: Do you have any evidence that in mice immunized with unfractionated soluble *Leishmania* antigen (SLA) plus IL-12 that the former molecule is the major antigen recognized?

Locksley: Mice that received LACK plus IL-12 were always slightly less resistant than mice that received SLA plus IL-12 or anti-IL-4. Two thirds of the former mice healed as well as with SLA plus IL-12, and one third were worse. This may be because SLA primes with greater number of antigens, so more T cells are primed. This may be required for an optimal response.

Lotze: You suggested that *Leishmania* may be able to target the compartment that contains class II molecules. What is the mechanism by which it enters that compartment? Is there a precise mechanism that targets it to the compartment or is that where it lives and multiplies?

Locksley: That's where it lives and multiplies but what targets it there is not known. When *Leishmania* infects a host, it persists for life. We know this from studies of HIV infection. Individuals can be infected asymptomatically with *Leishmania* and then five to 10 years later after subsequent infection with HIV, they can suffer reactivation of *Leishmania* with the appearance of clinical disease. There is little evidence that sterile immunity occurs in humans infected with *Leishmania*.

Lotze: Does *Leishmania* persist in a latent state in the host?

Locksley: Yes.

References

Asherson GL, Dieli F, Gautman Y, Siew LK, Zembala M 1990 Major histocompatibility complex regulation of the class of the immune response: the H-2^d haplotype determines poor interferon-gamma response to several antigens. Eur J Immunol 20:1305–1310

Hsieh C-S, Macatonia SE, O'Garra A, Murphy KM 1995 T cell genetic background determines default T helper phenotype development *in vivo*. J Exp Med 181:713–721

Kaye J, Hsu M-L, Sauron M-E, Jameson SC, Gascoigne NRJ, Hedrick SM 1989 Selective development of $CD4^+$ T cells in transgenic mice expressing a class II-restricted antigen receptor. Nature 341:746–749

Kearney ER, Pape KA, Loh DY, Jenkins MK 1994 Visualization of peptide-specific T cell immunity and peripheral tolerance induction *in vivo*. Immunity 1:327–339

Mougneau E, Altare F, Wakil AE et al 1995 Expression cloning of a protective *Leishmania* antigen. Science 268:563–566

Schorle H, Holtschke T, Hünig T, Schimpl A, Horak I 1991 Development and function of T cells in mice rendered interleukin 2 deficient by gene targeting. Nature 352:621–624

Scott B, Liblau R, Degermann S et al 1994 A role for non-MHC genetic polymorphism in susceptibility to spontaneous autoimmunity. Immunity 1:73–82

Shankar AH, Titus RG 1995 T cell and non-T cell compartments can independently determine resistance to *Leishmania major*. J Exp Med 181:845–855

Suzuki H, Kundig TM, Furlonger C et al 1995 Deregulated T cell activation and autoimmunity in mice lacking interleukin 2 receptor β. Science 268:1472–1475

Via CS, Shearer GM 1988 T-cell interactions in autoimmunity: insights from a murine model of graft-versus-host disease. Immunol Today 9:207–213

T cells and cytokines in intracellular bacterial infections: experiences with *Mycobacterium bovis* BCG

Stefan H. E. Kaufmann*, Ch. H. Ladel and Inge E. A. Flesch

*Department of Immunology, University of Ulm, Albert-Einstein-Allee 11, D-89070 Ulm, and *Max-Planck-Institute for Infectious Biology, Monbijoustrasse 2, 10117 Berlin, Germany*

Abstract. Intracellular bacteria reside in mononuclear phagocytes, and protective immunity is dominated by T lymphocytes. *Mycobacterium bovis* bacillus Calmette–Guérin (BCG) infection of mice represents an excellent model for studying immune mechanisms involved in defence against persistent intracellular bacteria that cause chronic disease. Gene disruption mutant mice include: $A\beta^{-/-}$, which lack conventional $CD4^+$ T cell receptor α/β (TCRα/β) T lymphocytes; β_2 microglobulin$^{-/-}$, which lack conventional $CD8^+$ TCRα/β lymphocytes; TCR$\beta^{-/-}$, which lack all TCRα/β lymphocytes; TCR$\delta^{-/-}$, which lack all TCRγ/δ lymphocytes; and RAG-1$^{-/-}$ mutants, which lack mature T and B lymphocytes. Studies of these mutants suggest that $CD4^+$ TCRα/β, $CD8^+$ TCRα/β and TCRγ/δ T lymphocytes all contribute to immunity against *M. bovis* BCG. Activation of antibacterial effector functions in macrophages by T helper 1 (Th1) cell-derived γ-interferon (IFN-γ) is central to protection. In contrast, Th2 cells are only marginally involved. Activation of Th1 and Th2 cells is regulated by interleukin 10 (IL-10) and IL-12, which are induced early in infection with *M. bovis* BCG. Although IL-12 is stimulated by *M. bovis* BCG in immunocompetent mice, studies with IFN-γ receptor-deficient and tumour necrosis factor α (TNF-α) receptor-deficient mutant mice suggest that *M. bovis* BCG-induced IL-12 secretion depends on IFN-γ and TNF-α. Hence, IL-12 cannot be the first cytokine produced during *M. bovis* BCG infection.

1995 T cell subsets in infectious and autoimmune diseases. Wiley, Chichester (Ciba Foundation Symposium 195) p 123–136

The control of infectious disease depends on specific T lymphocytes. Four paradigmatic types of antimicrobial defence can be distinguished. (1) T helper 1 (Th1) cells, which activate antimicrobial functions in professional phagocytes via certain cytokines (in particular γ-interferon [IFN-γ]), determine protection against intracellular bacteria and protozoa (Kaufmann 1993, Scott & Sher 1993). (2) Cytotoxic T lymphocytes (CTLs), which lyse infected target cells, are central to protection against many viral infections (Zinkernagel 1993). Differentiation of CTLs is regulated by Th1 cells through interleukin 2 (IL-2) and perhaps other cytokines. (3) Th2 cells activate other effector cells which control, for example, helminth infections (Scott & Sher 1993). (4) Antibodies secreted by B lymphocytes neutralize microbial toxins and certain viruses, and thus mobilize humoral effector mechanisms that protect against many extracellular bacteria (Gotschlich 1993). B cell maturation into antibody-secreting plasma cells is controlled by Th2 cells through the secretion of IL-4 and IL-5.

Th1 and Th2 cells serve as central regulators in all four types of antimicrobial immunity. They express a T cell receptor that is composed of an α chain and a β chain (TCRα/β), and they also express the co-receptor CD4 (Cardell et al 1994). CTLs express TCRα/β together with the CD8 co-receptor, and TCRγ/δ lymphocytes characteristically lack both the CD4 and the CD8 molecule but express a second TCR composed of a γ chain and a δ chain (Raulet 1994, Haas et al 1993). In contrast, B cells bear a completely different receptor for the antigen, namely surface immunoglobulin. These three lymphocyte populations can be distinguished phenotypically from Th1 and Th2 cells because Th1 and Th2 cells cannot be differentiated by specific cell surface markers, rather their distinction follows their characteristic cytokine patterns, i.e. IFN-γ and IL-2 for Th1 cells, and IL-4 and IL-5 for Th2 cells (Seeder & Paul 1994).

Successful control of infectious disease depends on the activation of a predominant T cell type. Early mechanisms that control the development of these T cells are, therefore, of extreme importance. Although other factors participate in the divergence into Th1 or Th2 cells, the decisive factor is the secretion of cytokines by cells of the innate immune system. The decision between CD4$^+$ and CD8$^+$ T cells is influenced by cytokines, but it is primarily determined by the two different antigen-processing pathways (McCoy 1994). Microbial peptides that reside in the cytosol provide the adequate recognition signal for CD8$^+$ T cells, whereas peptides that exist in the endosome supply the appropriate signal for recognition by CD4$^+$ T cells. CD4$^+$ T cell and CD8$^+$ T cell activation is regulated primarily by class II and class I major histocompatibility complex (MHC) molecules, respectively.

The four immune responses outlined above are not activated in an exclusive way in a given infection. Rather, the immune response to a certain pathogen is a complex network comprising several types of T cells and innate immune cells that interact with each other. The outcome of the host response must, therefore, be viewed as a labile system encompassing both protective and pathogenic avenues. In this paper we will focus on the immunity to intracellular bacteria that is dominated by Th1 cells. These cells activate macrophages via IFN-γ. It is likely that other T cell subsets are activated (to a lesser degree) in addition to the activation of Th1 cells. Some of these cells contribute to protection, and others contribute to pathology. We will discuss recent results from experimental infection of mice with *Mycobacterium bovis* bacillus Calmette–Guérin (BCG). We will focus on the effector phase, i.e. on the cells needed for optimum protection, and on the initiation phase, i.e. on parameters involved in the induction of these cells.

The effector phase *in vitro*: cytokines involved in activation of antimicrobial macrophages

The attenuated strain of *M. bovis* BCG is used widely as a vaccine against tuberculosis, and it has varying efficacy. Systemic infection with *M. bovis* BCG causes chronic infection in mice. The microorganisms remain in the phagosome after they are engulfed. Protection is considered to rest exclusively on MHC class II-restricted $CD4^+$ T cells, which produce IFN-γ (Kaufmann 1993). Consistent with this notion, *in vitro* activation of murine bone marrow-derived macrophages (BMM) with recombinant IFN-γ induced potent growth inhibition of *M. bovis* (Flesch & Kaufmann 1987). This antibacterial effect is due to the activation of nitric oxide synthase, which results in the potent production of reactive nitrogen intermediates (RNI) (Flesch & Kaufmann 1991). Infection of recombinant IFN-γ-stimulated BMM with *M. bovis* also induced synthesis of certain cytokines, including tumour necrosis factor α (TNF-α) and IL-10 (Flesch et al 1994). TNF-α participates in various Th1-controlled immune reactions, whereas IL-10 is a marker cytokine of Th2-type immunity (Seeder & Paul 1994).

IFN-γ stimulation of BMM alone failed to induce RNI production and required co-stimulation with recombinant TNF-α (Flesch et al 1994). Moreover, addition of anti-TNF-α antiserum to recombinant IFN-γ-stimulated and *M. bovis*-infected BMM interfered markedly with RNI production and mycobacterial growth inhibition. This suggests that macrophages, stimulated with recombinant IFN-γ and infected with *M. bovis*, produce TNF-α, and that endogenous or autocrine TNF-α mediates antimycobacterial activities. Addition of recombinant IL-10, either concomitantly or after stimulation with recombinant IFN-γ, had no influence on macrophage activation, as measured by mycobacterial growth inhibition, TNF-α production and RNI synthesis. However, recombinant IL-10, when added prior to recombinant IFN-γ stimulation and *M. bovis* infection, inhibited these activities markedly. We conclude that IL-10 antagonizes the IFN-γ-induced antimycobacterial macrophage functions. Although our conclusions are based on *in vitro* analyses, it is interesting that recombinant IL-10 was only inhibitory when its addition preceded that of recombinant IFN-γ. This time dependency could explain why potent activation of antimycobacterial functions by IFN-γ occurs in macrophages despite autocrine IL-10 production. Certain macrophages that encounter IL-10 before IFN-γ would be inactivated (Fig. 1). In this way,

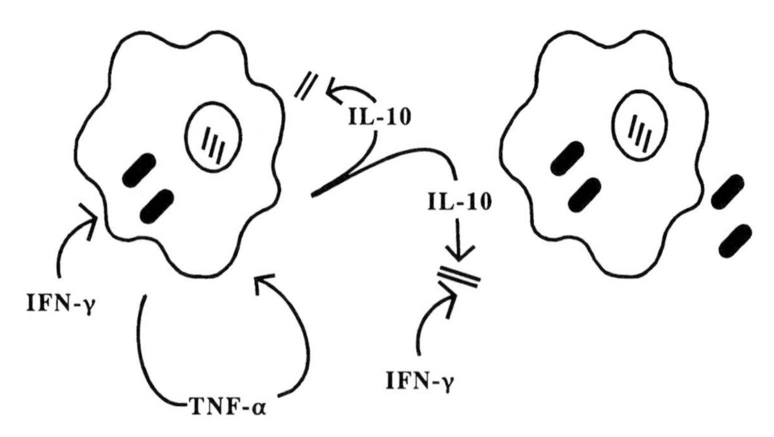

FIG. 1. Effect of recombinant γ-interferon (IFN-γ) with either tumour necrosis factor α (TNF-α) (synergistic effect) or interleukin 10 (IL-10) (antagonistic effect) on macrophage activation. Endogenous TNF-α and IL-10 are induced in IFN-γ-stimulated, *Mycobacterium bovis* bacillus Calmette–Guérin (BCG)-infected macrophages. Endogenous TNF-α is required as a co-stimulator for activation of antimycobacterial activities in macrophages by IFN-γ. In contrast, IL-10 inhibits IFN-γ stimulation, but only in naive macrophages that have not previously been exposed to IFN-γ and *M. bovis* BCG. Compiled from data by Flesch et al (1994).

macrophage activation could be prevented from becoming too extensive and, therefore, detrimental to the host organism. It is not known if this mechanism contributes to the containment of macrophage activation to the site of mycobacterial implantation.

The effector phase *in vivo*: *Mycobacterium bovis* BCG infection of T cell-deficient mutant mouse strains

The availability of gene deletion mutant mice with defined defects in their immune system has provided new approaches towards the analysis of antimicrobial host responses *in vivo* (Kaufmann & Ladel 1994). We have studied experimental infection with *M. bovis* BCG in the following gene deletion mutants:

(1) $A\beta^{-/-}$ mutant mice, which lack surface-expressed MHC class II molecules and are, therefore, devoid of functional $CD4^+$ T cells (Ladel et al 1995a);
(2) β_2 microglobulin-deficient mutant mice, which lack surface-expressed MHC class I molecules and are, therefore, devoid of thymus-dependent $CD8^+$ T cells (Ladel et al 1995a);
(3) $TCR\beta^{-/-}$ mutant mice, which lack all $TCR\alpha/\beta$ cells (Ladel et al 1995b);
(4) $TCR\delta^{-/-}$ mutant mice, which are devoid of all γ/δ T cells (Ladel et al 1995b);

(5) RAG-$1^{-/-}$ mutant mice, which lack all mature T and B lymphocytes (Ladel et al 1995b).

The RAG-$1^{-/-}$, TCR$\beta^{-/-}$ and A$\beta^{-/-}$ mutants suffered from exacerbated infection and ultimately died (Table 1). The TCR$\delta^{-/-}$ mutants controlled infection equally as well as their control littermates over long periods of time; however, bacterial numbers were elevated slightly. This increase started at around four months and was statistically significant. Although the β_2 microglobulin$^{-/-}$ mutants were able to control lower doses of *M. bovis* BCG infection, they failed to achieve control of higher inocula. Thus, half of the β_2 microglobulin$^{-/-}$ mice, but none of the β_2 microglobulin$^{+/-}$ controls, died within six months of infection with 3×10^6 *M. bovis* BCG organisms. We conclude from these findings that: (1) TCRα/β CD4$^+$ T cells are central to protection against *M. bovis* BCG; (2) γ/δ T cells play an auxiliary role that is manifested at later stages of infection and; (3) TCRα/β CD8$^+$ T lymphocytes play a more important role than generally assumed. Hence, our results challenge the dogma that *M. bovis* BCG vaccination fails to activate CD8$^+$ T lymphocytes.

IFN-γ production by T cells from *Mycobacterium bovis* BCG-infected T cell-deficient knockout mice

IFN-γ production by T lymphocytes from the first four mouse mutants described above was analysed after *in vitro* restimulation with mycobacterial antigens. Only spleen cells from β_2 microglobulin$^{-/-}$ knockout mice produced appreciable concentrations of IFN-γ. The failure of spleen cells from A$\beta^{-/-}$ and TCR$\beta^{-/-}$ mutants is in agreement with the notion that TCRα/β CD4$^+$ T cells are the major source of IFN-γ in mycobacterial infections. Low IFN-γ production by spleen cells from TCR$\delta^{-/-}$ mice, however, was unexpected and suggested an essential role of γ/δ T lymphocytes in IFN-γ production. A mixing experiment, using spleen cells from TCR$\beta^{-/-}$ and TCR$\delta^{-/-}$ mutants, led to the reconstitution of IFN-γ production in these cell mixtures (Table 2). IFN-γ secretion was not only re-established by mixing spleen cells from infected mice, but was also achieved when one of the two cell populations had been obtained from uninfected mice. We assume that the competence for IFN-γ production was induced in both α/β and γ/δ T cells independently from the other subset *in vivo*. Maximum IFN-γ production *in vitro*, however, not only involved antigen-specific T cell stimulation, but also regulatory interactions between the two T cell subsets. Support for such a bidirectional crosstalk has been obtained previously (Kaufmann et al 1993).

The initiation phase: IL-12 production depends on IFN-γ and tumour necrosis factor α

The development of Th1 and Th2 cells is regulated by cytokines at two different stages of maturation. Immediately after infection, cytokines produced

TABLE 1 Susceptibility to *Mycobacterium bovis* infection of mutant mouse strains

Mutant	*Susceptibility*
RAG-1$^{-/-}$	+ + +
TCR$\beta^{-/-}$	+ + +
A$\beta^{-/-}$	+ + +
β_2 microglobulin$^{-/-}$	+ +
TCR$\delta^{-/-}$	+

Mutant mice were infected with *Mycobacterium bovis* and survival recorded. In addition, bacterial growth in infected organs of surviving mice were determined at various time points. + + +, death; + +, death at higher inoculum; +, slight increase in bacterial growth at later time points. Compiled from data by Ladel et al (1995a,b).

by cells of the innate immune system determine maturation into Th1 or Th2 cells (Seeder & Paul 1994). Later, cytokines produced by the respective Th subset sustain this initial development. At both stages, regulation is complex, involving both positive and negative signals, many of which are not understood. Results from various models suggest that IL-12 is central to Th1 cell maturation (Trinchieri 1993). In contrast, IL-4 and IL-10 are considered essential for the development of Th2 cells. Immunocompetent mice were infected with *M. bovis* BCG, and splenic macrophages were restimulated *in vitro* with mycobacteria. These macrophages produced appreciable concentrations of both IL-10 and IL-12 (Flesch et al 1995). Thus, consistent with findings in other models, macrophages are good candidates for the early production of these two cytokines. To analyse the signals required for IL-12 production, we used BMM as a source of quiescent mononuclear phagocytes. Stimulation of these BMM with recombinant IFN-γ and infection with *M. bovis* BCG caused concomitant production of IL-12, IL-10 and TNF-α (see above and Flesch et al 1994, 1995). Addition, of anti-TNF-α antiserum to these cultures inhibited IL-12 production, although recombinant TNF-α failed to induce IL-12 synthesis by itself. These findings suggest that IL-12 synthesis by macrophages in mycobacterial infections is dependent on both IFN-γ and TNF-α. To analyse this issue in more depth, we used BMM from knockout mice that either lacked the IFN-γ receptor (IFN-γR$^{-/-}$) or the TNF-α receptor 1 (TNFR1$^{-/-}$) (Flesch et al 1995). Macrophages from both mutants failed to produce IL-12 in response to recombinant IFN-γ stimulation and *M. bovis* BCG infection. Moreover, spleen cells from *M. bovis* BCG-infected IFN-γR$^{-/-}$ or TNFR1$^{-/-}$ mice were unable to generate IL-12 mRNA and protein. These findings suggest a central role of IFN-γ and TNF-α in the regulation of IL-12 production during *M. bovis* BCG infection (Fig. 2). Hence, IL-12 cannot be the first cytokine produced in response to infection with *M. bovis* BCG. Although macrophages can serve as a source of TNF-α, these cells are unlikely to produce IFN-γ and other cells must be involved. Natural killer (NK) cells are potent IFN-γ producers but their activation depends on signals from other cells. In particular, IL-12 has been shown to activate NK cells and recent findings from this laboratory suggest that γ/δ T cells are involved in the control of NK cells (Ch. H. Ladel, C. Blum, I. E. A. Flesch, S. Tonegawa & S. H. E. Kaufmann, unpublished observations). Also, γ/δ T lymphocytes can produce IFN-γ; although, as discussed above, their activation depends on α/β T cells. Consequently, either an unknown cell is involved or a known cell is activated by an unknown mechanism. One possibility is that concomitant activation of α/β T cells and γ/δ T cells is mediated by superantigen-like molecules. In the human system conventional superantigens that stimulate α/β T cells have been identified in mycobacteria and low molecular weight fractions composed of phosphate and carbohydrates rapidly activate γ/δ T lymphocytes (Ohmen et al 1994, Schoel et al 1994).

Concluding remarks

This report describes our recent investigations on cytokine and T cell interactions in the Th1-dominated host response to the intracellular bacterium *M. bovis* BCG. The results illustrate how IFN-γ acts in concert with other

TABLE 2 Requirement for both α/β T cells and γ/δ T cells for optimum IFN-γ secretion during *Mycobacterium bovis* infection

T lymphocyte		
Infected	*Non-infected*	*IFN-γ*
Control	–	+++
α/β	–	+
γ/β	–	+
$\alpha/\beta+\gamma/\delta$	–	+++
α/β	γ/δ	+++
γ/δ	α/β	+++
–	$\alpha/\beta+\gamma/\delta$	–
None	–	–

Spleen cells from *Mycobacterium bovis*-infected or non-infected mice were restimulated *in vitro* with mycobacterial antigens. Pure α/β T cells were obtained from TCR$\delta^{-/-}$ mutants, pure γ/δ T cells from TCR$\beta^{-/-}$ mutants. γ-interferon (IFN-γ) production was measured by ELISA. +++, high IFN-γ concentration; ++, intermediate IFN-γ concentration; +, low-IFN-γ concentration; –, IFN-γ not detectable. Compiled from data by Ladel et al (1995a).

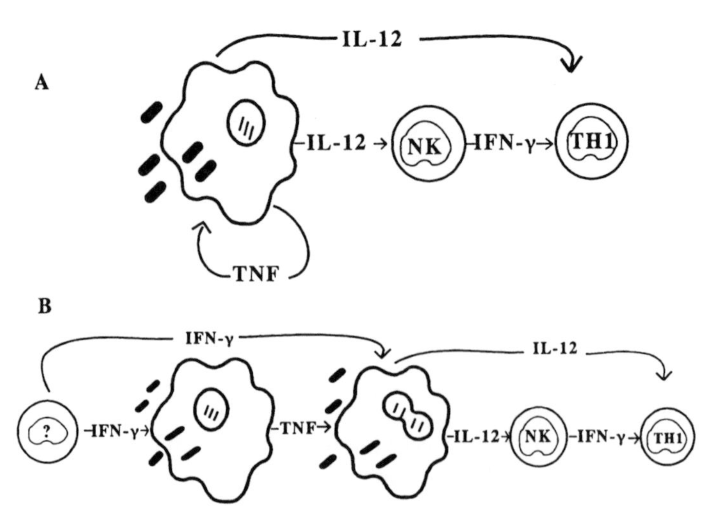

FIG. 2. Dependency of interleukin 12 (IL-12) production by *Mycobacterium bovis* bacillus Calmette–Guérin (BCG)-infected macrophages on γ-interferon (IFN-γ) and tumour necrosis factor (TNF). (A) According to current dogma, IL-12 is the first cytokine produced in infections with intracellular bacteria. (B) *M. bovis* BCG-infected macrophages from IFN-γR$^{-/-}$ or TNFR1$^{-/-}$ mutant mice failed to produce IL-12 in response to *M. bovis* BCG infection. These findings suggest a dependency on IFN-γ and TNF of IL-12 production. Although TNF could be produced by activated macrophages, the cellular source of IFN-γ remains to be identified. NK, natural killer cells; TH1, T helper 1 cell. Compiled from data by Flesch et al (1995).

cytokines. This is true for both the effector phase and the induction phase. In the effector phase, appropriate stimulation of antimycobacterial macrophage functions by IFN-γ is controlled by two macrophage-derived cytokines, TNF-α and IL-10. These have a positive and a negative input, respectively. In the induction phase, development of IFN-γ-producing T cells depends on the cytokine IL-12. In addition, our results suggest a dependency of IL-12 induction on IFN-γ and TNF-α in mycobacterial infections. Hence, these bacteria must not only be able to trigger IL-12 and TNF-α production in macrophages, but also possess the capacity to induce IFN-γ production in an as yet undefined cell population immediately after infection. Strong IL-10 production induced concomitantly by *M. bovis* BCG infection suggests that IL-12 effects dominate over IL-10 effects. This is consistent with published results obtained in an *in vitro* model by Hsieh et al (1993), who described the following dominance: IL-4 over IL-12 over IL-10. Recently, early IL-4 production has been described (Yoshimoto & Paul 1994). These IL-4 producers must be ignored or even inhibited in infections with intracellular bacteria such as *M. bovis* BCG.

Acknowledgements

This work received financial support from the Sonderforschungsbereich 322 'Lympho-Hämopoese', the German Science Foundation, project Ka 573/3-1 and the BMFT Verbundprojekt 'Mykobakterielle Infektionen'. We thank R. Mahmoudi for excellent secretarial help.

References

Cardell S, Merkenschlager M, Bodmer H et al 1994 The immune system of mice lacking conventional MHC class II molecules. Adv Immunol 55:423–440

Flesch I, Kaufmann SHE 1987 Mycobacterial growth inhibition by interferon-γ-activated bone marrow macrophages and differential susceptibility among strains of *Mycobacterium tuberculosis*. J Immunol 138:4408–4413

Flesch IEA, Kaufmann SHE 1991 Mechanisms involved in mycobacterial growth inhibition by gamma-interferon activated bone marrow macrophages: role of reactive nitrogen intermediates. Infect Immun 59:3213–3218

Flesch IEA, Hess JH, Oswald IP, Kaufmann SHE 1994 Growth inhibition of *Mycobacterium bovis* by IFN-γ stimulated macrophages: regulation by endogenous tumor necrosis factor-α and IL-10. Int Immunol 6:693–700

Flesch IEA, Hess JH, Huang S et al 1995 Early interleukin-12 production by macrophages in response to mycobacterial infection depends on interferon-γ and tumor necrosis factor α. J Exp Med 181:1615–1622

Gotschlich EC 1993 Immunity to extracellular bacteria. In: Paul WE (ed) Fundamental immunology. Raven Press, New York, p 1287–1308

Haas W, Pereira P, Tonegawa S 1993 γ/δ T-cells. Annu Rev Immunol 11:637–685

Hsieh C-S, Macatonia SE, Tripp CS, Wolf SF, O'Garra A, Murphy KM 1993 Development of T_H1 CD4$^+$ T cells through IL-12 produced by *Listeria*-induced macrophages. Science 260:547–549

Kaufmann SHE 1993 Immunity to intracellular bacteria. In: Paul WE (ed) Fundamental immunology. Raven Press, New York, p 1251–1286

Kaufmann SHE, Ladel CH 1994 Application of knock-out mice to the experimental analysis of the infections with bacteria and protozoa. Trends Microbiol 2:235–242

Kaufmann SHE, Blum C, Yamamoto S 1993 Crosstalk between a/β T cells and γ/δ T cells *in vivo*: activation of a/β T cell responses after γ/δ T cell modulation with the monoclonal antibody GL3. Proc Natl Acad Sci USA 90:9620–9624

Ladel CH, Daugelat S, Kaufmann SHE 1995a Immune response to *Mycobacterium bovis* bacille Calmette–Guérin infection in major histocompatibility complex I and II deficient mutant mice. Eur J Immunol 25:377–384

Ladel CH, Hess J, Daugelat S, Mombaerts P, Tonegawa S, Kaufmann SHE 1995b Contribution of α/β and γ/δ T lymphocytes to immunity against *Mycobacterium bovis* BCG infection: studies with T cell receptor-deficient knockout mice. Eur J Immunol 25:838–846

McCoy KL 1994 Mechanisms responsible for the processing and presentation of antigens to T cells. In: Snow EC (ed) Handbook of B and T lymphocytes. Academic Press, San Diego, CA, p 117–142

Ohmen JD, Barnes PF, Grisso CL, Bloom BR, Modlin RL 1994 Evidence for a superantigen in human tuberculosis. Immunity 1:35–43

Raulet DH 1994 MHC class I-deficient mice. Adv Immunol 55:381–421

Schoel B, Sprenger S, Kaufmann SHE 1994 Phosphate is essential for stimulation of Vγ9Vδ2 T lymphocytes by mycobacterial low molecular weight ligand. Eur J Immunol 24:1886–1892

Scott PA, Sher A 1993 Immunoparasitology. In: Paul WE (ed) Fundamental immunology. Raven Press, New York, p 1179–1210
Seder RA, Paul WE 1994 Acquisition of lymphokine-producing phenotype by $CD4^+$ T cells. Annu Rev Immunol 12:635–673
Trinchieri G 1993 Interleukin-12 and its role in the generation of T_H1 cells. Immunol Today 14:335–338
Yoshimoto T, Paul WE 1994 $CD4^{pos}$, $NK1.1^{pos}$ T cells promptly produce interleukin 4 in response to *in vivo* challenge with anti-CD3. J Exp Med 179:1285–1295
Zinkernagel RM 1993 Immunity to viruses. In: Paul WE (ed) Fundamental immunology. Raven Press, New York, p 1211–1250

DISCUSSION

Kaufmann: It is well known that intracellular bacteria, including *Mycobacterium bovis* bacillus Calmette–Guérin (BCG) and *Listeria monocytogenes*, are strong T helper 1 (Th1) cell inducers. Th1 cell development, of course, is favoured by early interleukin 12 (IL-12) production by macrophages infected with these intracellular pathogens. In addition, we have evidence for active curtailment of early IL-4-producing cells. We found that the early IL-4 producers, which are $CD4^+$ $NK1.1^+$ liver lymphocytes, disappear soon after infection with intracellular bacteria (M. Emoto & S. H. E. Kaufmann, unpublished results). Therefore, it appears that the active suppression of cells of the innate system which favours Th2 cell development takes place in intracellular bacterial infections. We think that IL-12 is involved in the disappearance of $CD4^+$ $NK1.1^+$ cells because administration of recombinant IL-12 causes this phenomenon (M. Emoto & S. H. E. Kaufmann, unpublished results). We assume that apoptosis of $CD4^+$ $NK1.1^+$ cells is responsible for this kind of suppression.

Liew: Is the mechanism of apoptosis purely speculation?

Kaufmann: We have not yet proven apoptosis.

Sher: I like this idea because we do not see significant Th2 cytokine production in *Toxoplasma*. Moreover, we also do not see a significant increase in anti-IL-12 antibody-treated mice or in γ-interferon (IFN-γ) knockout mice (Gazzinelli et al 1991, 1994). This may be because there are control mechanisms for suppressing Th2 responses (e.g. superantigen stimulation) that are independent of cytokine cross-regulation.

Kaufmann: Last year we published that treatment of mice with anti-NK1.1 monoclonal antibody improves resistance of mice against *L. monocytogenes* (Teixeira & Kaufmann 1994). These results were in contrast to experiments using SCID (severe combined immunodeficiency) mice. Initially, we believed that the anti-NK1.1 monoclonal antibody treatment affected only conventional natural killer (NK) cells. Now we believe that the treatment also affected unconventional lymphocytes, which have the $CD4^+$ $NK1.1^+$ phenotype, and that down-modulation of these lymphocytes improves resistance against listeriosis. Moreover, treatment with anti-IL-4 monoclonal antibody improved resistance against *L. monocytogenes* (Teixeira & Kaufmann 1994). Even if this antibody is given the day before infection, effects are observed at later time points (Day 4 postinfection). This suggests that IL-4 performs regulatory functions from the outset of infection. Finally, PCR analysis suggests that IL-4 mRNA levels are higher prior to *L. monocytogenes* infection than after infection (J. Hess & S. H. E. Kaufmann, unpublished results). Therefore, these results suggest that IL-4-producing $NK1.1^+$ cells are not conventional NK cells. Conventional NK cells contribute to antibacterial protection, whereas the unconventional $NK1.1^+$ $CD4^+$ liver lymphocytes seem to interfere with resistance.

Mitchison: Are the $NK1.1^+$ cells in the liver from a separate lineage or are they just T cells which happen to have migrated to the liver?

Kaufmann: Although these $CD4^+$ $NK1.1^+$ cells have been described in the thymus, bone marrow and liver of major histocompatibility complex (MHC) I-deficient mutant mice (Bendelac et al 1994, Coles & Raulet 1994, Ohteki & MacDonald 1994), we believe that they are unconventional T cells which develop separately from classical T cells.

Ramshaw: Exogenous IL-4 delivered by a recombinant virus increases the pathogenesis of *Listeria* dramatically (I. A. Ramshaw, unpublished results), which agrees with your results.

Flavell: Most of the cells in the liver are dying. Is it likely that the liver is just a site where $NK1.1^+$ cells go to die, as is the case for $CD8^+$ cells?

Kaufmann: During listeriosis, $CD4^+$ $NK1.1^+$ cells vanish from the liver, whereas in control mice they remain. This finding suggests that bacterial infection has a specific influence on these cells.

Flavell: It is possible that this just represents an acceleration of the natural process.

Kaufmann: Yes, this is possible. Five days after infection with *L. monocytogenes*, we observe influx of $CD8^+$ T cells into the liver, and these T cells express potent cytolytic activities *in vitro*.

Flavell: These $CD8^+$ cells may be important, but how do you know that their function is not performed elsewhere, and that the liver is not relevant?

Kaufmann: The $CD4^+$ $NK1.1^+$ cells are most abundant in the liver. Moreover, listeriosis manifests itself in the liver. This is the reason why we studied effects of bacterial infections on $CD4^+$ $NK1.1^+$ cells using lymphocytes isolated from the liver.

Coffman: One of the striking aspects of the peripheral $NK1.1^+$ cell population is that it secretes cytokines more rapidly than normal T cells. This suggests that these cells could be involved in the earliest IL-4 production. Yoshimoto & Paul (1994) showed that the peak of IL-4 production by these cells is about 90 min, both *in vivo* and *in vitro* (Yoshimoto & Paul 1994). Is this early cytokine burst also prevented by *Listeria* infection, or do you observe that the first burst occurs but nothing thereafter?

Kaufmann: We do not see such a rapid burst. The IL-4-producing $CD4^+$ $NK1.1^+$ liver lymphocytes disappear in a matter of days. These cells also disappear in *L. monocytogenes*-infected, MHC class II-deficient mice (M. Emoto & S. H. E. Kaufmann, unpublished results). This finding argues against the involvement of listerial superantigens, which generally have a MHC class II-dependent activity.

Locksley: Does the IL-4 burst depend on the presence of T cell receptor γ/δ (TCRγ/δ) lymphocytes? You could show this by seeing whether the burst also occurs in TCRγ/δ knockout mice.

Kaufmann: I don't know. We have not yet done these experiments.

Sher: Why are you convinced that IL-12 is responsible for the suppression of these cells?

Kaufmann: We observed that recombinant IL-12 curtails the IL-4-producing $CD4^+$ $NK1.1^+$ liver lymphocytes and that treatment of *L. monocytogenes*-infected mice with anti-IL-12 monoclonal antibody interferes with *Listeria*-induced reduction (M. Emoto & S. H. E. Kaufmann, unpublished results).

Lotze: But what does IL-12 do to these cells?

Kaufmann: It may cause apoptosis, but we have not studied this in sufficient detail.

Lotze: Do these cells produce IFN-γ?

Kaufmann: They produce IL-4 but not IFN-γ.

Mitchison: How do you gain access to these cells? Do you have to make a cell suspension out of the liver?

Kaufmann: Yes.

Mitchison: Are results using these liver cell suspensions applicable to human infections?

Kaufmann: S. Abrignani's group (personal communication) has studied human liver lymphocytes, and they have shown that these cells produce unique patterns of cytokines.

Romagnani: Also, $CD4^+$ $CD57^+$ (HNK-1^+) T cells with a large granular morphology have been described by Carlo Grossi in the germinal centres of human tonsils and lymph nodes (Velardi et al 1986).

Locksley: Most mouse $NK1.1^+$ $CD4^+$ T cells require IL-7 in addition to IL-2 for efficient growth *in vitro*.

Mitchison: Are these cells from human liver biopsies?

Locksley: No, these are mouse cells.

Abbas: Has anyone crossed β_2 microglobulin knockout mice with BALB/c mice to see if the IL-4 responses are ablated? Because a consistent observation is that β_2 microglobulin knockout mice lack these cells (Bendelac et al 1995).

Locksley: That's a good experiment. We've never examined this in the BALB/c background. We should start to think about the genetics of this.

Coffman: I would like to comment briefly on the prediction that *Leishmania* infection does not trigger this pathway. In the mouse model of visceral leishmaniasis using *Leishmania donovani*, the primary infection is in the liver. However, this leads to a Th1-like response and it has been virtually impossible to convert this to a Th2 response using various cytokines or anti-cytokine antibodies. This suggests that the situation may be more complicated than that for *Listeria* (Kaye et al 1991, Miralles et al 1994).

Kaufmann: We have constructed strains of *Salmonella typhimurium* that express listerial antigens either in secreted or in somatic form (J. Hess & S. H. E. Kaufmann, unpublished results). We have found that a single antigen provides good protection against subsequent infection with *Listeria* when presented in a secreted form. In contrast, the same antigen in somatic form is virtually inactive.

Allen: I would like to ask a question about these secreted antigens. Is it possible that any antigen that is secreted will make a good vaccine? Perhaps a key factor in making a good vaccine is to deliver the antigen to the right place.

Kaufmann: I assume that several antigens are protective in intracellular bacterial infections. It is probably important to mimic the natural situation, so the antigens we used are secreted during natural infection. Hence, the *S. typhimurium* constructs secreting listerial antigens mimicked natural infection, whereas those displaying these antigens in somatic form failed to do so.

Allen: In *Mycobacterium* the fibronectin-binding protein is secreted before it binds to the bacterial cell surface and it is readily seen by the class II compartment. This suggests that any antigen will do the trick because there is no antigen selection.

References

Bendelac A, Killeen N, Littman DR, Schwartz RH 1994 A subset of CD4$^+$ thymocytes selected by MHC class I molecules. Science 263:1774–1778

Bendelac A, Lantz O, Quimby ME, Yewdell JW, Bennink JR, Brutkiewicz RR 1995 CD1 recognition by mouse NK1$^+$ T lymphocytes. Science 268:863–865

Coles MC, Raulet DH 1994 Class I dependence of the development of CD4$^+$ CD8$^-$ NK1.1$^+$ thymocytes. J Exp Med 180:395–399

Gazzinelli RT, Hakim FT, Hieny S, Shearer GM, Sher A 1991 Synergistic role of CD4$^+$ and CD8$^+$ T lymphocytes in IFN-γ production and protective immunity induced by an attenuated *Toxoplasma gondii* vaccine. J Immunol 146:286–292

Gazzinelli RT, Wysocka M, Hayashi S et al 1994 Parasite-induced IL-12 stimulates early IFN-γ synthesis and resistance during acute infection with *Toxoplasma gondii*. J Immunol 153:2533–2543

Kaye PM, Curry AJ, Blackwell JM 1991 Differential production of Th1-derived and Th2-derived cytokines does not determine the genetically controlled or vaccine-induced rate of cure in murine visceral leishmaniasis. J Immunol 146:2763–2770

Miralles GD, Stoeckle MY, McDermott DF, Finkelman FD, Murray HW 1994 Th1 and Th2 cell-associated cytokines in experimental visceral leishmaniasis. Infect Immun 62:1058–1063

Ohteki T, MacDonald HR 1994 Major histocompatibility complex class I related molecules control the development of $CD4^+8^-$ and $CD4^-8^-$ subsets of natural killer 1.1^+ T cell receptor-α/β^+ cells in the liver of mice. J Exp Med 180:699–704
Teixeira HC, Kaufmann SHE 1994 Role of $NK1.1^+$ cells in experimental listeriosis. $NK1^+$ cells are early IFN-γ producers but impair resistance to *Listeria monocytogenes* infection. J Immunol 152:1873–1882
Velardi A, Tilden AB, Millo R, Grossi CE 1986 Isolation and characterization of Leu 7^+ germinal center cells with the T helper-cell phenotype and granular lymphocyte morphology. J Clin Immunol 6:205–215
Yoshimoto T, Paul WE 1994 $CD4^{pos}$, $NK1.1^{pos}$ T cells promptly produce interleukin 4 in response to *in vivo* challenge with anti-CD3. J Exp Med 179:1285–1295

General discussion III

Allergy

Lachmann: I would like to report some work from my unit (McHugh et al 1995) on the manipulation of the T helper 1 (Th1)/Th2 balance in humans, using insect venom anaphylaxis as a model. In contrast to other allergies, reactions to insect venom occur without a genetic predisposition. Bee venom allergy is largely a problem of beekeepers, their relatives and their neighbours. Therefore, it occurs in people who have been stung frequently and who have had considerable previous experience of the antigen. In contrast, anyone is at risk of wasp venom allergy, and it usually occurs in those who have been stung infrequently. Therefore, bee venom patients usually have a high titre of IgG antibodies in their blood and wasp venom patients usually have a low titre.

Bee venom anaphylaxis can occur at any stage of a beekeeper's career. Probably the commonest pattern of immune response is to go through a phase of local 'IgE-type' hyper-reactivity (which usually does not progress to anaphylaxis) followed by desensitization. It is probably, therefore, not true that the Th2 response is the inevitable end product of prolonged antigen stimulation. Instead, with bee venom, the IgE response may occur early and may disappear again with further antigenic stimulation. Those who have anaphylactic reactions have IgE antibodies to venom. Venom-sensitive beekeepers also have a high level of IgG antibodies. The cytokine response of allergic beekeepers' peripheral blood to bee venom shows a Th2-type response with high interleukin 4 (IL-4) and low γ-interferon (IFN-γ) levels. Similarly, wasp venom-sensitive patients' peripheral blood lymphocytes (PBL) show a Th2 cytokine response to wasp venom.

Traditionally, anaphylactic sensitivity has been treated by desensitizing the patients with regular injections of increasing amounts of antigen. This was believed to generate high levels of IgG, which compete with IgE for antigen so that cross-linking by antigen of IgE on mast cells is prevented. However, this is not a complete explanation. We have been experimenting with 'rush' desensitization, another traditional technique that has come back into fashion. This involves injecting patients, not every week over several months, but every half hour for a day. We have found that by the end of one day these patients can tolerate substantial amounts of venom. The traditional explanation for rush desensitization is that the mast cells are slowly degranulated so that, at the end of the day, there are no granules left to degranulate. We discovered that this explanation was not necessarily true when, in response to the last injection of the day, our first patient produced a full-blown anaphylactic reaction, having tolerated a relatively large amount of venom beforehand. By the next morning, the peripheral blood of patients undergoing this treatment showed a loss of Th2 response, in that there was a significantly lower level of IL-4 from the cells remaining in the circulation. This change in cytokine pattern, therefore, occurred rapidly. This is not comparable to the *in vitro* situation because *in vivo* there is a redistribution of cells from the circulation to the numerous injection sites. Therefore, a different pool of cells is sampled after the injections. Nevertheless, the IL-4 response disappears quickly, and at that stage the patients are no longer clinically sensitive. If these patients are subsequently injected at weekly intervals, the IFN-γ levels also increase, and the patients develop a fairly typical Th1 response. The patients then make antibodies and, typically, over a period of months the IgG1 and IgG4 anti-venom levels rise and the IgE levels gradually fall.

Romagnani: How do you measure this Th1 response?

Lachmann: We have used IL-4 and IFN-γ as the classic markers for Th2 and Th1 cytokines, respectively. We have investigated: (1) the ability of the responding T cells to produce cytokines by stimulating with phytohaemagglutinin (PHA), a T cell mitogen; (2) Th1 responses induced by the bacterial recall antigens streptokinase:streptodornase and purified protein derivative (PPD), to which there is a strong IFN-γ response and a low, if any, IL-4 response; (3) allergen stimulation, to which there is an IL-4 response and a low, if any, IFN-γ response; and (4) no stimulation.

Romagnani: Does the production of cytokines in the peripheral blood change after stimulation with PHA or recall antigens?

Lachmann: No. Only the response to venom is changed. PHA and recall antigens are used as controls for the specificity of the observed changes.

Abbas: Do you convert from IL-4 dominance to IFN-γ dominance within 24 h?

Lachmann: No. IL-4 disappears early and IFN-γ levels increase only later. However, the PBL tested are not a uniform cell sample. Some of the Th2 cells may have simply left the circulation and moved to the antigen sites.

Desensitization seems to involve a number of mechanisms, none of which we understand perfectly. The early clinical desensitization, which is seen after just one day of repeated antigenic stimulation, must be an effect on mast cells. At the time of antigen challenge, the anaphylactic reaction, which is now prevented, can take place within a minute or two, so there can be no question of T cells making any contribution during this time interval. Some change has taken place in the mast cells during the desensitization procedure that prevents them from releasing mediators. Therefore, the effector cell population (possibly T cells secreting cytokines) can be produced within 24 h. Many cytokines produced by T cells affect mast cells but we don't know which, if any, is the critical one.

Mitchison: If the rush injections are given at separate sites, is your hypothesis that something is moving from one site in the skin to another site?

Lachmann: No. It just represents one way of giving increasing doses of antigen.

Flavell: Do you use any adjuvant?

Lachmann: No, although we have thought about using it. For example, we have thought of coupling venom to PPD, which is a good way of getting a Th1 response, but we have never had the courage to try it in humans.

Flavell: That may not be the best approach because it may induce apoptotic death of Th2 cells. More cells may be cycling if they are exposed to a continuous antigenic stimulus. This may be mediated by IL-2. However, this doesn't explain how a Th1 response is obtained. A Th2 response also has to be eliminated.

Lachmann: That's right, and reduction in the Th2 response seems to be what happens first. We do not know whether the cells just leave the circulation or whether they die. In the next phase of desensitization (in the early week, after rush desensitization) there is still no obvious change in antibody levels but the patients are clinically non-sensitive and have a Th1 response in their peripheral blood. IgE levels tend to rise in the first stages of desensitization, which is followed by an increase in IgG levels. After maintenance injections for two to three years, (when it is usual to stop the regular injections) most of the patients have lost much of their IgE, while maintaining their level of IgG. We hope that patients will then remain desensitized, but the necessary long-term studies are not available.

Dutton: Could one determine whether or not the mast cells are degranulated at the end of the 24 h?

Lachmann: One could look at skin biopsies but it would be difficult to look in the real target areas.

Abbas: Have you taken a skin biopsy to see if IL-4-producing cells home in to the site of antigen administration?

Lachmann: Not yet. But such studies are about to be done.

Mosmann: Do you observe an immediate reaction at every injection site or do you observe a late reaction at any or all of them?

Lachmann: I can't give you a statistical answer. We observe immediate (and late) reactions but they become less frequent in conventional desensitization.

Mosmann: The late-phase reactions are interesting because of the production of cytokines like tumour necrosis factor later on. Do you know anything about the late-phase reaction?

Lachmann: In the skin, large and exaggerated swelling reactions come up after 12–24 h. These are, to some extent, believed to be IgE dependent. The exaggerated nature of the reaction is due, in part, to the fact that bee venom and wasp venom contain phospholipase A2, which generates the local production of arachidonic acid. The allergic reactions attract polymorphs that have a high level of prostaglandin synthetase. Consequently, large amounts of eicosanoids are produced. We have treated these reactions with modest success by giving non-steroidal, anti-inflammatory drugs in addition to antihistamines.

Lamb: There is another approach to desensitization, which has been studied by S. R. Durham (personal communication), that uses pollen to desensitize over a period of time. He's looked at patients two years after desensitization, and he has found that there's essentially only a minor reduction in the levels of IgE. These patients, as far as they're concerned, are cured because they don't take any medication. They have an elevated level of IgG and they have an immediate skin test reactivity, although they've lost the late-phase skin reactivity. This seems like a different immunological profile in some respects to what Peter Lachmann described, yet the end result is the same in these patients, i.e. they are cured. Do we really know what we want to achieve immunologically by desensitizing an allergic response? Do we want to shift the balance of Th1 and Th2 cytokines or do we want to try and tolerate Th2 responses?

Mitchison: One point that may not be accepted by everybody is Lichtenstein's view that the last thing that we want to achieve is to eliminate all 'non-specific' IgE (Lichtenstein 1993). His view is that the lower the levels of IgE, the more sites are available on mast cells for absorption of allergy-related IgE. Mexican immigrants in the United States have mast cells that are loaded with IgE, which protects against allergy. This argues against aiming at cytokine-based conversion of the whole immune system away from IgE production.

Lachmann: Lichtenstein may be right about atopics, but in insect venom allergy most of the IgE antibody is to insect venom.

Mitchison: But they might be better off with a lot of IgE.

Mitchison: Does anyone know if cytokine studies have been done in conjunction with CATVAX (Immunologic, MA), which is designed to desensitize against allergy to cat fur?

Abbas: These studies have reached phase II clinical trials, but the results can't be released yet because it's a double-blind phase II trial.

Lachmann: Desensitization against allergens encountered on mucosal surfaces is less well established than desensitization against insect venom, which travels to target organs in the circulation. It's also more dangerous, particularly in severe asthmatics. There have been some deaths during attempts to desensitize severe asthmatics, so this therapy is now allowed only where full resuscitation facilities are available.

Ramshaw: There is some similar work with IL-6 knockout mice, which are deficient in mucosal antibodies. If the mucosa of control mice is stained with IgA, there is a good IgA response, whereas this response is absent in the IL-6 knockout mice. If the knockout mice are challenged with antigen, they don't produce any mucosal immunity. We have measured antibody-secreting cells in the lung and challenged these cells with a vaccinia virus encoding a foreign gene, namely the haemagglutinin gene of influenza, and looked at the antibody response to the haemagglutinin gene (Ramsay et al 1994). A good response is observed only in the littermate controls, where IL-6 stimulates IgA and IgG at the mucosa in response to the local expression of the haemagglutinin gene by the virus (Ramsay et al 1994). If the knockout mice are challenged with the same virus, they do not make mucosal IgA or IgG antibodies. However, if they are challenged with the vaccinia virus that encodes the IL-6 gene, the IL-6 mucosal knockout mice are reconstituted and they recover fully. In fact they produce a much better response than the normal mice. Therefore, IL-6 is absolutely crucial for the immune response at the mucosa.

Romagnani: Is this effect at the level of T cells?

Ramshaw: Probably not, in the sense that IL-6 is a B cell differentiation factor in these circumstances.

References

Lichtenstein LM 1993 Allergy and the immune system. Sci Am 269:84–93

McHugh SM, Deighton J, Stewart AG, Lachmann PJ, Ewan PW 1995 Bee venom immunotherapy induces a shift in cytokine responses from a T_H2 to a T_H1 dominant pattern: comparison of rush and conventional therapy. Clin Exp Allergy 25:828–838

Ramsay AJ, Husband AJ, Ramshaw IA et al 1994 The role of interleukin-6 in mucosal IgA responses *in vivo*. Science 264:561–563

Cytokines in immune regulation/pathogenesis in HIV infection

Gene M. Shearer, Mario Clerici*, Apurva Sarin, Jay A. Berzofsky† and Pierre A. Henkart

*Experimental Immunology and †Metabolism Branches, National Cancer Institute, National Institutes of Health, Bethesda, MD 20892, USA and *Cattedra di Immunologia, Universita degli Studi, Milano, Italy*

Abstract. Two hallmarks of immunopathogenesis in the progression of HIV-infected individuals to AIDS are the loss of T helper (Th) cell function in response to antigens and the critical reduction in $CD4^+$ T cell numbers. It is probable that these two phenomena are related. We observed that: (1) the failure to detect antigen-stimulated Th cell responses *in vitro* correlates with increased pokeweed mitogen/staphylococcal enterotoxin B (P/S)-stimulated and antigen-stimulated T cell death; and (2) both of these events are similarly modulated by immunoregulatory cytokines. Interleukin 2 (IL-2) and IL-12 (Th1-type cytokines), as well as antibodies to IL-4 and IL-10 (which are Th2-type cytokines) restore *in vitro* Th cell responses to recall antigens such as influenza virus and HIV envelope synthetic peptides (*env*). P/S-induced T cell death affects both $CD4^+$ and $CD8^+$ T cell subsets, whereas death induced by stimulation with *env* affects only $CD4^+$ T cells. In both examples, Th1-type cytokines and antibodies to Th2-type cytokines protect against T cell death. In contrast, IL-4 and IL-10 do not protect against death, and anti-IL-12 antibody can enhance T cell death. Our findings indicate that the loss of Th cell function and increased T cell death seen *in vitro* are correlated, and that *in vivo* HIV infection gives rise to inappropriate cytokines resulting in immune dysfunction and immunopathogenesis.

1995 T cell subsets in infectious and autoimmune diseases. Wiley, Chichester (Ciba Foundation Symposium 195) p 142–153

The most widely known immunological feature resulting from HIV infection and associated with progression to AIDS is the decline in $CD4^+$T cells (Levy 1993). However, the more subtle loss of *in vitro* T helper (Th) cell function precedes the drop in CD4 count (Lane et al 1985, Shearer et al 1986, Giorgi et al 1987, Miedema et al 1988, Clerici et al 1989a,b), and it is predictive of the loss of $CD4^+$ cells (Lucey et al 1991) and the time to AIDS diagnosis and death (Dolan et al 1995). Because the decreased CD4 count is a reflection of events that have already taken place by the time it is detected, it is reasonable to consider that investigation of immunological changes that occur prior to detection of this benchmark surrogate marker will be informative for understanding the immunopathogenesis and, ultimately, for selecting effective therapy for AIDS. A second observation that is associated with the decline in CD4 count is a cell death phenomenon frequently referred to as apoptosis or programmed cell death (Amiesen & Capron 1991, Gougeon & Montagnier 1993, Meyaard et al 1992).

For the past several years, our respective laboratories have focused on HIV-induced dysregulation of Th cell function, HIV-specific T cell responses and mechanisms of cell death. We have combined our experiences to gain a better understanding of the events that contribute to the destruction of the immune system. To this end, we have studied the loss of Th cell function to specific antigens, the changes in the cytokine profile associated with such loss, and mitogen-induced and antigen-induced T lymphocyte cell death detected in HIV-infected individuals.

Results

The loss of Th cell function resulting from HIV infection is associated with changes in immunoregulatory cytokine profiles: interleukin 2 (IL-2) (Clerici et al 1989a,b) and γ-interferon (IFN-γ) production (Maggi et al 1987, Clerici et al 1993a) are reduced, and IL-4 and IL-10 (Clerici et al 1994a) are increased (Clerici et al 1993b, Clerici et al 1994b, Barcellini et al 1994, Meyaard et al 1994a). Furthermore, the addition of IL-12 (Clerici et al 1993a), as well as antibodies specific for IL-4 (Clerici et al 1993b) or IL-10 (Clerici et al 1994a) restored Th cell function to cultures of peripheral blood mononuclear cells (PBMC) from HIV-infected (HIV$^+$) patients. Because of these changes in cytokine profiles, and the earlier observations that the immunological characteristics of AIDS progression include both loss of Th cell function and hypergammaglobulinemia (Mildvan et al 1982), we suggested that a switch from a dominant Th1-like to a dominant Th2-like state is characteristic of AIDS progression (Clerici & Shearer 1993). We refined the hypothesis (Clerici & Shearer 1994) to include the fact that immunoregulatory cytokines are produced not only by CD4$^+$ T cells but also by monocytes, B cells and CD8$^+$ T cells (Trinchieri 1993, Erard et al 1994, Romagnani et al 1994). Thus, immune function in the patient involves paracrine as well as autocrine regulation, and it is much more complex than that limited to the autocrine regulation detected in long-term cultures and clones of CD4$^+$ T cells.

It has been suggested that T cell death involving an apoptotic mechanism contributes significantly to both the loss of Th cell function and the CD4$^+$ T cells in the progression to AIDS (Amiesen & Capron 1991). It has been reported that both loss of Th cell function and T cell death is increased in peripheral blood samples from HIV$^+$ individuals (Meyaard et al 1992), and that such death is increased upon mitogen stimulation (Clerici et al 1994b).

TABLE 1 Effect of cytokines and anti-cytokine antibodies on T cell function and on antigen-stimulated or mitogen-stimulated T cell death

Cytokine or antibody	*T cell function*	*T cell death*
IL-2	enhances	protects
IL-12	enhances	protects
IFN-γ	enhances	protects
Anti-IL-4 + anti-IL-10	enhances	protects
IL-4	suppresses	does not protect[a]
IL-10	suppresses	does not protect[a]
Anti-IL-12	suppresses	does not protect[a]

[a]can exacerbate T cell death.

This T cell death is observed in both the $CD4^+$ and $CD8^+$ subsets (Clerici et al 1994b). However, we have recently observed that stimulation with nominal antigens such as infectious influenza A virus, as well as synthetic peptides of HIV envelope (*env*) results in the selective death of $CD4^+$ T cells (Clerici et al 1995). Mitogen-stimulated or antigen-stimulated T cell death was not observed in PBMC from HIV^- donors. Nevertheless, activation of T cells from HIV^- donors by pan-T cell receptor stimulation appears to 'prime' them for T cell death, such that restimulation results in extensive killing of these T cells.

It should be noted that the same cytokines that regulate Th cell function in HIV^+ individuals also affect mitogen-stimulated T cell death. Thus, IL-2, IL-12, IFN-γ and antibodies to IL-4 and IL-10 block T cell death, whereas IL-4, IL-10 and antibody to IL-12 do not inhibit death and can enhance T cell death (Clerici et al 1994c). Table 1 summarizes the effects of these cytokines and their antibodies on T cell function and on T cell death.

Discussion

The changes in cytokine profiles observed in progression of HIV^+ individuals to AIDS suggest a shift from a Th1-like to a Th2-like cytokine dominance (Clerici & Shearer 1993, 1994). These changes have been proposed to account for the reduced cellular and increased humoral arms of the immune system resulting from HIV infection. The T cells of HIV^+ patients have been paradoxically described as being anergic to *in vitro* stimulation, and also as being in an activated state (Wachter et al 1986). It is possible that both observations are descriptively accurate because such activated T cells could be primed to undergo T cell death as the result of stimulation. *In vitro* antigen stimulation of T cell proliferation, IL-2 production and cytotoxic T lymphocyte generation might not be observed, due to antigen-induced T cell death in the cultures. It would be interesting to know whether the *in vivo* lack of delayed type hypersensitivity reported in HIV^+ individuals (Blatt et al 1993, Markowitz et al 1993) is actually due to anergy or to localized T cell death in the skin at the site of antigenic challenge.

Our finding that $CD4^+$ but not $CD8^+$ T cells from HIV^+ individuals die when stimulated with specific antigen, including infectious influenza A virus or non-infectious *env*, suggests one possible mechanism for the selective decline in CD4 counts in HIV^+ patients. This contrasts with mitogen-stimulated T cell death observed for both $CD4^+$ and $CD8^+$ T cells in the same PBMC preparations (Clerici et al 1994b). Thus, it is possible that chronic antigenic stimulation *in vivo* contributes significantly to $CD4^+$ T cell depletion over the lengthy course of HIV disease. A recent report suggested that there is no correlation between apoptosis and progression to AIDS (Meyaard et al 1994b). In contrast to our studies, that report assessed death in unstimulated cultures. It is our opinion that antigenic stimulation is an important aspect of the *in vivo* T cell death model. Furthermore, the report of Meyaard et al (1994b) may have involved patients in whom *in vivo* T cell death had already been extensive prior to *in vitro* testing, thereby leaving fewer $CD4^+$ T cells available to undergo apoptosis in the experiment.

If the findings that loss of T cell function, increased T cell death and the reversal of both phenomena by Th1-type cytokines such as IL-2, IL-12, or antibodies to IL-4 and IL-10 are relevant to AIDS progression, a high priority should be given to cytokine-based therapy for reversing both Th cell function and the $CD4^+$ T cell decline characteristic of AIDS progression. We suggest that such a therapy should be started before the CD4 count drops to a critical level.

Summary

We have investigated two immunological events that are predictive for progression to AIDS: loss of Th cell function; and decline in the number of $CD4^+$ T cells. We suggest that both are closely associated with antigen-stimulated T cell death, and that this death may be the consequence of an earlier event in the HIV^+ individual's response to the virus, which involved an aberrant T cell activation signal and resulted in T cell death following a second antigenic stimulus. Because loss of Th cell function and antigen-stimulated T cell death can be modulated by cytokines, we also suggest that a Th1-type cytokine-based therapy that enhances and/or restores cellular immunity should be used to stabilize asymptomatic HIV^+ individuals and to reverse T cell decline in progressive patients.

References

Amiesen JC, Capron A 1991 Cell dysfunction and depletion in AIDS: programmed cell death hypothesis. Immunol Today 12:102–105

Barcellini W, Rizzardi GP, Borghi MO, Fain C, Lazzarin A, Meroni PL 1994 Th1 and Th2 cytokine production by peripheral blood mononuclear cells from HIV-infected patients. AIDS 8:757–762

Blatt SP, Hendrix CW, Butzin CA et al 1993 Delayed type hypersensitivity skin testing predicts progression to AIDS in HIVB-infected patients. Ann Intern Med 119:177–184

Clerici M, Shearer GM 1993 A T_H1–T_H2 switch is a critical step in the etiology of HIV infection. Immunol Today 14:107–110

Clerici M, Shearer GM 1994 The T_H1–T_H2 hypothesis of HIV infection: new insights. Immunol Today 15:575–581

Clerici M, Stocks NI, Zajac RA et al 1989a Interleukin-2 production used to detect antigenic peptide recognition by T helper lymphocytes from asymptomatic, HIV seropositive individuals. Nature 339:383–385

Clerici M, Stocks NI, Zajac RA et al 1989b Detection of three distinct patterns of T helper cell dysfunction in asymptomatic, HIV-seropositive patients: independence of $CD4^+$ cell numbers and clinical staging. J Clin Invest 84:1892–1899

Clerici M, Lucey DR, Berzofsky JA et al 1993a Restoration of HIV-specific cell-mediated immune responses by interleukin-12 *in vitro*. Science 262:1721–1724

Clerici M, Hakim FT, Venzon DJ et al 1993b Changes in interleukin-2 and interleukin-4 production in asymptomatic, human immunodeficiency virus-seropositive individuals. J Clin Invest 91:759–765

Clerici M, Wynn TA, Berzofsky JA et al 1994a Role of interleukin-10 in T helper cell dysfunction in asymptomatic individuals infected with the human immunodeficiency virus (HIV-1). J Clin Invest 93:768–775

Clerici M, Sarin A, Coffman RL et al 1994b Type 1/type 2 cytokine modulation of T cell programmed cell death as a model for HIV pathogenesis. Proc Natl Acad Sci USA 91:11811–11815

Clerici M, Sarin A, Berzofsky JA et al 1995 Regulation of antigen-stimulated $CD4^+$ T cell death in HIV infection: role of immunoregulatory cytokines and lymphotoxin, submitted

Dolan MJ, Clerici M, Blatt SP et al 1995 *In vitro* T cell function, delayed type hypersensitivity skin testing, and $CD4^+$ T cell subset phenotyping independently predict survival time in patients infected with human immunodeficiency virus. J Infect Dis 172:79–87

Erard F, Dunbar PR, Le Gros G 1994 The IL4-induced switch of $CD8^+$ T cells to a T_H2 phenotype and its possible relationship to the onset of AIDS. Res Immunol 145: 643–646

Giorgi JV, Fahey JL, Smith DC et al 1987 Early effects of HIV on CD4 lymphocytes *in vivo*. J Immunol 138:3725–3730

Gougeon ML, Montagnier L 1993 Apoptosis in AIDS. Science 260:1269–1270

Lane HC, Depper JM, Greene WC, Whalen G, Waldmann TA, Fauci AS 1985 Qualitative analysis of immune function in patients with the acquired immunodeficiency syndrome: evidence for selective defect in soluble antigen recognition. N Engl J Med 313:79–84

Levy JA 1993 Pathogenesis of human immunodeficiency virus infection. Microbiol Rev 57:183–289

Lucey DR, Melcher GP, Hendrix CW et al 1991 Human immunodeficiency virus infection in the US Air Force: seroconversion, clinical staging, and assessment of T helper cell functional assay to predict change in $CD4^+$ T cell counts. J Infect Dis 164:631–637

Maggi E, Macchia D, Parronchi P et al 1987 Reduced production of interleukin 2 and interferon gamma and enhanced helper activity for IgG synthesis by cloned CD4 T cells from patients with AIDS. Eur J Immunol 17:1685–1690
Markowitz N, Hansen NI, Wilcosky TC et al 1993 Tuberculin and anergy testing in HIV-seropositive and HIV-seronegative persons. Ann Intern Med 119:185–193
Meyaard L, Otto SA, Jonker RR, Mijnster MJ, Keet RPM, Miedema F 1992 Programmed death of T cells in HIV-1 infection. Science 257:217–219
Meyaard L, Otto SA, Keet IPM, Roos MLT, Miedema F 1994b Programmed cell death of T cells in human immunodeficiency virus infection: no correlation with progression to disease. J Clin Invest 93:982–988
Meyaard L, Otto SA, Keet IPM, van Lier RAW, Miedema F 1994a Changes in cytokine secretion patterns of $CD4^+$ T-cell clones in human immunodeficiency virus infection. Blood 84:4262–4268
Miedema F, Petit AJC, Terpstra FG et al 1988 Immunologic abnormalities in human immunodeficiency virus (HIV)-infected asymptomatic homosexual men. HIV affects the immune system before $CD4^+$ T helper cell depletion occurs. J Clin Invest 82:1908–1914
Mildvan D, Mathur U, Enlow RW et al 1982 Opportunistic infections and immune deficiency in homosexual men. Ann Intern Med 96:700–704
Romagnani S, Maggi E, Del Prete G 1994 HIV can induce a Th1 to Th0 shift and preferentially replicates in $CD4^+$ T-cell clones producing Th2-type cytokines. Res Immunol 145:611–618
Shearer GM, Bernstein DC, Tung KSK et al 1986 A model for the selective loss of major histocompatibility complex self-restricted T cell immune responses during the development of acquired immune deficiency syndrome (AIDS). J Immunol 137:2514–2521
Trinchieri G 1993 Interleukin-12 and its role in the generation of Th1 cells. Immunol Today 14:335–338
Wachter H, Fuchs D, Hausen A et al 1986 Are conditions linked with T-cell stimulation necessary for progressive HTLV-III infection? Lancet I:97

DISCUSSION

Locksley: You took a sample of the peripheral blood of infected patients, so presumably few infected cells are present, and you then challenged these cells with specific antigens. There are only about one in 10 000 responder cells present, but these must be making enough lymphotoxin to kill about 60% of the cells in the assay. I don't understand the mechanism of this amplification system.

Shearer: I don't understand the mechanism either. We don't know how much lymphotoxin is actually produced by these cells. We're now adding exogenous lymphotoxin to determine whether it will generate the same effect. The critical point is that many activated or primed cells are present which are not specific for influenza virus antigens or synthetic HIV envelope peptides (*env*). These activated T cells may be set to undergo T cell death depending on whether they encounter T helper 1 (Th1)-type or Th2-type cytokines. However, if they are exposed to lymphotoxin, they undergo apoptosis. This is our interpretation but we do not have proof for some aspects of this model.

Locksley: This is a powerful and dangerous system. Why don't many T cells die following infection with the influenza virus?

Shearer: Our assays are closed systems, i.e. we do not add extra cells. Some people believe that the immune system has the dynamic potential to regenerate $CD4^+$ cells; however, this could just represent a redistribution from the lymph nodes. Therefore, the human body is a relatively open system, unlike the situation *in vitro*.

Mitchison: Why is there a high level of apoptosis at the start of the assay, even before anything happens?

Shearer: Meyaard et al (1992) have cultured unstimulated cells for 48 h and they observe apoptosis which, they claim, occurs more frequently than in HIV^- controls. We haven't seen such a marked difference. This may depend on whether fetal calf serum is added to the cultures or whether a non-stimulating human plasma is used.

Mitchison: But even at the start, apoptosis was as high as 20% in some individuals.

Shearer: Yes. It is occasionally 20%, and sometimes it is even higher. These cases may represent HIV^+ patients who have an ongoing infection. Therefore, in some cases the cells may already be stimulated *in vivo*. One could ask why these cells aren't stimulated all the time by the HIV.

Mosmann: If lymphotoxin is the key to this, why does it occur at a higher level in samples from HIV^+ patients? Because that's the opposite of what you would expect from the bias away from the Th1 cytokine pattern in HIV^+ patients. It is possible that lymphotoxin is the key to this, or there may be something else that requires lymphotoxin action. Have you looked at the levels of lymphotoxin in similar cultures to those in which you've measured the other cytokines?

Shearer: No. That needs to be done. The prediction that the levels of lymphotoxin would be lower, assumes that the non-response observed *in vitro* after seven days represents anergy. I now believe that it represents apoptosis. If the cells die within 48–72 h and $[^3H]$thymidine is added, little $[^3H]$thymidine uptake would be observed six days later.

Mosmann: The timing of this is interesting.

Shearer: Yes. But the effect of lymphotoxin appears to depend on the other cytokines present, namely interleukin 10 (IL-10), IL-4, IL-2 and γ-interferon (IFN-γ).

Abbas: If apoptosis is involved, which is what most of us would predict at this stage, then the cytokine effects that you observe are almost exactly the opposite of what I would predict you would see, i.e. I predict that activation-induced apoptosis with staphylococcal enterotoxin B would be inhibited by adding IL-4. The reason for this is that IL-4 drives the T cells to become Th2 cells. Th2 cells express lower levels of Fas ligand and they don't undergo apoptosis (Ramsdell et al 1994, Ettinger et al 1995). Therefore, the Th2-type cytokines are protective and the Th1-type cytokines exacerbate cell death. Every model of activation-induced T cell death that has been looked at in humans or in mice is either largely or entirely Fas ligand dependent (Brunner et al 1995, Dhien et al 1995, Ju et al 1995). The easiest way of showing this is to use the Fas–Fc fusion protein, which is a conventional blocking reagent. I predict that the effects that you observe will be Fas ligand dependent, and I do not understand your results.

Shearer: We are not the only people who have observed these results. For example, Terri Finkel in Denver and Jean-Claude Amiesen in Lille, France (Amiesen & Capron 1991, Finkel et al 1995) have also observed this. The Fas and Fas ligand experiments that you suggested are being done, and I believe that Peter Krammer in Heidelberg, and Jean-Claude Amiesen (personal communications) have results which support the idea that this cell death is Fas dependent.

Flavell: Pippa Marrack (personal communication) has looked at the role of lipopolysaccharide in preventing antigen-stimulated apoptosis. She has found that that protection can be blocked by anti-tumour necrosis factor α (TNF-α) antibody, which is again exactly the opposite of your results.

Shearer: We have also blocked apoptosis with anti-TNF-α antibody (A. Sarin, G. M. Shearer & P. A. Henkart, unpublished observations).

Abbas: But Pippa Marrack blocked the protection with anti-TNF-α or with anti-IFN-γ antibodies. However, many of us have tried to change the sensitivity of Fas^+ cells to apoptosis by simply adding cytokines, but we have not observed anything.

Swain: If the production of cytokines is actually inducing apoptosis, one should observe that cytokines are produced early, and that the cells then die. Also, if the reason that one does not see proliferation is that the cells die, then one should see proliferation early on. Have you observed proliferation early on?

Shearer: Yes. The kinetics of both proliferation and apoptosis are different, depending on whether specific antigen, mitogen or anti-CD3 antibody is used. Proliferation in response to a specific antigen is observed after five or six days, which is when apoptosis is observed; whereas proliferation in response to mitogen is observed after two or three days, which is again when apoptosis is observed.

Swain: What happens on Day 1?

Shearer: We haven't looked at cell death on Day 1. We have looked at proliferation, but I can't give you the numbers.

Dutton: In your model there are cells that are half way along the road to apoptosis. Are there any cell surface markers that identify these cells?

Shearer: There are claims that CD38 and major histocompatibility complex class II antigens may be such markers.

Dutton: Are these markers polyclonal on all the cells?

Shearer: Not on all the cells, but on a reasonable proportion.

Mitchison: The Tunel staining technique could be used as an assay for this.

Shearer: The Tunel technique has already been used (Clerici et al 1994), and these results have also been confirmed with DNA ladders of 200 bp.

Sher: Do you have any evidence that Th1 and Th2 cells undergo selective apoptosis?

Shearer: No. We have proposed that only Th1 cells or Th1-like cells are killed selectively. In contrast, Sergio Romagnani has shown that Th2 cells or Th2-like cells are more likely to be infected and possibly killed by a virus (Maggi et al 1994a). It is possible that there are two ways to kill $CD4^+$ T cells. One way, involving Th1-like cells, may be more susceptible to apoptosis; and the other way, involving Th2-like cells, may be more susceptible to death by infection.

Sher: Do you observe virus-infected cells in the group of cells that are being killed?

Shearer: These cells may be killed by the influenza virus, but they may not necessarily be influenza virus-specific clones. One experiment that our laboratories are currently doing is to determine the different effects of stimulation with influenza virus, *env* or both in T cells from the same patient. If we observed 40% T cell death with all three stimuli, we would conclude that the apoptosis is not clonal. However, if we observed 80% T cell death when both stimuli are added, then we may actually be killing specific clones. It is unlikely that specific clones are differentially infected or killed.

Lachmann: Do the cells undergoing apoptosis have gp120 bound to their CD4?

Shearer: I don't know. In my opinion, the strong association between gp120 and the CD4 receptor is a unique feature of AIDS. I wouldn't call this the kiss of death, but it could be the activation event. However, this does not explain why $CD8^+$ cells are activated and killed if they are stimulated via the T cell receptor.

Lachmann: Have you added gp120 to the cells of HIV^- people?

Shearer: This experiment has been done. The cells become sensitized to apoptosis, and the addition of an anti-gp120 antibody/antigen complex induces apoptosis (Gougeon et al 1993). It is not known whether this phenomenon is modulated by cytokines.

Mitchison: A contrasting paper was published last summer (Graziosi et al 1994). Some people believe that there is a change in the Th1/Th2 balance in AIDS, whereas others do not. Gene Shearer, what is your opinion?

Shearer: The expression 'Th1/Th2 balance' should be reserved for clones. We can learn a lot from clones but patients are not clones.

Mitchison: But this paper did not describe work on clones. Most of the experiments involved using PCR analysis of peripheral blood lymphocytes that came directly out of the patient and were not stimulated.

Shearer: Most of their experiments were performed without stimulation because they felt that one should measure what comes directly out of the patient. However, what happens in the patient is probably under the influence of antigenic stimulation, which is why we stimulate *in vitro*. Therefore, differences between their results and our results could be accounted for by differences in stimulation and other technical aspects. We have verified our results by PCR analysis of IL-12 and IL-10, and we have shown that the level of IL-12 decreases and the level of IL-10 increases in HIV^+ patients. Part of the problem also lies in the classic definition of Th1 and Th2 cytokines. IFN-γ and IL-4 are made by $CD4^+$ T cells, but they are also made by other cells. At what point do you call a cell either a Th1 cell or a Th2 cell? If you clone cells for six months, there is a good chance that you throw away other cells that are making important immunoregulatory cytokines. These cytokines are going to play a role in the regulation of Th1-type responses. We need to think about therapy and prophylactic vaccines for the patient, and not just design experiments on the basis of T cell clones.

Romagnani: We have looked extensively at the Th1/Th2 cell balance in HIV^+ patients (Maggi et al 1994a). Firstly, we assessed the production of cytokines following stimulation of peripheral blood mononuclear cells with phytohaemagglutinin or TPA (12-O-tetradecanoylphorbol 13-acetate) plus Ca^{2+} ionophore. We did not find increased IL-4 production in these patients, rather a decrease in comparison with HIV^- controls. We then looked at $CD4^+$ T cell clones from HIV^+ patients, which were generated using three different techniques. (1) When we stimulated single T cells from HIV^+ patients with phytohaemagglutinin, we obtained a smaller proportion of IL-4-producing T cell clones than when we stimulated cells from HIV^- controls. However, the clonal efficiency was significantly lower in HIV^+ patients than in HIV^- controls, suggesting a higher degree of mortality in the cloned T cells of the former. (2) We also generated clones from skin biopsies of HIV^+ patients that were stimulated with IL-2. We found an increase in the number of IL-4-producing T cell clones in comparison to controls, but no difference in the number of IFN-γ-producing clones. In other words, under this experimental condition, there was a shift from Th1 cells to Th0 cells. (3) We also obtained the same results when we established antigen-specific clones from the blood of HIV^+ patients, i.e. a shift from Th1 cells to Th0 cells in comparison with control clones (Maggi et al 1994a). One problem is that the conclusions drawn from these experiments are limited by the small number of patients that one can reasonably test by using cloning techniques.

Finally, in some HIV^+ patients, particularly those with high levels of IgE and hypereosinophilia, we obtained numerous clones with a $CD8^+$ phenotype that was shifted towards a clear-cut Th2 cell profile (Maggi et al 1994b). We conclude that the Th2 cell shift occurs in some but not all patients and, therefore, it is not a general phenomenon.

I would also like to mention that, in another model based on the *in vitro* infection with HIV of T cell clones that have an established Th1, Th2 or Th0 phenotype, we found that Th2-type clones were able to replicate the virus more efficiently than Th1-type clones (Maggi et al 1994a). It is possible that Th2 clones express CD30 and that the CD30 ligand/CD30 interaction may favour enhanced viral replication. The difficulty of finding clear-cut Th2 cells may be due to the fact that when Th cells are stimulated by an antigen under conditions inducing a Th2-type response (so that they express CD30), enhanced HIV replication and early death of HIV^+ $CD4^+$ T cells are favoured because of the interaction between Th2 cells and B cells (or other cells) expressing the CD30 ligand.

Mitchison: What is your opinion of trying therapy, presumably with IL-12, to encourage the development of Th1 cells?

Romagnani: I am not sure that IL-12 would be beneficial. We have to determine the effect of this cytokine on, for example, the expression of CD30 and HIV replication.

Ramshaw: Infection with HIV leads to the development of cytotoxic T cells that are highly effective at controlling the levels of the virus. These cells are much more effective than AZT (3′-azido-3′-deoxythymidine). This may be due to particular cytokines expressed under conditions of cell-mediated immunity that control the virus. I know of no evidence where antibodies play any role in the control of HIV replication *in vivo*. The question is not whether a strong cell-mediated response occurs but how can it be maintained.

References

Amiesen JC, Capron A 1991 Cell dysfunction and depletion in AIDS: programmed cell death hypothesis. Immunol Today 12:102–105

Brunner T, Mogil RJ, LaFace D 1995 Cell-autonomous Fas (CD95)/Fas-ligand interaction mediates activation-induced apoptosis in T cell hybridomas. Nature 373:441–444

Clerici M, Sarin A, Coffman RL et al 1994 Type 1/2 cytokine modulation of T cell programmed cell death as a model for HIV pathogenesis. Proc Natl Acad Sci USA 91:11811–11815

Dhein J, Walczak H, Baumler C, Debatin K-M, Krammer PH 1995 Autocrine T-cell suicide mediated by APO-1(Fas/CD95). Nature 373:438–441

Ettinger R, Panka DJ, Wang J, Stanger BZ, Ju S-T, Marshak-Rothstein A 1995 Fas ligand mediated cytotoxicity is directly responsible for apoptosis of normal $CD4^+$ T cells responding to a bacterial superantigen. J Immunol 154:4302–4308

Finkel TH, Tudor-Williams G, Banda NK et al 1995 Apoptosis occurs predominantly in bystander cells and not in productively infected cells of HIV- and SIV-infected lymph nodes. Nat Med 1:129–134

Gougeon ML, Garcia S, Meeney J 1993 Programmed cell death in AIDS-related HIV and SIV infections. AIDS Res Hum Retroviruses 9:553–563

Graziosis C, Pantaleo G, Gantt KR et al 1994 Lack of evidence for the dichotomy of T_H1 and T_H2 predominance in HIV-infected individuals. Science 265:248–252
Ju S-T, Panka DJ, Cui H et al 1995 Fas (CD95)/FasL interactions required for programmed cell death after T-cell activation. Nature 373:444–448
Lichtman AH, Chin J, Schmidt JA, Abbas AK 1988 The role of interleukin-1 in the activation of T lymphocytes. Proc Nat Acad Sci USA 85:9699–9703
Maggi E, Mazzetti M, Ravina A et al 1994a Ability of HIV to promote a Th1 to Th0 shift and to replicate preferentially in Th2 and Th0 cells. Science 265:244–248
Maggi E, Giudizi M-G, Biagotti R et al 1994b Th2-like $CD8^+$ T cells showing B cell helper function and reduced cytolytic activity in human immunodeficiency virus type 1 infection. J Exp Med 180:489–495
Meyaard L, Otto SA, Jonker RR, Mijnster MJ, Keet RPM, Miedema F 1992 Programmed death of T cells in HIV-1 infection. Science 257:217–219
Ramsdell F, Seaman MS, Miller RE, Picha KS, Kennedy MK, Lynch DH 1994 Differential ability of Th1 and Th2 T cells to express Fas ligand and to undergo activation-induced cell death. Int Immunol 6:1545–1553

Regulation of CD4+ T cell differentiation[1]

Tetsuo Nakamura*†, Mercedes Rincón*[2], Yumiko Kamogawa*†[3]
and Richard A. Flavell*†

**Yale University School of Medicine, New Haven, CT 06520, †Howard Hughes Medical Institute, Yale University School of Medicine, 310 Cedar Street, New Haven, CT 06520-8011, USA*

Abstract. Naive T cells can be induced to differentiate from an uncommitted precursor to T helper 1 (Th1) and Th2 cells. During this differentiation, genes for transcription factors are activated, and transcription factors such as AP-1 accumulate. To study this activation, we have developed reporter transgenic mice for a number of factors, including AP-1. Naive T cells require two signals to activate AP-1. However, upon becoming effector cells, activation through diacylglycerol analogues is sufficient. Rested effector cells lose accumulated AP-1, and the induction of AP-1 synthesis requires both Ca^{2+} diacylglycerol signals, but not co-stimulation.

1995 T cell subsets in autoimmune and infectious diseases. Wiley, Chichester (Ciba Foundation Symposium 195) p 154–172

Upon the completion of development in the thymus, T cells exit to the periphery. CD4+ T cells recognize major histocompatibility complex (MHC) class II peptides, and therefore survey the antigenic space of the extracellular compartments and the vacuoles that potentially harbour intracellular parasites. They also play a crucial role both in assisting B cells to make antibodies and in the activation of macrophages. CD8+ T cells are specialized in their ability to recognize MHC class I molecules, which present cytoplasmic peptides. These cells are therefore ideally suited to combat cytoplasmic pathogens, such as viruses and certain bacteria (e.g. *Rickettsia*) (Janeway & Travers 1994). The divergent role of CD4+ T cells in the activation of macrophages on the one hand, and activation of B cells to make antibodies on the other hand, was clarified by the observations of Bottomly (1989), who showed that certain cloned T cells showed either one of these properties but not both. A molecular basis for this dichotomy was provided by Mosmann & Coffman (1989), who

[1]This paper was presented by R. Flavell.
[2]T. Nakamura & M. Rincón share equally in experimentation.
[3]Present address: Shinjuku-Ku, Institute of Gastroenterology, Tokyo, Japan.

showed that the differentiation of these clones was caused by the cytokines that they produce. Thus, T cell clones with macrophage-activating properties secrete the cytokines γ-interferon (IFN-γ) and tumour necrosis factor. These cytokines then activate macrophages and, therefore, comprise the effector function of these T cells, which are known as T helper 1 (Th1) or T inflammatory cells (Cher & Mosmann 1987, Stout & Bottomly 1989). Likewise, Th2 CD4$^+$ T cells secrete the cytokines interleukin 4 (IL-4), IL-5, IL-6 and IL-13. These molecules assist the proliferation and differentiation of B cells into antibody-secreting plasma cells (Kim et al 1985, Boom et al 1988, Killar et al 1987, Cherwinski et al 1987, Mosmann et al 1986, Mosmann & Coffman 1989). Th1 and Th2 cells are derived from thymocytes which, on exit to the periphery, are presumably naive because the antigen for T cells that escapes selection is not normally present within the thymus. It is not clear how the differentiated cell types are derived from their thymic precursors. A priori, it could be imagined that precursor cells are already committed to either the Th1 or Th2 lineage in the thymus. Alternatively, committed cells could develop in the periphery after T cell activation. We have been interested for some time in the mechanisms whereby differentiated, specialized T effector cells are derived from naive T cells; how memory T cells are ultimately obtained from this source; and whether such cells are truly committed, or whether they can be induced to 'change their spots' and revert from a Th1 to a Th2 phenotype. In this article we will discuss the evidence that supports the notion that naive T cells precursors to Th1 and Th2 cells are uncommitted and can be induced selectively to become Th1 or Th2 cells during differentiation. Moreover, we will discuss the molecular basis for the difference in signalling requirements for effector cells, naive T cells and potentially memory T cells.

Production of Th1 and Th2 cells can be elicited *in vitro* by manipulating the cytokine environment

Swain et al (1990, 1991) have pioneered an approach whereby naive T cells can be induced to differentiate into effector cells by stimulation with a polyclonal activator in the presence of appropriate cytokines. They showed that T cells, cultured in the presence of concanavalin A, IL-12 and IL-4, differentiate into a population that secretes Th2 cytokines but not Th1 cytokines. Parallel experiments suggested that culture of T cells in IFN-γ reduces the number of Th2 cells, and in some cases elicits significant numbers of Th1 cells (Swain et al 1990, 1991, Kamogawa et al 1993). However, it was recently found that inclusion of IL-12 into the culture medium containing concanavalin A and IL-2 mediates the production of a large number of Th1 cells, and essentially no Th2 cells (see Kamogawa et al 1993). We have used such a culture system to produce populations of essentially exclusively Th1 or Th2 cells (Fig. 1).

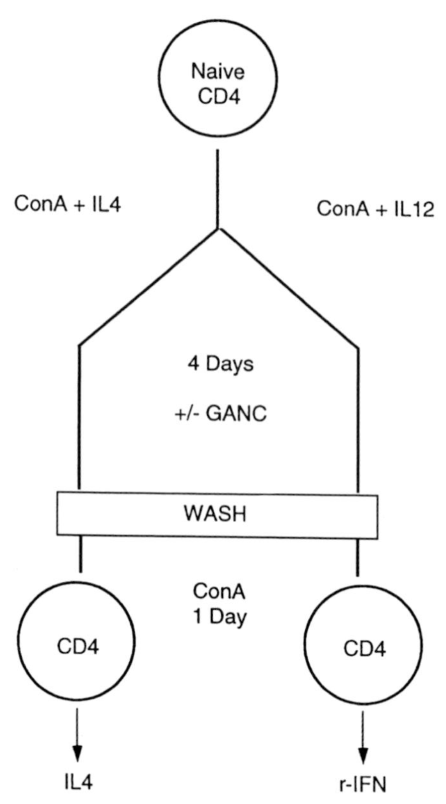

FIG. 1. Experimental strategy to generate interleukin (IL-4)-producing- or γ-interferon (IFN-γ)-producing $CD4^+$ T cells. Sorted naive ($CD45RB^{high}$ and $Pgp\text{-}1^{low}$) $CD4^+$ T cells or total $CD4^+$ T cells (5×10^5) were cultured with 2.5×10^5 antigen-presenting cells and incubated with 2.5 μg/ml concanavalin A (ConA), 10 U/ml IL-2 and either 1000 U/ml IL-4, 1000 U/ml IFN-γ or 3.5 ng/ml IL-12 for four days in the presence or absence of 10 mM ganciclovir (GANC). Cells were recovered and restimulated with 2.5 μg/ml concanavalin A alone for 20 h. Culture supernatants were measured for cytokines by bioassay or ELISA, and cells were harvested and used for herpes simplex virus thymidine kinase assays and *in situ* hybridization. r-IFN, recombinant IFN-γ.

Cytokine mRNA synthesis during T cell differentiation

At some point during the stimulation of naive $CD4^+$ T cells, differentiated cytokines are synthesized and secreted in large amounts. We used the competitive reverse transcriptase PCR (RT-PCR) method of Reiner et al (1993) to determine the timing of cytokine mRNA production under these conditions. Immediately after activation, T cells produce low levels of IL-4 mRNA. Inclusion of IL-4 in the culture medium leads to the accumulation of substantial levels of IL-4 mRNA, which starts approximately one day after culture and peaks after about four days in culture. Restimulation of effector cells elicited in this environment leads to the rapid and accumulated synthesis of IL-4 mRNA, which is detectable 6–8 h after synthesis. Similarly, inclusion of IL-12 in the culture medium leads to the synthesis and preferential accumulation of IFN-γ mRNA, starting within approximately one to two days after stimulation of naive T cells, and leads to low levels of IL-4 mRNA synthesis. (However, we do not detect significant levels of IFN-γ mRNA after stimulation of naive T cells.)

Lineage ablation of CD4$^+$ T cells mediated by herpes simplex virus thymidine kinase and ganciclovir

Over the last few years, in collaboration with the laboratory of Kim Bottomly, we have developed a useful technique for the analysis of T cell lineages (Minasi et al 1993, Kamogawa et al 1993). This technique relies on the early observations of Borrelli et al (1988), who showed that cells expressing the gene encoding herpes simplex virus thymidine kinase (HSV-TK) are selectively susceptible to killing by the anti-herpes drug ganciclovir. We adapted this technique to study cytokine gene expression by generating mice in which the IL-2, IL-4 or IFN-γ promoters were used to direct the synthesis of HSV-TK in transgenic mice (Minasi et al 1993, Kamogawa et al 1993, T. Nakamura & R. A. Flavell, unpublished results). These transgenic mice express the gene encoding HSV-TK selectively in cells that also express these cytokines, as shown by *in situ* hybridization experiments or by the tissue specificity of these promoters. Further experiments (described below) testified that this system directs the expression of transgene-encoded RNA specifically and apparently exclusively in those cells that express the relevant cytokine mRNAs. Stimulation of naive CD4$^+$ T cells leads to the synthesis of HSV-TK, which can readily be detected by an enzymatic assay (Minasi et al 1993, Kamogawa et al 1993). Utilizing this assay, we have shown that HSV-TK is synthesized with similar kinetics to the synthesis of the cytokines, although HSV-TK persists for longer, presumably because it is a more stable protein. The specificity of the system allows the study of T cell lineages in the following way. If a given CD4$^+$ T cell expresses a cytokine (either IL-2, IL-4 or IFN-γ) then T cells, derived from transgenic mice in which the respective cytokine promoter drives HSV-TK mRNA synthesis, will accumulate intracellular HSV-TK upon activation. If ganciclovir is included in the culture medium, these cells die (Minasi et al 1993, Kamogawa et al 1993). The efficiency of this system is high, and in the case of IL-2 HSV-TK transgenic mice, approximately 80% of T cells can be eliminated under the appropriate circumstances. Similarly, T cells from the IL-4 HSV-TK transgenic mice are almost entirely eliminated by this process.

Before describing the results obtained with this system, certain caveats should be mentioned. First, in this system we are studying the promoter activity of cytokine genes and not the accumulation of cytokines themselves. Thus, it is possible in all our studies that the cells express cytokine mRNA but not cytokine protein, and the cells that we examine and score as positive for HSV-TK expression may not be synthesizing or secreting significant levels of cytokines. Second, it is difficult to determine the level of sensitivity of this system, and it is possible that the threshold of HSV-TK expression required for the killing of cells is not the same as the threshold of detection of cytokine mRNA or protein that one can develop in more traditional assays. Therefore, it is possible that not all the observations of these transcription-based studies will completely parallel the studies of cytokine proteins. By using this technique, we were able to show that T cells from IL-4 HSV-TK transgenic mice can be readily killed upon activation by ganciclovir. T cells were induced to differentiate in the presence of IL-4 which, as stated above, leads to a high proportion of Th2-like effector cells as a final differentiation product. These cells were completely ablated by culturing with ganciclovir. We were surprised, however, to find that cells derived from the IL-4 HSV-TK transgenic mice were also eliminated when they were induced to differentiate in the presence of IL-12 (Fig. 2). Numerous control experiments showed that this was not an artefact of the transgene because, as described above, specific expression of the gene could be noted. This effect was not due to non-specific cytotoxicity because the transgene-negative littermates do not exhibit any cytotoxicity (Kamogawa et al 1993). An alternative explanation is that Th1 effector cells require a gene product from Th2 cells in order to differentiate effectively. This gene product could not be IL-4 itself because addition of IL-4 to the culture does not prevent the ablation of the low level of Th1 cells produced under those circumstances. It is possible that additional cytokines, such as IL-5, IL-6, IL-13 or a previously undescribed cytokine, could mediate this effect. To test this possibility, we performed ablation experiments with cells cultured in the presence of IL-2 and IL-12, supplemented with a cytokine supernatant generated by the activation of Th2 populations. These supernatants should contain these hypothetical 'helper' materials. However, lineage ablation was unaffected when we performed this experiment. We conclude that Th1 precursors derived from naive $CD4^+$ T cells express the IL-4 transgene and, therefore, the gene encoding IL-4. As described above, this is consistent with our observations that IL-4 mRNA can be detected after stimulation of naive $CD4^+$ T cells at low levels immediately after activation. Our interpretation of these experiments is that the expression of cytokine promoters is initially uncommitted. Therefore, we proposed that precursors of Th1 effector cells are uncommitted and have a Th0 phenotype, namely a phenotype in which the genes responsible for both the Th1 and Th2 phenotypes are expressed (Kamogawa et al 1993). Our studies do not measure cytokine secretion, so one could argue that these cells do not reflect true Th0 cells (defined as cloned T cells that secrete mixed cytokines). Therefore, we propose to call these cells Th precursor (Thp) cells.

Th precursor cells become committed to a particular cytokine gene expression pattern after activation

If Thp cells are not committed to a pathway resulting in the expression of specific cytokine genes, do they eventually become committed and express only the cytokine genes that are specific to a given Th1 or Th2 lineage? To test this idea, we cultured $CD4^+$ T cells in the presence of IL-2 and IL-4 for four days in the absence of ganciclovir. This gives sufficient time for the differentiation of these cells into effector cells. As in our standard assay, restimulation of these cells leads to the rapid secretion of IL-4 by Th2 effector cells. This second stimulation was performed by removing the cytokines from the initial stimulation and restimulating with concanavalin A alone. We showed that the cells secreting IL-4 following this second stimulation were completely ablated by the inclusion of ganciclovir. This experiment shows, not surprisingly, that Th2 effector cells transcribe the IL-4 gene at levels high enough to generate sufficient HSV-TK to mediate lineage ablation by ganciclovir killing. However, in T cells that have been stimulated in the presence of IL-2 and IL-12, and then restimulated with concanavalin A (in the absence of IL-2 and IL-12) in the presence of ganciclovir, no ablation of IFN-γ-producing cells is obtained. This is in stark contrast to the results described above, where Thp cells, which are destined to become exclusively Th1 cells, are rapidly and completely eliminated by this treatment. These results support the contention that Thp cells activate the IL-4 promoter and are initially uncommitted but become committed after a particular stage has been reached.

At what stage does this commitment event occur? To determine this, we cultured T cells in the presence of cytokines to induce either Th1 or Th2 effector cells, and we added ganciclovir 24 h or 48 h after primary stimulation rather than at the beginning of the culture. As described above, substantial levels of cytokine mRNAs are synthesized after 24 h. Co-culture with IL-12 results in the production of IFN-γ mRNA and a small amount of IL-4 mRNA. Co-culture with IL-2 plus IL-4 results in the production of IL-4 mRNA but not IFN-γ mRNA. Inclusion of ganciclovir 24 h after culturing the cells results in the complete ablation of all Th1 and Th2 precursor cells. However, the inclusion of ganciclovir into cultures after 48 h eliminates Th2 effector cell production completely, but it does not effect the production of Th1 effector cells (T. Nakamura & R. A. Flavell, unpublished results). These results suggest that a commitment event occurs between 24 h and 48 h, at least in the case of precursors that become Th1, during which the inappropriate cytokine promoters are switched off (or at least expression is reduced to below the threshold level that makes killing by ganciclovir possible). We interpret these

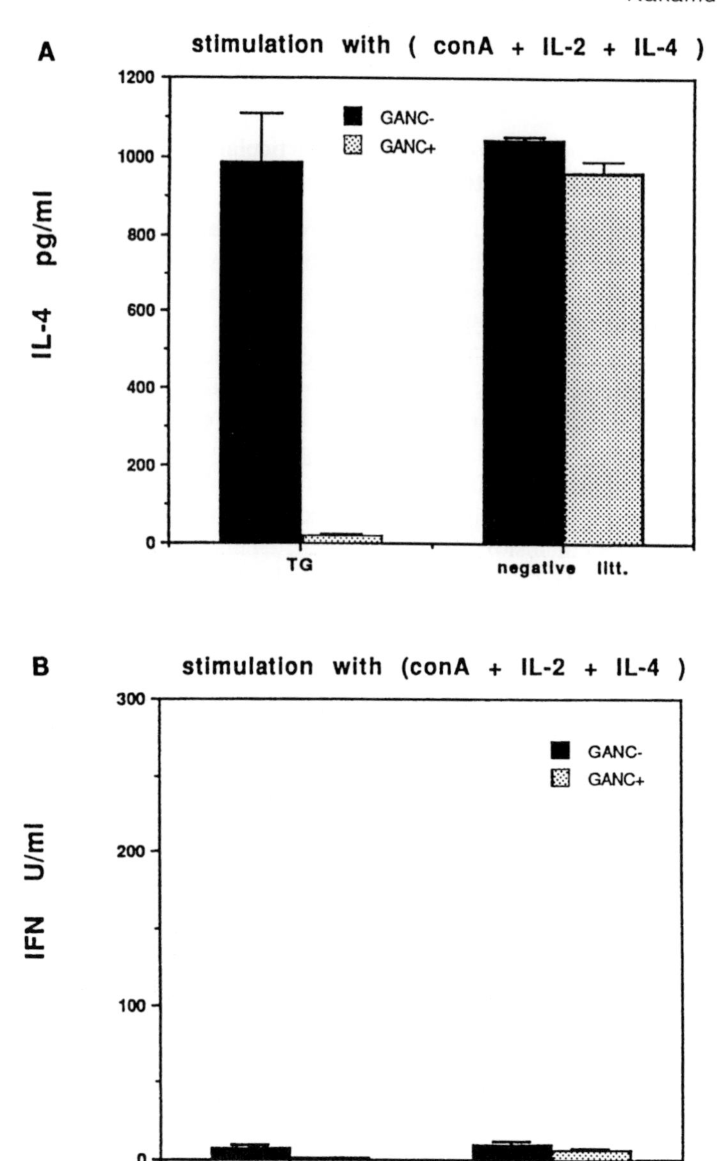

FIG. 2. Ganciclovir (GANC) ablates both interleukin 4 (IL-4) and γ-interferon (IFN-γ) production by naive T cells from IL-4–thymidine kinase transgenic (TG) mice. (A and B) Sorted naive CD4 T cells from transgenic mice (three mice per experiment, F5 generation) and negative littermates (litt) (three mice per experiment) were stimulated with concanavalin A (conA), IL-2 and IL-4, in the presence (hatched bar) or the absence

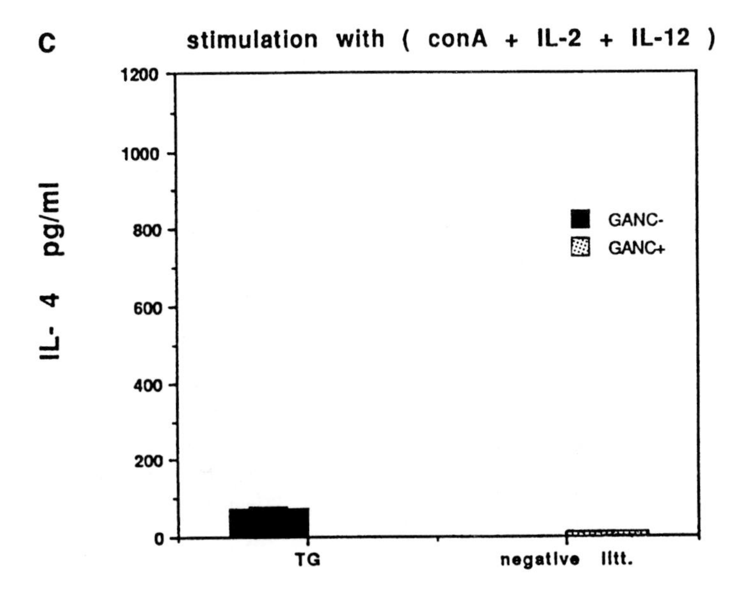

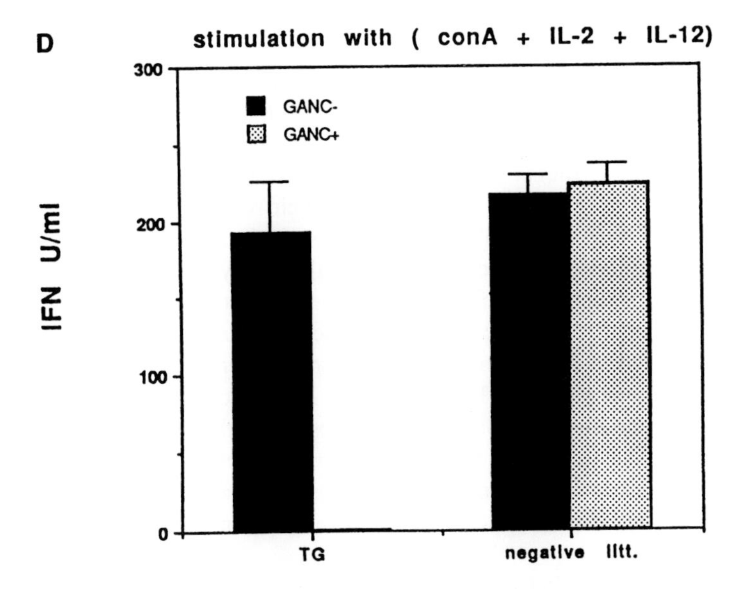

(closed bar) of ganciclovir. IL-4 release (A) and IFN-γ release (B) after restimulation with concanavalin A. (C and D) Stimulation with concanavalin A, IL-2 and IL-12 induces cells to make IFN-γ upon restimulation. IL-4 release (C) and IFN-γ release (D) from concanavalin A-stimulated cells. Note that the scale for IFN-γ in B and D is different.

results to mean that the key events that mediate the differentiation of $CD4^+$ T cells into Th1 cells *in vitro* occurs sometime between 24 h and 48 h after stimulation. In the case of $CD4^+$ T cells that differentiate into Th2 cells, the commitment event may occur within 24 h after primary stimulation because preferential IL-4 transcription was already observed after 24 h in culture.

The molecular regulation of cytokine gene expression during T cell differentiation

The molecular basis for the regulation of cytokine gene expression has been studied by extensive promoter mapping experimentation, and the promoter region of the IL-2 gene is the most well characterized (Crabtree 1989, Ullman et al 1990). Several inducible transcription factors bind to this region. The AP-1 transcription factor is a complex mixture of homodimers and heterodimers of different members of the Fos and Jun family proteins (Cohen & Curran 1988, Halazonetis et al 1988, Nakabeppu et al 1988, Hirai et al 1989, Ryder et al 1989, Zerial et al 1989). The NFAT (nuclear factor of activated T cells) complex consists of a cytoplasmic component encoded by the NFATp or NFATc genes and a nuclear component which, at least under certain circumstances, seems to contain members of the Fos and Jun families (Flanagan et al 1991, Northrop et al 1994, Jain et al 1993, McCaffrey et al 1993). The Oct-1-associated protein (OAP) that binds to the NFIL-2 (nuclear factor of IL-2) activating site also contains some of the Fos and Jun family members (Ullman et al 1990). Binding sites for NFκB have also been described (Hoyos et al 1989). Mutation of all of these regulatory sites shows that they play an important role *in vivo* and that all of them are required for full transcription of the IL-2 gene. In addition to their role in the regulation of IL-2 gene expression, they are involved in the transcription of other cytokine genes. To study the regulation of the transcriptional activity of these nuclear factors and their role during T cell differentiation, we generated novel transgenic mice in which binding sites specific for each of the given transcription factors have been oligomerized and then linked to a basal promoter that drives the synthesis of the reporter gene encoding luciferase (Fig. 3) (Rincón & Flavell 1994). Thus, although the gene encoding luciferase is present in the genome of every cell, functionally active luciferase activity will only be detected when a particular transcription factor is present. We have used T cells from these mice to elucidate the activation requirements of transcription factors in naive T cells and rested T cells.

The studies with AP-1 luciferase transgenic mice have been particularly informative. It is widely believed that AP-1 is activated exclusively by the stimulation of protein kinase C (PKC) because phorbol esters such as 12-*O*-tetradecanoylphorbol 13-acetate (TPA) can activate AP-1 DNA binding and transcriptional activity in numerous human and mouse cell lines. However, in

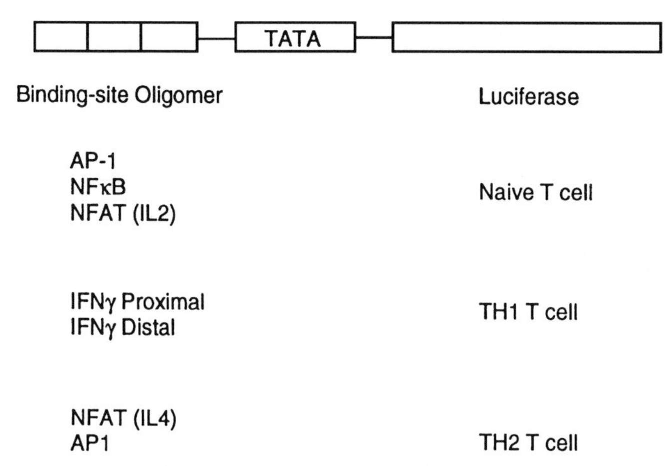

FIG. 3. Constructs for reporter transgenes. The gene encoding luciferase is driven by a minimal promoter transactivated by oligomerized binding sites for the factors described. Transgenic mice bearing these constructs were generated by standard procedures (see Rincón & Flavell 1994). IFNγ, γ-interferon; IL2, interleukin 2; NFAT, nuclear factor of activated T cells; TH, T helper.

primary T cells isolated from transgenic mice this is not the case. Although PKC stimulation is sufficient to induce DNA binding, AP-1 transcriptional activity requires PKC stimulation as well as a Ca^{2+} flux mediated by the antigen receptor (or its surrogates TPA and ionomycin) in conjunction with a second signal, which can be provided by co-stimulatory molecules such as CD28 (Fig. 4) (Rincón & Flavell 1994). We believe that post-translational modification of AP-1 is required in order to provide transcriptional activation and that this is mediated by the Jun kinase family (JNK 1–3) (Hibi et al 1993, Derijard et al 1994, Su et al 1994, Rincón & Flavell 1994). When freshly isolated $CD4^+$ T cells from T cell receptor (TCR) transgenic mice are stimulated in the presence of polyclonal activators, such as concanavalin A, or specific antigen in the presence of antigen-presenting cells (as a source of co-stimulation) and IL-4 to induce further differentiation to Th2 phenotype, cells accumulate AP-1 complexes. The transcription of AP-1 peaks at approximately Day 2–3 but decreases sharply at Day 4 to a low or undetectable level (although cells contain high levels of AP-1 complex in the nucleus). It is unusual for preformed transcription factors to be detectable in cells that are not utilizing such molecules to direct transcription. However, upon restimulation of these effector cells, a high AP-1 transcriptional activity is rapidly induced, which is at least an order of magnitude more than that obtained under similar conditions by naive $CD4^+$ T cells. In these effector cells, AP-1 transcriptional activity can be induced by TPA alone without the

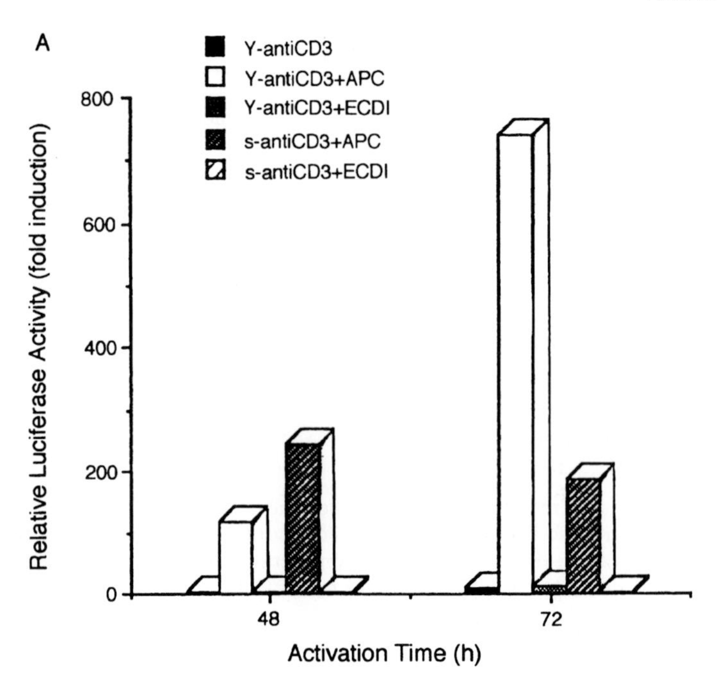
A
Y-antiCD3
Y-antiCD3+APC
Y-antiCD3+ECDI
s-antiCD3+APC
s-antiCD3+ECDI
Relative Luciferase Activity (fold induction)
800
600
400
200
0
48
72
Activation Time (h)

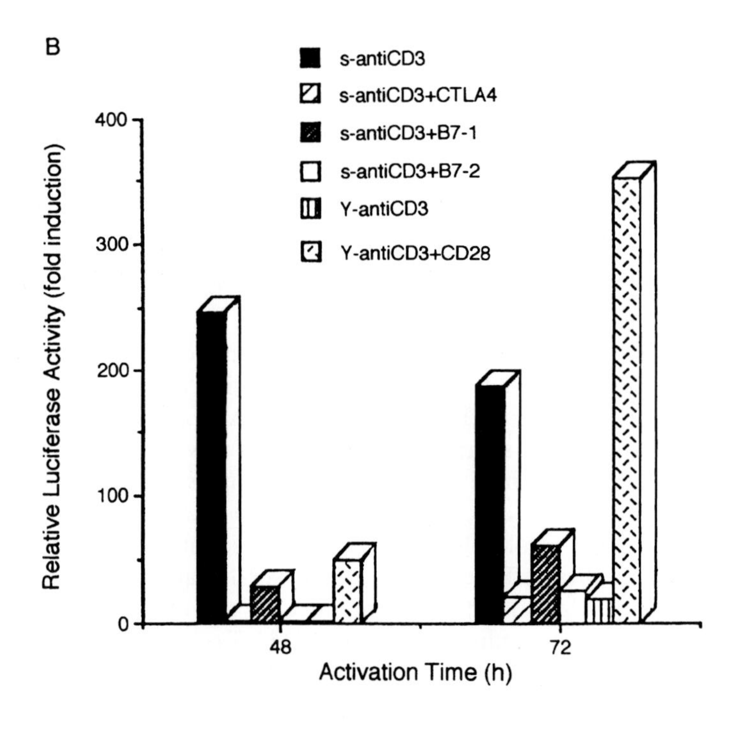
B
s-antiCD3
s-antiCD3+CTLA4
s-antiCD3+B7-1
s-antiCD3+B7-2
Y-antiCD3
Y-antiCD3+CD28
Relative Luciferase Activity (fold induction)
400
300
200
100
0
48
72
Activation Time (h)

requirement for Ca^{2+}. Moreover, signalling mediated by TCR alone in the absence of co-stimulation can also induce AP-1 transcriptional activity. These results are in contrast to what is observed for freshly isolated (presumably mostly naive) T cells from TCR transgenic mice, indicating that a change in the activation requirements for AP-1 occurs during differentiation to effector cells. A high proportion of these T cells can be recovered if they are rested for two days, and during this resting phase the preformed AP-1 complexes disappear almost entirely from the nucleus (M. Rincón & R. A. Flavell, unpublished results). AP-1 DNA binding is induced in these cells when they are treated with TPA alone. A Ca^{2+} signal is also required to induce AP-1 transcriptional activity, which is similar to the activation requirements of freshly isolated T cells. However, a crucial difference between the activation parameters of rested cells when compared with naive T cells is that co-stimulation does not appear to play an important role.

These results show that a key factor, which promotes the transcription of IL-2 and probably IL-4, responds differently to extracellular signals, depending on the state of differentiation of the T cells in which this factor is present. Thus, naive T cells require both stimulation through the TCR and the co-stimulator CD28 in order to proliferate and synthesize cytokines. In addition, as we show here, the activation requirements for a transcription factor that regulates cytokine expression are the same as those of the cell itself. We propose that the stimulation of naive T cells has stringent activation requirements to ensure that only the appropriate activation leads to clonal expansion of these cells. However, during clonal expansion and accompanying differentiation, these T cells accumulate high levels of preformed transcription complexes that render the cell able to produce high levels of cytokines at short notice. Cells that differentiate to Th2 cells accumulate high levels of preformed AP-1 complex in the nucleus, which mediate transcription within a few hours of activation. This enables the cell to deliver its effector function promptly and efficiently in the absence of co-stimulators, as would be required of an effector cell. After resting, this population

FIG. 4. A co-stimulatory signal is required for AP-1 transcriptional activity in purified resting T cells. T lymphocytes purified from spleen and lymph nodes from transgenic mice were stimulated with (A) immobilized anti-CD3 monoclonal antibody (mAb) (125 μl/well at 5 μg/ml) (Y-antiCD3) or 1 μg/ml soluble anti-CD3 mAb (s-antiCD3) in the presence or absence of mitomycin C-treated cells (antigen-presenting cells, APCs) or 1-ethyl-3-(3-dimethylaminopropyl)-carbodiimide (ECDI)-fixed spleen cells from non-transgenic B7 mice, or (B) 1 μg/ml soluble anti-CD3 mAb plus APC in the presence or absence of 10 μg/ml soluble CTLA4–Ig (CTLA4), 10 μg/ml anti-B7-1 mAb (B7-1) or 10 μg/ml anti-B7-2 mAb (B7-2) or immobilized anti-CD3 mAb without APC in the presence or absence of 1 μg/ml soluble anti-CD28 mAb (CD28). Luciferase activity was analysed 48 h or 72 h following activation. Similar results were obtained in two other individual experiments.

gives rise to quiescent cells that can be reactivated to express AP-1. In this case, the kinetics and level of the protein is analogous to that seen for naive T cells. However, the activation requirements for this process appear to be different. These results suggest that we may have generated the *in vitro* equivalent of memory T cells which show an enhanced ability to respond.

Finally, we have generated transgenic mice expressing reporters specific for end-stage cytokines of both the Th1 and Th2 pathways. Specifically, reporters utilizing the NFAT site of the IL-4 promoter (Todd et al 1993) will provide a potent marker for the Th2 lineage, and the proximal and distal enhancer elements of IFN-γ promoter markers (Penix et al 1993) will provide markers for Th1 differentiation. These mice will be analysed in the near future, and they should hopefully provide molecular insights into the way in which the differentiation to Th1 and Th2 cells occurs.

Acknowledgements

T. N. is, and Y. K. was, an associate of the Howard Hughes Medical Institute. R. A. F. is an investigator of the Howard Hughes Medical Institute, which supported this work, in part. M. R. was supported by a grant from the National Institutes of Health (AI29902).

References

Boom WH, Liano D, Abbas AK 1988 Heterogeneity of helper/inducer T lymphocytes. II. Effects of interleukin 4- and interleukin 2-producing T cell clones on resting B lymphocytes. J Exp Med 168:1350–1363

Borelli E, Heyman R, Hsi M, Evans RM 1988 Targeting of an inducible toxic phenotype in animal cells. Proc Natl Acad Sci USA 85:1–5

Bottomly K 1989 Subsets of CD4 T cells and B cell activation. Semin Immunol 1:21–31

Cher DJ, Mosmann TR 1987 Two types of murine helper T cell clone. II. Delayed-type hypersensitivity is mediated by Th1 clones. J Immunol 138:3688–3694

Cherwinski HM, Schumacher JH, Brown KD, Mosmann TR 1987 Two types of mouse helper T cell clone. III. Further differences in lymphokine synthesis between Th1 and Th2 clones revealed by RNA hybridization, functionally monospecific bioassays, and monoclonal antibodies. J Exp Med 166:1229–1244

Cohen DR, Curran T 1988 fra-1: a serum-inducible, cellular immediate–early gene that encodes a Fos-related antigen. Mol Cell Biol 8:2063–2069

Crabtree GR 1989 Contingent genetic regulatory events in T lymphocyte activation. Science 243:355–361

Dérijard B, Hibi M, Wu I-H et al 1994 JNK1: a protein kinase stimulated by UV light and Ha-Ras that binds and phosphorylates the c-Jun activation domain. Cell 76:1025–1037

Flanagan WM, Corthesy B, Bram RJ, Crabtree GR 1991 Nuclear association of a T-cell transcription factor blocked by FK-506 and cyclosporin A. Nature 352:803–807

Halazonetis TD, Georgopoulos K, Greenberg ME, Leder P 1988 c-Jun dimerizes with itself and with c-Fos forming complexes of different DNA binding affinities. Cell 55:917–924

Hibi M, Lin A, Smeal T, Minden A, Karin M 1993 Identification of an oncoprotein- and UV-responsive protein kinase that binds and potentiates the c-Jun activation domain. Genes & Dev 7:2135–2148

Hirai S-I, Ryseck R-P, Mechta F, Bravo R, Yaniv M 1989 Characterization of JunD: a new member of the *jun* proto-oncogene family. EMBO J 8:1433–1439
Hoyos B, Ballard DW, Böhlein E, Siekevitz M, Greene WC 1989 Kappa B–specific DNA binding proteins: role in the regulation of human interleukin-2 gene expression. Science 244:457–460
Jain J, McCaffrey PG, Miner A et al 1993 The T-cell transcription factor NFATp is a substrate for calcineurin and interacts with Fos and Jun. Nature 365:352–355
Janeway CA Jr, Travers P (eds) 1994 Immunobiology. Current Biology, London
Kamogawa Y, Minasi L-aE, Carding SR, Bottomly K, Flavell RA 1993 The relationship of IL-4- and IFNγ-producing T cells studied by lineage ablation of IL-4-producing cells. Cell 75:985–995
Killar L, MacDonald G, West J, Wood A, Bottomly K 1987 Cloned, Ia-restricted T cells that do not produce interleukin 4 (IL-4)/B cell stimulatory factor 1 (BSF-1) fail to help antigen-specific B cells. J Immunol 138:1674–1679
Kim J, Woods A, Becker-Dunn E, Bottomly K 1985 Distinct functional phenotypes of cloned Ia-restricted helper T cells. J Exp Med 162:188–201
McCaffrey PG, Luo C, Kerpolla TK et al 1993 Isolation of cyclosporin-sensitive T cell transcription factor NFATp. Science 262:750–754
Minasi L-aE, Kamogawa Y, Carding S, Bottomly K, Flavell RA 1993 The selective ablation of interleukin 2-producing cells isolated from transgenic mice. J Exp Med 177:1451–1459
Mosmann TR, Coffman RL 1989 Th1 and Th2 cells: different patterns of lymphokine secretion lead to different functional properties. Annu Rev Immunol 7:145–173
Mosmann TR, Cherwinski H, Bond MW, Giedlin MA, Coffman RL 1986 Two types of murine helper T cell clone. I. Definition according to profiles of lymphokine activities and secreted proteins. J Immunol 136:2348–2357
Nakabeppu Y, Ryder K, Nathans D 1988 DNA binding activities of three murine jun proteins: stimulation by fos. Cell 55:907–915
Northrop JP, Ho SN, Chen L et al 1994 NF-AT components define a family of transcription factors targeted in T-cell activation. Nature 369:497–502
Penix L, Weaver WM, Pang Y, Young HA, Wilson CB 1993 Two essential regulatory elements in the human interferon γ promoter confer activation specific expression in T cells. J Exp Med 178:1483–1496
Reiner SL, Zheng S, Corry DB, Locksley RM 1993 Constructing polycompetitor cDNAs for quantitative PCR. J Immunol Methods 165:37–46
Rincón M, Flavell RA 1994 AP-1 transcriptional activity requires both T-cell receptor-mediated and co-stimulatory signals in primary T lymphocytes. EMBO J 13:4370–4381
Ryder K, Lanahan A, Perez-Albuerne E, Nathans D 1989 Jun-D: a 3rd member of the Jun gene family. Proc Natl Acad Sci USA 86:1500–1503
Stout RD, Bottomly K 1989 Antigen-specific activation of effector macrophages by IFN-γ producing (Th1) T cell clones: failure of IL-4 producing (Th2) T cell clones to activate effector function in macrophages. J Immunol 142:760–765
Su B, Jacinto E, Hibi M, Kallunki T, Karin M, Ben-Neriah Y 1994 JNK is involved in signal integration during costimulation of T lymphocytes. Cell 77:727–736
Swain SL, Weinberg AD, English M, Huston G 1990 IL-4 directs the development of Th2-like helper effectors. J Immunol 145:3796–3806
Swain SL, Huston G, Tonkonogy S, Weinberg A 1991 Transforming growth factor-β and IL-4 cause helper T cell precursors to develop into distinct effector helper cells that differ in lymphokine secretion pattern and cell surface phenotype. J Immunol 147:2991–3000

Todd MD, Grusby MJ, Ledere JA, Lacy E, Lichtman AH, Glimcher LH 1993 Transcription of the interleukin 4 gene is regulated by multiple promoter elements. J Exp Med 177:1663–1674

Ullman KS, Northrop JP, Verweij CL, Crabtree GR 1990 Transmission of signals from the T lymphocyte antigen receptor to the genes responsible for cell proliferation and immune function. Annu Rev Immunol 8:421–452

Zerial M, Toschi L, Ryseck R-P, Schuermann M, Muller R, Bravo R 1989 The product of a novel growth factor activated gene, *fosB*, interacts with JUN proteins enhancing their DNA binding activity. EMBO J 8:805–813

DISCUSSION

Mitchison: Immunologists are familiar with the idea that inappropriate transcripts are made during the process of differentiation. For example, in acute lymphoblastic leukaemia diagnostic samples and cell lines with unequivocal B cell precursor or T cell precursor immunophenotypes, there is inappropriate IgM or T cell receptor (TCR) β chain rearrangement in approximately 25% of the cases (Furley et al 1987). Is this a universal phenomenon in cell differentiation?

Flavell: We looked at whether the promoter was expressed in the wrong places. We found that in mice it was expressed in T cells (but not in B cells) in the brain, the foot or the elbow; i.e. it was expressed only in the places where it should be expressed. However, this does not mean that its expression in a T helper precursor (Thp) cell is physiologically important.

Mitchison: Are immunologists the only people doing these experiments? Do people studying the differentiation of neural cells, for example, see inappropriate transcription?

Flavell: It's pretty standard in development that if there are a series of choices, all the choices are expressed first and then a decision is made to express only the preferred choice and switch off the rest.

Locksley: In your studies of the AP-1 transcription factor did you only look at Th2 effectors or did you also look at Th1 effectors?

Flavell: We've looked at non-specific effectors in the presence of concanavalin A, and we observe the same result—a high level of AP-1 if we look at Th2 effectors, and a low level of AP-1 if we look at Th1 effectors. Therefore, there is both an interesting dichotomy in the expression and a definite mechanism of regulation that we have not finished studying.

Locksley: It is possible that other transcription factors are specific for Th1 effectors.

Flavell: Yes. We've made transgenic mice containing γ-interferon (IFN-γ) response elements and we know that they express IFN-γ but we haven't done enough with them to determine whether expression is lineage specific.

Trinchieri: Does methylation or chromatin structure affect transcription from the promoter?

Flavell: We haven't looked at that.

Mosmann: Your mRNA studies showed that interleukin 4 (IL-4) was expressed equally and early on in both the Th1 and Th2 pathways, whereas IFN-γ showed a strong preference for expression only in the Th1 differentiation pathway.

Flavell: That appears to be the case. We've made herpes simplex virus thymidine kinase (HSV-TK) transgenic mice with the IFN-γ construct, which are problematic because they only breed from females. When we did that experiment, we obtained nearly all males, and the one female we did get, did not ablate very well. We made these transgenic mice again recently but we haven't had time to analyse them yet.

Mosmann: So there is a real difference between the levels of IL-4 and IFN-γ mRNAs.

Flavell: Yes, this is a typical finding. Has anyone else looked at the production of IFN-γ in cells stimulated with concanavalin A and IL-4?

Abbas: We have. We did not find that IFN-γ mRNA was produced in the presence of IL-4, but then we did not observe the production of IL-4 mRNA in cells stimulated with IL-12 either.

Mitchison: Do you have any evidence that the proteins are being made?

Flavell: I wouldn't like to comment on the proteins because there are many steps beyond transcription, for example mRNA stability, that may also be important. The HSV-TK protein is stable and I presume that if IL-4 protein was being made, the genes encoding the two proteins would be transcribed at a similar rate. Therefore, there may be a problem with stability.

Swain: We've noticed that in Th2 polarized effector populations there is a transient increase in the mRNA encoding IL-2 but there is no IL-2 protein present (X. Zhang & S. L. Swain, unpublished observations). Do you have any results with the IL-2 knockout mice? Can you generate the effectors from them without drugs and analyse them in the same way as the IL-2 HSV-TK mice?

Flavell: We haven't looked at it in depth, so I don't know the answer. It's an interesting point.

Swain: The question addresses whether IL-2 is turned off permanently in Th2 cells. When Th2 cells are restimulated, they make a small amount of IL-2 mRNA which then disappears. You might find that you could kill the effector function of these Th2 cells with ganciclovir if they re-express IL-2.

Flavell: We've not looked at this. We first have to repeat the assays measuring the levels of IL-2 mRNA. Abul Abbas, have you looked at the levels of IL-2 in Th2 cells?

Abbas: Yes. It is virtually undetectable in Th2 cells. There is also an interesting aside to this story. Some established clones make certain mRNAs but not the proteins (Lederer et al 1994). For example, CDC25 is a well known

Th2 cell clone that makes mRNA encoding IL-2, but the protein cannot be detected. Also, AE7 is a well established Th1 cell clone that makes mRNA encoding IL-4, but again IL-4 protein is not detected. Therefore, the level of mRNA does not always reflect the level of protein.

Flavell: I would prefer to look at populations rather than clones because clones are generated from selected cells, which may have strange characteristics.

Allen: Have you looked at anergized cells *in vivo* in cytochrome c TCR transgenic mice crossed with AP-1 luciferase transgenic mice that have had the antigen delivered intravenously?

Flavell: This is the reason we did this experiment in the first place. We wanted to look at anergy but all we get is death. We do not have an *in vivo* model in which we have generated sufficiently large numbers of anergized cells. We have a couple of rather bizarre systems in which we could do this, for example in one of our T antigen models, but we have not been able to anergize cells *in vitro*.

Mitchison: You have not examined any antigen-specific responses. I expect that you would have to cross your TCR transgenic mice with mice transgenic for the antigen which they recognize.

Flavell: But most of these experiments were performed on a TCR background using antigen stimulation. Even the concanavalin A experiments were performed on the TCR transgenic mouse.

Lotze: If you immunize transgenic or normal mice *in vivo* with antigen, can you return at a later date and ask questions regarding the requirement for co-stimulation?

Flavell: We would need to do adoptive transfer experiments to answer this.

Lotze: You could use non-transgenic mice and look at the effect of a specific antigen.

Mitchison: Why can't you immunize your mice with cytochrome c, then come back six months later and see if they're still active?

Abbas: Because if you take the cells out even only two weeks later, you find that they do not have the properties of cells that have previously encountered the antigen. The reason for this may be that a large number of T cells are stimulated at the same time. Therefore, many activated cells may be deleted, there may be a constant replenishment of cells with new immigrants from the thymus and naive populations may dominate in the immunized intact TCR transgenic mouse. This is why Jenkins' approach of adoptive transfer is used (Kearney et al 1994). But the problem with this is that you lose some of the power of transgenic mice by having to dilute the specific T cells.

Lotze: Why can this not be done in a non-transgenic mouse?

Flavell: Because we have to be able to look at the low levels of luciferase produced by the cells.

Lotze: Could you use alloantigen to broadly stimulate those cells recognizing the major histocompatibility complex-associated peptide?

Flavell: But then you would not know whether you were looking at recall responses of previously immunized cells.

Cantrell: The initial expression of AP-1 is slow and peaks after three days, which is reminiscent of autocrine stimulation. Are you sure that CD28 is required for the regulation of AP-1 transcription? Or could CD28 be required for the production of other cytokines? All the kinetics of c-Jun expression in primary cells suggest that induction occurs within minutes or hours. This implies that there is a limiting step, which may be the production of another cytokine. Tumour necrosis factor α (TNF-α) may be involved in this. Have you looked at whether anti-TNF-α antibody suppresses induction?

Flavell: No, we have not looked at that. Why are you suggesting TNF-α?

Cantrell: Because it's a powerful inducer of these stress-activated kinases, which may be the crucial kinases that mediate this phosphorylation.

Flavell: Another approach would be to look at direct Jun kinase activation by other stimuli, such as TNF.

Cantrell: But you need to determine whether there are any indirect effects because the effect of CD28 on Jun kinase, for example, occurs within minutes. Therefore, you need to explain this three-day delay.

Flavell: I agree, it's a puzzling result and we don't know the mechanism.

Mitchison: It is important to establish whether an activated cell can be detected by its enlarged internal store of AP-1. This would represent a characteristic that is independent of surface markers, and it would be missing in anergized T cells.

Allen: Lenardo has shown that there is a reduced level of AP-1 in anergized cells (Kang et al 1992).

Flavell: His experiments may not be relevant to these cells.

Mosmann: Presumably, during this three-day period the cells are differentiating into states that can make IL-4. Does the ability to make IL-4 occur simultaneously with the appearance of the AP-1 activity?

Flavell: We've done this experiment with and without IL-4 in the pre-culture, but we do not see a difference in AP-1 induction. Therefore, the induction does not require IL-4 but it does occur simultaneously.

Mosmann: Does it gain the ability to make IL-4 spontaneously in the absence of IL-4?

Flavell: Yes.

Cantrell: Does IL-4 itself up-regulate transcription of the gene encoding IL-4, and are there any IL-4 response elements, such as STAT (signal transducer and activator of transcription) response elements, in the IL-4 promoter?

Abbas: We looked for an IL-4 STAT using gel shift assays with an oligonucleotide from the IL-4 promoter (J. A. Lederer, V. L. Perez & A. K. Abbas, unpublished results). We are now trying to see whether particular transfections and reporter constructs can be used to define how IL-4 stimulates endogenous IL-4 gene expression. These are not easy experiments to do because we have to transfect growth factor-dependent cell lines rather than immortalized cell lines. So far, it looks as though IL-4 by itself up-regulates endogenous transcription of the gene encoding IL-4.

Trinchieri: If transcription of IL-4 is directly regulated by IL-4 STAT, it is surprising that the expression of IL-4 in culture is delayed. This suggests that the molecular mechanisms involved are more complex.

References

Furley AJ, Chain LC, Mizutani S et al 1987 Lineage specificity of rearrangement and expression of genes encoding the T cell receptor–T3 complex and immunoglobulin heavy chain in leukaemia. Leukaemia 1:644–652

Kang S-M, Beverly B, Tran A-C, Brorson K, Schwartz RH, Lenardo MJ 1992 Transactivation by AP-1 is a molecular target of T cell clonal anergy. Science 257:1134–1138

Kearney ER, Pape KA, Loh DY, Jenkins MK 1994 Visualization of peptide-specific T-cell immunity and peripheral tolerance induction in vivo. Immunity 1:327–339

Lederer JA, Liou JS, Todd MD, Glimcher LH, Lichtman AH 1994 Regulation of cytokine gene expression in T helper cell subsets. J Immunol 152:77–86

The role of subsets of CD4$^+$ T cells in autoimmunity[1]

D. Fowell[2], F. Powrie[3], A. Saoudi, B. Seddon, V. Heath and D. Mason

MRC Cellular Immunology Unit, Sir William Dunn School of Pathology, University of Oxford, Oxford OX1 3RE, UK

Abstract. It is generally considered that T cells which are reactive with self-antigens are effectively eliminated by two processes: clonal deletion and the induction of T cell anergy. More recently, it has been shown that some potentially autoreactive T cells remain unactivated because the self-antigens for which they are specific are not presented on competent antigen-presenting cells. All these mechanisms of self-tolerance may be regarded as passive in the sense that the autoreactive cells are either deleted or are intrinsically non-responsive. If this view of self-tolerance is adopted, then one would predict that rendering animals relatively lymphopoenic should not give rise to autoimmune disease. This prediction is not verified by experiment. Rats rendered relatively lymphopoenic by adult thymectomy followed by repeated low dose γ-irradiation develop a high incidence of autoimmune diabetes. Furthermore, it has been shown that the reconstitution of these rats with a specific subset of CD4$^+$ T cells from syngeneic donors prevents the development of this disease. The protective cells have the CD45RC^{low} phenotype, they are resistant to adult thymectomy and the majority of them appear to be non-activated in the donor rats. In contrast, the CD45RC^{high} CD4$^+$ subset does not provide protection from diabetes. Instead, on injection into athymic rats, it gives rise to pathological changes in a variety of organs: stomach, pancreas, liver, thyroid and lung. In addition, the CD45RC^{low} CD4$^+$ subset prevents these manifestations of autoimmunity in these circumstances. Recently, we have shown that CD4$^+$ CD8$^-$ thymocytes are a highly potent source of cells that have the ability to control autoimmune diabetes in rats. It appears that the thymus has three distinct functions: positive selection; negative selection; and the generation of a population of cells that seem specialized for the control of autoimmunity.

1995 T cell subsets in infectious and autoimmune diseases. Wiley, Chichester (Ciba Foundation Symposium 195) p 173–188

---

[1]This paper was presented by D. Mason.
[2]Present address: Department of Medicine, Infectious Diseases Divison, University of California, San Francisco, USA.
[3]Present address: DNAX Research Institute of Molecular and Cellular Biology Inc., Palo Alto, California, USA.

One of the most impressive characteristics of the adaptive immune system is its ability to distinguish self from non-self. Although much is known about the mechanisms that make this distinction, many questions are unanswered. Indeed, the pathogenesis is not fully understood for any autoimmune disease.

Mechanisms that maintain self-tolerance can be classified into two broad types: those that depend on a functional absence of autoreactive cells or the presentation of self-antigens in a non-immunogenic form; and those that involve a positive activity of the immune system, specifically by the action of T cells with some regulatory function. For the purposes of this paper, the first group of mechanisms will be termed 'passive' and the second 'active', although it is recognized that this terminology signifies only the final state of the tolerogenic mechanisms, and not how these states are achieved.

The passive mechanisms, i.e. clonal deletion (Kappler et al 1987), clonal anergy (Morahan et al 1989) and T cell indifference (Ohashi et al 1991) are well described with regard to self-tolerance among T cells. Until recently, it was commonly assumed that between them they accounted fully for self–non-self discrimination. If this were so, however, an attenuation of the immune system by γ-irradiation or the transfer of a selected subset of T cells from a self-tolerant donor rat to a T cell deficient recipient would not give rise to the development of autoimmune diseases. However, this prediction is not borne out in practice, and the remainder of this paper will describe experiments that illustrate this point.

The major conclusions from these studies can be summarized as follows: (1) normal healthy rats possess T cells with the potential to cause organ-specific autoimmune diseases; (2) this potential is not normally realized because other T cells, with an identifiable phenotype, prevent it; and (3) the mechanism underlying this protective function is not yet understood. The cells that mediate this function will be termed regulatory T cells.

$CD4^+$ T cell subsets defined on the basis of their expression of various isoforms of CD45

Before discussing the evidence for an active role of T cells in the maintenance of self-tolerance, it is necessary to digress briefly to describe the functional heterogeneity of rat $CD4^+$ T cells that is revealed when they are fractionated on the basis of CD45 isoforms which contain the C exon product of the CD45 gene.

The CD45 molecule, which is expressed at a high level on the surface of all leukocytes, exists in a number of different isoforms. These isoforms result from differential usage of the three exons, A, B and C, which encode the portion of CD45 close to the extracellular NH_2 terminus (Thomas 1989). One isoform expresses none of these three exons and is designated CD45RO. In principle, eight different isoforms of CD45 can be generated and it is apparent that they all exist. The functional significance of the complexity of the extracellular portion of the molecule is unknown, but the intracellular portion, which is common to all, has phosphotyrosine phosphatase activity. Its exact role *in vivo* is unknown, but it plays an essential part in signal transduction in T cells via its activation of the kinase $p56^{lck}$ (Schraven et al 1992). Different subsets of $CD4^+$ T cells in rats, mice and humans express different isoforms of CD45 and these subsets have different functional characteristics (Mason 1992). In particular, depending on the level of expression of the C exon of CD45, $CD4^+$ T cells in the rat can be resolved into two subsets, $CD45RC^{low}$ and $CD45RC^{high}$. Both of these subsets are themselves heterogeneous because they contain cells at different stages of maturation and activation (Mason 1992). However, the separation into $CD45RC^{low}$ and $CD45RC^{high}$ is valuable in that it reveals functional heterogeneity among $CD4^+$ cells. In particular, the $CD45RC^{low}$ subset contains T cells that produce predominantly T helper 2 (Th2) cytokines on activation, whereas the $CD45RC^{high}$ subset yields more γ-interferon (IFN-γ) and little interleukin 4 (IL-4). These differences in cytokine synthesis are reflected in the *in vivo* attributes of these cells because the $CD45RC^{low}$ subset provides the majority of B cell help in secondary immune responses, whereas the $CD45RC^{high}$ subset responds in assays for cell-mediated immunity (Fowell et al 1991, Powrie & Mason 1990a). The following sections of this paper illustrate that the two subsets play very different roles in the induction and control of autoimmunity.

Studies with congenitally athymic rats

Athymic rats, like their mouse equivalent, are essentially hairless and have all the immunological deficiencies that are anticipated from the lack of a thymus. If these rats are injected intravenously with $CD45RC^{high}$ $CD4^+$ T cells, they develop a fatal wasting disease with mononuclear cell infiltrates in the liver, stomach, pancreas and thyroid. In addition, the lungs are severely affected with large numbers of lymphocytes and macrophages throughout the tissue. In a minority of rats anti-thyroglobin antibodies are found in the serum (Powrie & Mason 1990b). In contrast, rats given the $CD45RC^{low}$ subset of $CD4^+$ T cells remain healthy and appear more resistant to infection than nude rats that are not injected with cells. Significantly, with respect to the pathogenesis of the wasting disease induced by transfer of $CD45RC^{high}$ $CD4^+$ cells, co-transfer with the $CD45RC^{low}$ subset prevented the wasting disease and multiorgan mononuclear cellular infiltration. Subsequent experiments using severe combined immunodeficiency (SCID) mice as recipients of subsets of mouse $CD4^+$ T cells that are separated on the basis of their levels of CD45RB expression have produced similar results (Powrie et al 1994), indicating that the phenomenon is not an unusual characteristic of rats or the strain of rat used in the earlier work. Although the mononuclear infiltrates observed in various organs when CD45RC^{high} CD4^{+} T cells were transferred into nude recipients may be interpreted as manifestations of autoimmunity, other explanations are possible. The transfers were made from euthymic congeneic donors but a residual degree of histoincompatibility, for tissue specific alloantigens, cannot be dismissed completely. However, this hypothesis does not readily account for the protective effect of the CD45RC^{low} subset. Alternatively, it is possible that the CD45RC^{high} CD4^{+} cells are making some abortive attempt to eliminate pathogens to which the nude rats themselves are unable to respond. This suggestion is less easy to dismiss, and it may indeed account for the massive macrophage infiltration seen in the lungs of rats with the wasting disease induced by the transfer of the CD45RC^{high} T cells. Similar considerations may apply to observations of colitis in various gene knockout mice (Sadlack et al 1993, Shull et al 1992, Kuhn et al 1993). It is evident that it is difficult, at a site of strong antigenic stimulation with foreign antigens, to distinguish dysregulated inflammatory responses evoked by autoantigens from those that are of xenogeneic origin. For these reasons a second system was adopted to study the role of CD4^{+} subsets in autoimmunity.

The development of autoimmune diabetes in rats rendered lymphopoenic by thymectomy and low dose irradiation

The BB rat, which has the RT1^{U} major histocompatibility complex (MHC) allotype, is lymphopoenic for some genetically determined, but unidentified, reason. It develops diabetes spontaneously with an equal incidence in both sexes (Rossini et al 1985). Lymphopoenia induced experimentally by adult thymectomy and repeated low dose γ-irradiation in a different rat strain (PVG.RT1^{C}) results in autoimmune thyroiditis and/or diabetes in a significant percentage of rats, and female rats are more frequently affected (Penhale et al 1990). We have used both of these observations to obtain a high incidence of autoimmune diabetes in PVG.RT1^{U} rats. These rats have the same MHC allotype as the BB rat, but they have the background genes of PVG. When PVG.RT1^{U} rats are thymectomized at six weeks of age and subjected to four doses of 250 rad ^{137}Cs γ-irradiation at two week intervals starting at eight weeks of age, virtually all males and about 70% of females develop fatal diabetes by nine to 10 weeks after the final dose of irradiation (Fowell & Mason 1993). This high incidence of the disease in males is valuable for studying the pathogenesis of the disease because highly significant results can be obtained using relatively small numbers of rats. The PVG.RT1^{U} rat strain is unremarkable, particularly because it shows no tendency to develop autoimmunity spontaneously.

Evidence that the disease was cell mediated rather than caused by some unexpected consequence of the irradiation protocol was obtained by treating the rats with an anti-CD8 monoclonal antibody (Fig. 1). This treatment

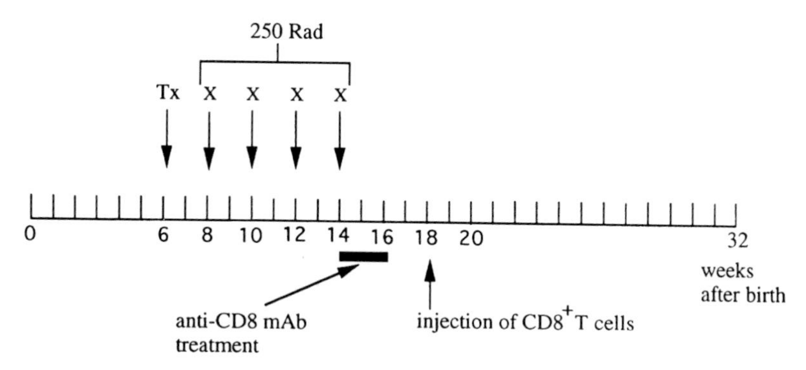

FIG. 1. Role of $CD8^+$ T cells in the development of diabetes in lymphopoenic rats. PVG.RT1^U rats were thymectomized (Tx) and irradiated (X) with four doses of 250 rad. They were then injected intraperitoneally with OX8 anti-CD8 monoclonal antibody on the day of last irradiation and three times weekly for two weeks. The control rats received an irrelevant monoclonal antibody, OX21, instead of OX8. $CD8^+$ T cells were negatively selected from thoracic duct lymphocytes of normal PVG.RT1^U rats by rosette depletion and 10^7 cells were administered intravenously to CD8 depleted Tx–X rats two weeks after cessation of antibody treatment.

depleted the rats of virtually all the $CD8^+$ cells that had survived the irradiation and completely protected them from diabetes. Diabetes was again observed if these antibody-treated rats were injected with purified $CD8^+$ T cells from normal donors after the anti-CD8 monoclonal antibody had been metabolized by the recipients (Fowell & Mason 1993). The most direct interpretation of these findings is that normal PVG.RT1^U rats have $CD8^+$ T cells with a diabetogenic potential, and that this potential is only revealed by radiation-induced lymphopoenia.

Prevention of diabetes by T cell transfer

The identity of the radiation-sensitive cell that prevented the spontaneous development of diabetes in non-lymphopoenic rats was determined by subjecting rats to the diabetes-inducing protocol and injecting them with purified populations of T cells from normal syngeneic donors to see whether any of these were protective. The results were similar to those obtained in the studies of the wasting disease in nude rats in that $CD45RC^{low}$ $CD4^+$ peripheral T cells from normal donors prevented the disease completely. These cells expressed the RT6 antigen but not the CDw90 (Thy-1) molecule. Cells that produce IL-4 but little IFN-γ on activation express this phenotype (McKnight et al 1991) and are generated on priming with exogenous antigens (Fowell et al 1991, Powrie & Mason 1990a,b). The phenotype analysis suggested that the regulatory T cells that prevented the diabetes were already primed in the donor rats, and doses as low as 5×10^6 could prevent the disease in 100% of recipients.

CD4$^+$ thymocytes also prevent diabetes

The peripheral CD4$^+$ T cell subset that prevented diabetes in thymectomized and irradiated recipients had a primed T cell phenotype. In addition, insulin-dependent diabetes is a tissue-specific autoimmune disease. Consequently, it was anticipated that the priming of the protective peripheral T cells took place in the periphery, possibly in the lymph nodes draining the pancreas. Such a hypothesis would predict that CD4$^+$ thymocytes, as a naive population, would be far less potent at controlling the diabetes than 'educated' peripheral T cells. However, when CD4$^+$ CD8$^-$ thymocytes from normal donors were assayed for their ability to prevent diabetes in the usual cell transfer experiments, they were more potent than CD45RC^{low} CD4$^+$ peripheral T cells. As few as 1.5×10^6 CD4$^+$ thymocytes protected the majority of recipients (Fig. 2), but at no dose of CD4$^+$ CD8$^-$ thymocytes did the prevention of diabetes reach 100%. Essentially the same result was obtained when peripheral CD4$^+$ T cells were tested for their ability to prevent diabetes: only when the CD45RC^{low} subset of CD4$^+$ peripheral T cells was isolated and transferred was the prevention of diabetes observed in all recipients. Given the earlier finding in nude rats that the CD45RC^{high} subset of CD4$^+$ peripheral T cells displayed autoimmune potential, the failure of unfractionated CD4$^+$ T cells to protect all recipients from diabetes has been ascribed to an antagonistic effect of the CD45RC^{high} subset in the inoculum on the protective action of the reciprocal subset. If this interpretation is extended to the observations of the inability of CD4$^+$ CD8$^-$ thymocytes to confer protection from diabetes in 100% of the rats, the inferences are that: (1) these thymocytes are, like peripheral CD4$^+$ T cells, functionally heterogeneous; and (2) the two subpopulations have antagonistic effects with respect to the control of diabetes. We are currently exploring this possibility, but it may be relevant to note that a functional heterogeneity has recently been described among mouse CD4$^+$ thymocytes, with a minority subset programmed to produce high levels of IL-4 on activation (Arase et al 1993). If a subset of CD4$^+$ CD8$^-$ thymocytes has the special function of controlling autoimmune disease in peripheral tissues, then the question arises as to how this subset is selected intrathymically.

The nature of the autoantigen in diabetic rats and a possible role for 'leaky' promoters

A major problem in studying the pathogenesis of autoimmune disease is that the autoantigens are unknown in the majority of cases. The situation is also complicated by the fact that the autoimmune destruction of a tissue may

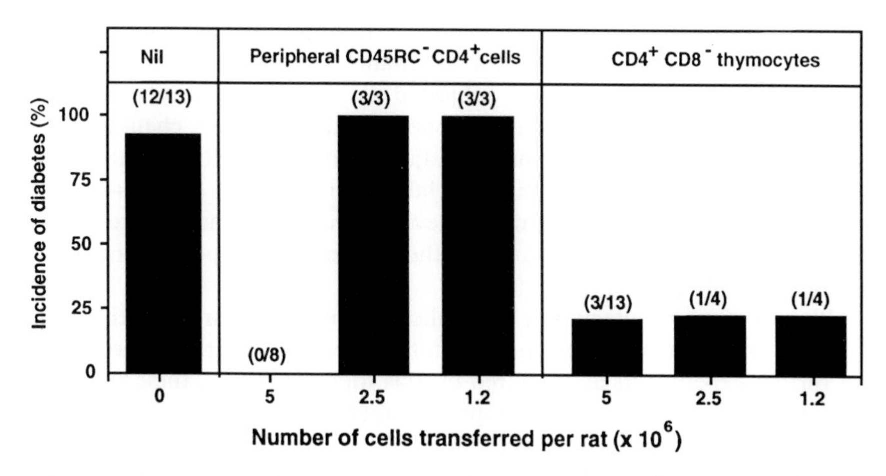

FIG. 2. Dose response for suppression of diabetes by $CD45RC^{low}$ $CD4^+$ peripheral T cells and $CD4^+$ $CD8^-$ thymocytes. T cell subsets were negatively selected from thoracic duct lymphocytes or thymus of normal $PVG.RT1^U$ rats by rosette depletion. They were then injected intravenously to Tx–X rats on the last irradiation day. The purity of isolated cells was >95%. Numbers in parentheses represent the number of rats in each group.

expose otherwise intracellular antigens to the immune system, resulting in the generation of an immune response. This possibility calls into question the attempts to identify autoantigens by examining humoral or cell-mediated responses to candidate autoantigens once the disease has developed. The problem is compounded by the fact that for some tissues, for example the β cells of the pancreas, substantial tissue destruction can occur before the appearance of clinical signs of disease. However, given that autoimmune diseases which target individual cell types are frequently ones that involve endocrine tissues, it is possible that it is the secretions of those tissues that

TABLE 1 Effects of neonatal treatment of male rats on the incidence of diabetes

Peptide injected	*Incidence*
A chain and B chain	2/13
B chain (residues 10–29) alone	0/3
B chain (residues 1–18) alone	0/2
A chain alone	4/4
ovalbumin	6/7

Statistics: inocula containing B chain peptides versus A chain peptides or ovalbumin $P < 3.3 \times 10^{-5}$.

contain the primary autoantigens. With this in mind, we tested the ability of peptides derived from the A and B chains of insulin to influence the development of diabetes in rats subjected to the thymectomy/γ-irradiation protocol. Rats were injected intraperitoneally with various A chain and/or B chain peptides in incomplete Freund's adjuvant at birth, and again at one and four weeks of age. They were then taken through the diabetes-inducing protocol in the usual way. A significant level of protection was observed only in rats given the B chain peptides, and not those given A chain peptides or control injections (Table 1).

Although these results appear to implicate insulin itself in the pathogenesis of insulin-dependent diabetes, they do not address the underlying mechanism, and further experiments are required on this topic. However, these results are interesting in another regard. Attempts to target transgenes to the β cells of the pancreas have resulted in the observation that the insulin promoter is 'leaky' in that transgene mRNA can be detected in the thymus (Heath et al 1992). A similar result was obtained when the promoter for the α1-anti-trypsin inhibitor gene was used to target gene expression to the liver (Bachmann et al 1994). These results suggest that mRNA for the endogenous genes under the control of these promoters should also be detectable in the thymus. Studies in both the mouse (Jolicoeur et al 1994) and the rat (V. Heath, unpublished results) confirm this prediction. If the insulin mRNA in the thymus is translated into protein, then peptides derived from it may play a role in the maintenance of tolerance to the hormone in the periphery. Two possibilities can be proposed: (1) that the intrathymic expression of insulin-derived peptides induces clonal deletion of insulin-reactive T cells and; (2) that insulin-specific regulatory T cells are generated which play an active role in tolerance. It is evident that these two hypotheses are not mutually exclusive.

Several reports indicate that the thymus plays an active role in self-tolerance. When a bird or mouse is engrafted with a xenogeneic or allogeneic thymic rudiment, respectively, before the recipient's immune system has matured to a functional level, the recipient, when mature, is tolerant of tissues from the thymus donor (Ohki et al 1987, Salaün et al 1990). The significant point is that the recipient's own thymus does not have to be removed to achieve this tolerant state. Although other interpretations are possible, these experiments suggest that recipient T cells maturing in the engrafted thymus have the capacity to inhibit the xenogeneic or allogeneic T cell response expected from the T cells maturing in the endogenous thymus (Coutinho et al 1993). The authors of that review put forward essentially the same hypothesis proposed in this paper, and the two experimental approaches support each other. The conclusion that emerges is that the thymus plays at least three different roles: it mediates both positive and negative selection, and it also generates a repertoire of regulatory T cells whose physiological role is to control autoimmunity to tissue-specific antigens.

Acknowledgements

We thank the members of the MRC Cellular Immunology Unit who have contributed to this work. In particular, the interactions with Neil Barclay and Andrew McKnight have been invaluable and the technical expertise of Mike Puklavec and Steve Simmonds is much appreciated.

References

Arase H, Arase N, Nakagawa K, Good RA, Onoe K 1993 NK1.1$^+$ CD4$^+$ CD8$^-$ thymocytes with specific lymphokine secretion. Eur J Immunol 23:307–310

Bachmann MF, Rohrer UH, Steinhoff U et al 1994 T helper cell unresponsiveness: rapid induction in antigen-transgenic and reversion in non-transgenic mice. Eur J Immunol 24:2966–2973

Coutinho A, Salaun J, Corbel C, Bandeira A, Le Douarin N 1993 The role of thymic epithelium in the establishment of transplantation tolerance. Immunol Rev 133:225–240

Fowell D, Mason DW 1993 Evidence that the T cell repertoire of normal rats contains cells with the potential to cause diabetes: characterization of the CD4$^+$ T cell subset that inhibits this autoimmune potential. J Exp Med 177:627–636

Fowell D, McKnight AJ, Powrie F, Dyke R, Mason D 1991 Subsets of CD4$^+$ T cells and their roles in the induction and prevention of autoimmunity. Immunol Rev 123:37–64

Heath WR, Allison J, Hoffmann MW et al 1992 Autoimmune diabetes as a consequence of locally produced interleukin-2. Nature 359:547–549

Jolicoeur C, Hanahan D, Smith KM 1994 T cell tolerance toward a transgenic beta-cell antigen and transcription of endogenous pancreatic genes in thymus. Proc Natl Acad Sci USA 91:6707–6711

Kappler JW, Roehm N, Marrack P 1987 T cell tolerance by clonal elimination in the thymus. Cell 49:273–280

Kuhn R, Lohler J, Rennick D, Rajewsky K, Muller W 1993 Interleukin-10-deficient mice develop chronic enterocolitis. Cell 75:263–274

Mason DW 1992 Subsets of CD4$^+$ T cells defined by their expression of different isoforms of the leucocyte-common antigen, CD45. Biochem Soc Trans 20:187–190

McKnight AJ, Barclay AN, Mason DW 1991 Molecular cloning of rat interleukin-4 cDNA and analysis of the cytokine repertoire of subsets of CD4$^+$ T cells. Eur J Immunol 21:1187–1194

Morahan G, Allison J, Miller FJAP 1989 Tolerance of class I histocompatibility antigens expressed extrathymically. Nature 339:622–624

Ohashi PS, Oehen S, Buerki K et al 1991 Ablation of tolerance and induction of diabetes by virus infection in viral antigen transgenic mice. Cell 65:305–317

Ohki H, Martin C, Corbel C, Coltey M, Le Douarin NM 1987 Tolerance induced by thymic epithelial grafts in birds. Science 237:1032–1035

Penhale WJ, Stumbles PA, Huxtable CR, Sutherland RJ, Pethick DW 1990 Induction of diabetes in PVG/c strain rats by manipulation of the immune system. Autoimmunity 7:169–179

Powrie F, Mason D 1990a Subsets of rat CD4$^+$ T cells defined by their differential expression of variants of the CD45 antigen: developmental relationships and *in vitro* and *in vivo* functions. Curr Top Microbiol Immunol 159:79–96

Powrie F, Mason D 1990b OX-22^{high} CD4^{+} T cells induce wasting disease with multiple organ pathology: prevention by the OX-22^{low} subset. J Exp Med 172: 1701–1708
Powrie F, Leach MW, Mauze S, Menon S, Caddle LB, Coffman RL 1994 Inhibition of Thl responses prevents inflammatory bowel disease in SCID mice reconstituted with CD45RB(HI) CD4^{+} T cells. Immunity 1:553–562
Rossini AA, Mordes JP, Like AA 1985 Immunology of insulin-dependent diabetes mellitus. Annu Rev Immunol 3:289–320
Sadlack B, Merz H, Schorle H, Schimpl A, Feller AC, Horak I 1993 Ulcerative colitis-like disease in mice with a disrupted interleukin-2 gene. Cell 75:253–261
Salaün J, Bandeira A, Khazaal I et al 1990 Thymic epithelium tolerizes for histocompatibility antigens. Science 247:1471–1474
Schraven B, Schirren A, Kirchgessner H, Siebert B, Meuer SC 1992 Four CD45/P56lck-associated phosphoproteins (pp29–pp32) undergo alterations in human T cell activation. Eur J Immunol 22:1857–1863
Shull MM, Ormsby I, Kier AB et al 1992 Targeted disruption of the mouse transforming growth factor-β1 gene results in multifocal inflammatory disease. Nature 359:693-699
Thomas ML 1989 The leukocyte common antigen family. Annu Rev Immunol 7:339–369

DISCUSSION

Shearer: Have you looked for these protective cells in the spleens of young rats, at approximately seven days of age, then put them into the rats that have been conditioned to see if they're protective?

Mason: No, we haven't done any experiments with spleen cells of rats that young.

Shearer: Where do these cells come from? Are they lymph node cells?

Mason: The cells in the periphery that we use in our experiments are thoracic duct lymphocytes, so they're a circulating population of cells. However, they're not recent thymic migrants because we can remove the thymus and still obtain these cells from rats seven months after thymectomy. Therefore, although the thymus contains these cells, when their progeny migrate to the periphery they live for a long time. This is what you would expect from the cell population that controls autoimmune disease.

Shearer: It would be worthwhile to determine whether there are dysregulatory cell populations in young rats because it is possible that young humans and young mice are not normal in terms of their T helper 1 (Th1)-type and Th2-type cytokine profiles.

Mason: Yes, there is some evidence that thymectomy of young mice leads to autoimmunity.

Shearer: There was one Canadian study which suggested that there's a high incidence of type 1 diabetes in human infants given cows' milk up to the age of three months, but not in infants over the age of three months (Karjalainen et al 1992). It is possible that autoreactive cells left the thymus but didn't die in the periphery until the infant was three months old.

Ramshaw: I'm not convinced that you have identified the antigen as insulin. Irradiation has multiple effects; for example, it may reactivate retroviruses in the pancreas. This creates a problem if you're looking at the control of immune responses and are trying to identify a self-antigen.

Mason: The original model developed by John Penhale was for thyroiditis and not for diabetes (Penhale et al 1973). Therefore, there are two endocrine-specific autoimmune diseases that occur spontaneously in humans, dogs, mice and rats. They can both be induced in rats by the mechanism that I presented, and they can be controlled by the cells that I described. My view is that if diabetes in our thymectomized and irradiated rats is a retrovirus-induced disease, then so is autoimmune diabetes in humans. Otherwise, one has to postulate entirely different mechanisms for the induction of diabetes (and thyroiditis) in rats and humans, although the histopathology in both species is indistinguishable.

Mosmann: You showed partial protection over a wide range of thymocyte numbers. Your interpretation was that the thymocytes in the thymus had already made some commitment to their future function. But an alternative explanation is that the thymocytes are simply precursors that can differentiate into a mixture of phenotypes in the periphery. The balance of that mixture may shift one way or the other, which may result in partial protection.

Mason: The potency of the cells in the thymus is as high if not higher than that of the peripheral cells. Therefore, it's difficult to believe that these cells come from an uncommitted population. If there is a precursor frequency for any one antigen in the naive population of one in 10^5, then, because we transfer only 10^6 thymocytes, the diabetes is going to be controlled by 10 cells specific for that antigen. Therefore, selection must take place in the thymus for these particular antigens. One could argue that positive selection in the thymus involves the recognition of many intrathymic self-antigens but that's a different situation in that such selection does not result in self-recognition in the periphery.

Mitchison: The numbers must be more impressive than that because, presumably, in that experiment all the double positives die quickly.

Mason: No. We isolated only the single CD4$^+$ thymocytes.

Lachmann: Colleagues of yours in Oxford (Bennett et al 1995) have demonstrated a genetic predisposition for autoimmune diabetes in the promoter region of the insulin gene. They suggested that hypoexpression of insulin was the predisposing event. Does this conflict with your model?

Mason: There is evidence that a low level of transcription of the insulin gene is associated with susceptibility to insulin-dependent diabetes (Bennett et al 1995, Kennedy et al 1995). The most direct interpretation of this result would implicate insulin itself as a relevant autoantigen in this disease. If, as our reverse transcriptase PCR results indicate, the insulin gene is expressed intrathymically then we would argue that it is here, rather than in the pancreas, that the genetic variation in expression influences the disease susceptibility. Of course, a high level of expression would be expected to be more effective in mediating a clonal deletion of insulin-reactive T cells. Whether it has any effect in the selection of putative regulatory T cells is impossible to say.

Lachmann: In both genetically predisposed humans and in non-obese diabetic (NOD) mice there are environmental differences that influence the severity of the diabetes. This has, at least by some, been interpreted as suggesting that an infectious cofactor, possibly a virus, also plays a part in producing the disease.

Mason: We haven't done any experiments with rats but in the NOD mouse the cleaner they are kept the higher is the incidence of diabetes. One could argue that the bacterial flora is being refined but no one can really answer that question.

Ramshaw: Diabetes in NOD mice can be prevented by injecting them with *Mycobacterium bovis* bacillus Calmette–Guérin. This is a non-specific effect (Shehadeh et al 1994).

Flavell: This is also true for tumour necrosis factor (TNF) (R. A. Flavell & I. Grewal, unpublished results). We were sceptical that this was a real physiological effect, rather than a systemic pharmacological effect. Therefore, we made transgenic NOD mice that expressed TNF on the islet. We expected that they would become more susceptible, but we found that they were protected instead. We also transferred cells from normal NOD mice to NOD severe combined immunodeficiency (SCID) recipients, and we found that they became diabetic within a few weeks. However, if cells from the NOD TNF transgenic mice were transferred, they did not become diabetic. Transfer of a mixture of cells also protects. Therefore, this is similar to your results. We don't yet have any results that may suggest a possible mechanism.

Lotze: Nora Sarvetnick showed that γ-interferon induces autoimmune diabetes and that interleukin 10 (IL-10) causes infiltration of the pancreas with $CD4^+$ cells and $CD8^+$ cells, and a loss of pancreatic exocrine cells, although the endocrine cells are intact. If the NOD mice expressed IL-10 under an insulin promoter, the disease was worse (Wogensen et al 1993, 1994).

Romagnani: IL-12 also accelerates the progression of diabetes in NOD mice. This has recently been shown by Trembleau et al (1995).

Mason: Polly Matzinger thymectomized B10 mice and transplanted two thymuses, one from CB mice and one from syngeneic B10 mice, into each mouse. These mice accepted skin grafts from CB donors although CB and B10 mice differ at several minor histocompatibility loci and normal B10 mice reject CB skin. In the mice with the two thymus grafts, both grafts were healthy and had a normal histological appearance. The question raised by the experiments was what prevented the T cells that matured in the B10 graft from rejecting the CB skin? The authors suggested that in some way the T cells that matured in the CB thymus graft suppressed the graft-rejecting capacity of these B10 thymus-derived cells. That is, tolerance to CB skin was a dominant characteristic (Zamoyska et al 1989). There are other similar examples in the literature, and they are not easy to explain unless one takes the view that tolerance is, in part, a dominant, active phenomenon and not a passive one that depends solely on the absence of self-reactive T cells.

Abbas: Do those thymocytes make IL-4?

Mason: If CD4$^+$ thymocytes are stimulated, they make IL-4.

Abbas: If we're trying to prove that suppression is a mechanism for inducing self-tolerance, then healthy people ought to have insulin-specific or islet-specific, IL-4-producing T cells. Also, a population of healthy people should have IL-4-producing T cells that are reactive against self-antigens, and a decline in the numbers of those cells should precede the development of autoimmunity. Is it feasible to do this study, either in rats or in a large human population?

Mason: I couldn't make a case for saying that insulin is the only self-antigen that is protective. We have not examined other candidates like glutamic acid decarboxylase.

Abbas: Do normal rats have islet-specific IL-4-producing T cells?

Mason: We have not examined this point. Islet-specific T cells would provide a number of options to look at with regard to possible target antigens to which the T cells that prevent diabetes respond. It is possible that several different T cell specificities are involved.

Mitchison: But it's a general rule that T cells reactive with disease-inducing autoantigens can be detected in normal individuals. This has been found for myelin basic protein and for acetylcholine receptors (Burns et al 1983, Salvetti et al 1991).

Abbas: Do you know what those T cells make?

Mitchison: No. There are other mechanisms of containment besides Th1 cells and Th2 cells; for example, the idiotypic network, or majority rule, within clusters (Brunner et al 1994).

Mason: There are difficulties in explaining how clonal deletion in the thymus can be achieved for more than a few thousands of self-peptides. If we say that an antigen-presenting cell has a million class II molecules on its cell surface, and that only a hundred of these are required to delete a T cell against a particular antigen, then only 10 000 antigens can be deleted.

Allen: If you bring the number down from 100 to 10, you can delete your entire self-reactive repertoire. We don't know how many receptors need to be ligated for negative selection.

Mason: It is true that we do not know how many receptors on a thymocyte need to be engaged to ensure clonal deletion. However, each dendritic cell in the thymus interacts with several T cells simultaneously and I find it difficult to believe that clonal deletion will be effective if only 10 major histocompatibility complex/peptide associations are present on an intrathymic dendritic cell for a particular peptide. It is possible that the high affinity T cells are deleted, especially if a thymocyte encounters many antigen presenting cells in the thymus before it is exported. In this case, the lower affinity cells will escape deletion and may possibly be activated in the periphery, especially if the self-antigen to which they can respond is presented at a high level and in the presence of a strong adjuvant, for example, pertussis toxin or Freund's adjuvant.

Lotze: Paul Allen, didn't you show this with cardiac muscle peptides? Did you use peptide and adjuvant to get them to respond?

Allen: Yes, and pertussis toxin (Smith & Allen 1992). These T cells may have survived challenge in the thymus. It is possible that the high affinity ones are deleted, whereas the low affinity ones escape deletion. These may migrate to the periphery, where they remain silent. If there is a genetic predisposition or a concomitant infection, then these T cells can become players.

Mason: I take your point. The protective T cell-mediated mechanism that I have described for diabetes may be necessary because low affinity self-reactive T cells that escape clonal deletion in the thymus can potentially be activated by cross-reactive foreign antigens and so become pathogenic. In general, such a phenomenon may be limited to certain self-antigens because clonal deletion will have eliminated T cells specific for self-antigens that are well presented intrathymically.

To test directly the validity of the central hypothesis of our paper, we will make a transgenic mouse with the gene for a foreign antigen under the control of a tissue-specific promoter and then determine whether the thymocytes from such a mouse, but not from a non-transgenic litter mate, can control a peripheral response to the antigen encoded by the transgene. Our current results imply that the thymus plays a major role in peripheral tolerance.

Mitchison: But that's different from autoimmunity being the result of a thymic defect. However, the observation that T cells occur which have two different receptors with different specificities may illustrate how a defect in the thymus could cause autoimmune disease.

Mason: It has not been shown that both of these receptors are positively selected. It is possible that the mechanism for generating the T cell receptor is involved (Mason 1994, Hardardotti et al 1995). One can explain the frequency of T cells with two α chains simply on the basis of one α chain being expressed, but failing to achieve positive selection. If the second α chain locus then rearranges and the α/β combination that results achieves positive selection, rearrangement stops. Although such a T cell does not exhibit allelic exclusion for its α chains, only one allele has been positively selected. Therefore, it is possible that one of those receptors is not functional.

Locksley: In the MBP transgenic mouse the incidence of spontaneous encephalomyelitis increases if T cells expressing endogenous α chains are deleted (Lafaille et al 1994). All T cells express the same β chain in these mice, so T cells expressing the endogenous α chains might see altered ligands. These

T cells may maintain the regulation of the transgene-expressing T cells. This raises the question of whether altered ligands maintain some of the regulatory role of the T cell peripheral repertoire.

Coffman: It is tempting to suggest that a Th2-like mechanism is involved in the regulation you described because the pathogenic T cells are Th1 cells, but that may not always be the case. Fiona Powrie transferred the CD45RB^{high} subset of splenic CD4$^+$ T cells to syngeneic SCID mice, and she observed the rapid development of an inflammatory bowel disease that was confined largely to the colon (Powrie et al 1994). Co-transfer of these cells with the remainder of the CD4$^+$ cell population, the CD45RB^{low} subset, prevented the development of the disease. The disease is caused by a strong, presumably dysregulated, Th1 response to antigens of the intestine, the intestinal flora or both. The CD45RB^{low} population appears to have a regulatory action that prevents the activation or expansion of this population, and it is reasonable to suggest that it is a Th2-like population. However, neither anti-IL-4 antibodies, anti-IL-10 antibodies or both can inhibit the regulatory activity of the CD45RB^{low} cells. Recently, Fiona has shown that anti-transforming growth factor β (TGF-β) antibody can block the regulation of the pathogenic effects of the CD45RB^{high} population by these cells (F. Powrie, J. Carlino & R. L. Coffman, unpublished results).

Mason: The cells in the thymus produce TGF-β. I didn't want to imply that IL-4 was solely responsible, I was just using it as a marker for the cells.

Coffman: One of the predictions of the view that the thymus is where the regulatory T cells prevent autoimmunity is that the BB thymus should have lower titres of such cells. Have you tested this?

Mason: We haven't done that. BB rats develop diabetes spontaneously, and they fail to make T cells in the periphery that express RT6, which is a lipid-anchored member of the immunoglobulin superfamily with an unknown function. However, cells that are protective in the periphery do express RT6, suggesting that it can be used as a marker for these cells with a protective function. We are trying to find markers on CD4$^+$ thymocytes that divide them into two populations.

Flavell: Have you transferred RT6$^+$ thymocytes?

Mason: RT6 is not expressed in the thymus. It is not expressed until the T cells have matured in the periphery.

Mitchison: Sergio Romagnani, how solid are the results on human autoimmune diseases or candidate autoimmune diseases, because few human diseases are really known to be autoimmune diseases. Are these candidate autoimmune diseases Th1-type diseases?

Romagnani: In organ-specific autoimmune diseases, such as type 1 diabetes, multiple sclerosis and Crohn's disease, there is a prevalence of Th1 cells. This has been shown either by PCR or clonal analysis. In contrast, a Th2 response, rather than a Th1 response, seems to be involved in systemic lupus erythematosus. High levels of proinflammatory cytokines are apparent in rheumatoid arthritis suggesting that a Th1-type response may be involved, but it is not yet clear which type of T cells are actually present in the synovia and synovial fluid—they may be Th0 cells or Th1 cells (Simon et al 1994).

References

Bennett ST, Lucassen AM, Gough SCL et al 1995 Susceptibility to human type 1 diabetes at IDDM2 is determined by tandem repeat variation at the insulin gene minisatellite locus. Nat Genet 9:284–292

Brunner MC, Mitchison NA, Schneider SC 1994 Immunoregulation mediated by T-cell clusters. Folia Biol (Prague) 40:359–369

Burns J, Rosenzweig A, Zweiman B, Lisak RP 1983 Isolation of myelin basic protein-reactive T cell lines from normal human blood. Cell Immunol 81:435–440

Hardardotti F, Baron JD, Janeway CA 1995 T cells with two functional antigen-specific receptors. Proc Natl Acad Sci USA 92:354–358

Karjalainen J, Martin JM, Knip M et al 1992 A bovine albumin peptide as a possible trigger of insulin-dependent diabetes mellitus. N Eng J Med 327:302–307

Kennedy GC, Germaine MS, Rutter W 1995 The minisatellite in the diabetes susceptibility locus IDDM2 regulates insulin transcription. Nat Genet 9:293–298

Lafaille JJ, Nagashima K, Katsuki M, Tonegawa S 1994 High incidence of spontaneous encephalomyelitis in immunodeficient anti-myelin basic protein T cell receptor transgenic mice. Cell 78:399–408

Mason D 1994 Allelic exclusion of α chains in TCRs. Int Immunol 6:881–885

Penhale WJ, Farmer A, McKenna RP, Irvine WJ 1973 Spontaneous thyroiditis in thymectomised irradiated Wistar rats. Clin Exp Immunol 15:225–229

Powrie F, Correa-Oliveira R, Mauze S, Coffman RL 1994 Regulatory interactions between $CD45RB^{high}$ and $CD45RB^{low}$ $CD4^+$ T cells are important for the balance between protective and pathogenic cell-mediated immunity. J Exp Med 179:589–600

Salvetti M, Jung S, Chang SF, Will H, Schalke BCG, Wekerle H 1991 Acetylcholine receptor-specific T lymphocyte clones in the normal human immune repertoire: target epitopes, HLA restriction and membrane phenotypes. Ann Neurol 19:508–516

Shehadeh NS, Calcinaro F, Bradley BJ, Brunchlim I, Vardi P, Lafferty KJ 1994 The effect of adjuvant therapy on development of diabetes in mouse and man. Lancet 343:706

Simon AK, Seipelt E, Sieper J 1994 Divergent T-cell cytokine patterns in inflammatory arthritis. Proc Natl Acad Sci USA 91:8562–8566

Smith SC, Allen PM 1992 Expression of myosin class II major histocompatibility complexes in the normal myocardium occurs before induction of autoimmune myocarditis. Proc Natl Acad Sci USA 89:9131–9135

Trembleau S, Penna G, Bosi E, Mortara A, Gately MK, Adorini L 1995 Interleukin 12 administration induces T helper type 1 cells and accelerates autoimmune disease in NOD mice. J Exp Med 181:817–821

Wogensen L, Huang X, Sarvetnick N 1993 Leukocyte extravasation into the pancreatic tissue in transgenic mice expressing interleukin 10 in the islets of Langerhans. J Exp Med 178:175–185

Wogensen L, Lee MS, Sarvetnick N 1994 Production of interleukin 10 by islet cells accelerates immune-mediated destruction of β cells in nonobese diabetic mice. J Exp Med 179:1379–1384

Zamoyska R, Waldmann H, Matzinger P 1989 Peripheral tolerance mechanisms prevent the development of autoreactive T cells in chimeras grafted with two minor incompatible thymuses. Eur J Immunol 19:111–117

Signalling events in the anergy induction of T helper 1 cells[1]

Joanne Sloan-Lancaster and Paul M. Allen

Department of Pathology, Washington University School of Medicine, Box 8118, 660 South Euclid Avenue, St. Louis, MO 63110-1093, USA

Abstract. T cells can interact productively with altered peptide ligands (APLs) resulting in different phenotypic outcomes. Stimulation of T helper 1 cells with an APL on live antigen-presenting cells results in the induction of anergy. We investigated the intracellular signalling events involved in generating this anergy by comparing protein tyrosine phosphorylation patterns after stimulation with the anergy-inducing APL or the immunogenic peptide. Stimulation by an APL resulted in a unique pattern of T cell receptor (TCR) phospho-ζ species, which was not observed with any dose of immunogenic peptide. This altered phospho-ζ pattern had a profound functional significance, in that the tyrosine kinase ZAP-70 was not activated. Thus, anergy can be induced by changing the constellation of intracellular signalling events in a T cell. These findings demonstrate that the TCR–CD3 complex can engage selective intracellular biochemical signalling pathways as a direct consequence of the nature of the ligand recognized and the initial phosphotyrosine pattern of the TCR–CD3 proteins. This then leads to different phenotypes.

1995 T cell subsets in infectious and autoimmune diseases. Wiley, Chichester (Ciba Foundation Symposium 195) p 189–202

The recognition of antigen by T cells is a seminal event in the induction of an immune response. The molecular basis for this recognition event has now been elucidated with the identification of the T cell receptor (TCR) (Hedrick et al 1984), the direct demonstration of peptide antigens binding to major histocompatibility complex (MHC) molecules (Babbitt et al 1985) and the solving of the crystal structure of MHC molecules (Bjorkman et al 1987). T cells, through their TCR, recognize a bimolecular ligand composed of a peptide antigen bound to a self-MHC molecule on the surface of an antigen-presenting cell (APC) or target cell (Germain 1994). The signalling events involved in T cell activation are starting to be identified (Weiss & Littman 1994). The TCR

[1]This paper was presented by P. M. Allen.

chains themselves do not possess any direct signalling capability, but they transmit signals through associated transmembrane molecules. The ligation of the TCR results in an initial and rapid induction of intracellular protein tyrosine phosphorylation events, followed by increased serine/threonine phosphorylation, phospholipid hydrolysis and increased intracellular Ca^{2+} levels. The general view of the TCR signalling complex has been that it functions as an 'on-or-off' switch, with its ligation resulting in subsequent activation of all downstream signalling events.

Biological effects of altered peptide ligands

Induction of selective T cell functions

Recent studies from ours and other laboratories have indicated that the binary view of T cell activation is too simplistic. It has now been definitively shown that the TCR can interact productively with less than optimal ligands (Evavold & Allen 1991, Sloan-Lancaster et al 1993, De Magistris et al 1992, Racioppi et al 1993). We have generated analogues of the immunogenic peptide Hb(64–76) by introducing amino acid substitutions in identified TCR contact residues (Lorenz & Allen 1988, Evavold et al 1992). The presentation of such altered peptide ligands (APLs) induced partial T cell activation in the form of cytokine production without proliferation (Evavold & Allen 1991), T cell cytolytic activity, or up-regulation of cell surface markers without proliferation or cytokine production (Evavold et al 1993a, Sloan-Lancaster et al 1993). We have defined APLs as analogue peptides in which a TCR contact residue has been changed (Evavold et al 1993b). The resulting peptide cannot induce a full stimulatory response, as measured by T cell proliferation; however, it binds to the MHC molecule as strongly as the immunogenic peptide. APLs can also act as TCR antagonists, which was originally shown by Sette et al (1994) and subsequently by several other groups (De Magistris et al 1992, Jameson et al 1993, Racioppi et al 1993, Spain et al 1994). A key observation was then made by the Bevan and Tonegawa laboratories, who showed that APLs were able to induce the positive selection of thymocytes in fetal thymic organ cultures (Hogquist et al 1994, Ashton-Rickardt et al 1994). These studies provided a molecular basis for how MHC-restricted T cells, which are able to recognize all potential pathogens, can develop in the absence of such pathogens. Thus, by subtly changing the TCR ligand, one can induce a wide range of selective T cell functions, suggesting that differential signalling upon TCR engagement can occur.

Anergy induction in Th1 and Th2 cells by altered peptide ligands

After our initial description of the activity of APLs, we made the important observation that both T helper 1 (Th1) and Th2 cells could be rendered anergic to subsequent stimulation by first interacting with an APL on live APCs (Sloan-Lancaster et al 1993, 1994a). For the Th1 clone, PL.17, interaction with the Ser70 APL, which contained an Ala to Ser substitution at position 70, resulted in the T cells failing to be stimulated to proliferate by the immunogenic peptide and fresh APC. This unresponsiveness occurred even in the presence of a co-stimulatory signal, which distinguished it from the anergy described by Jenkins and Schwartz (Jenkins & Schwartz 1987). We obtained similar results for Th2 clones, in that interaction with an APL resulted in the inability to proliferate (Sloan-Lancaster et al 1994a). However, differences in the anergic phenotype between Th1 and Th2 cells was observed. For Th1 clones, the anergic cells did not make any of the autocrine growth factor, interleukin 2 (IL-2), whereas the anergized Th2 clones made normal levels of their growth factor, IL-4. Thus, anergized Th1 and Th2 clones both failed to proliferate upon restimulation, and they differed in the production of autocrine growth factors. However, T cell unresponsiveness was induced in both Th1 and Th2 cells after a less than optimal TCR engagement. These findings led us to propose that some, but not all, signal transduction pathways linked directly to the TCR were being activated, and that this resulted in T cell anergy (Evavold et al 1993b).

T cell signalling events induced by altered peptide ligands

Altered peptide ligands induce protein phosphorylation

To investigate the initial signalling events involved in the induction of anergy, we first examined protein tyrosine phosphorylation (Sloan-Lancaster et al 1994b). The Th1 clone, PL.17, was stimulated using live APCs and: (1) the immunogenic peptide, Hb(64–76); (2) the anergy inducing APL, Ser70 APL; or (3) a control peptide. Whole cell lysates were generated after 2, 10 or 20 min of stimulation and used for phosphotyrosine Western blot analysis. The results from these studies clearly indicated that the APLs were inducing signalling events in the T cells, as indicated by the appearance of antigen-specific phosphotyrosine protein bands. The pattern induced by the Ser70 APL was unique compared to that induced by the agonist Hb(64–76) or the control peptide. A strikingly different tyrosine phosphorylation pattern was identified in two species at 18 kDa and 20–21 kDa. Stimulation with Hb(64–76) resulted in the appearance of strong bands at both molecular masses, whereas the Ser70 APL stimulated a strong band at 18 kDa and a weak band at 20–21 kDa. The molecular masses of the 18 kDa and the 20–21 kDa species correlated with the tyrosine phosphorylated forms of the TCR ζ chain. This identity was confirmed by anti-ζ immunoprecipitation and Western blotting of the immunoprecipitated species. Thus, TCR engagement by an anergy-inducing

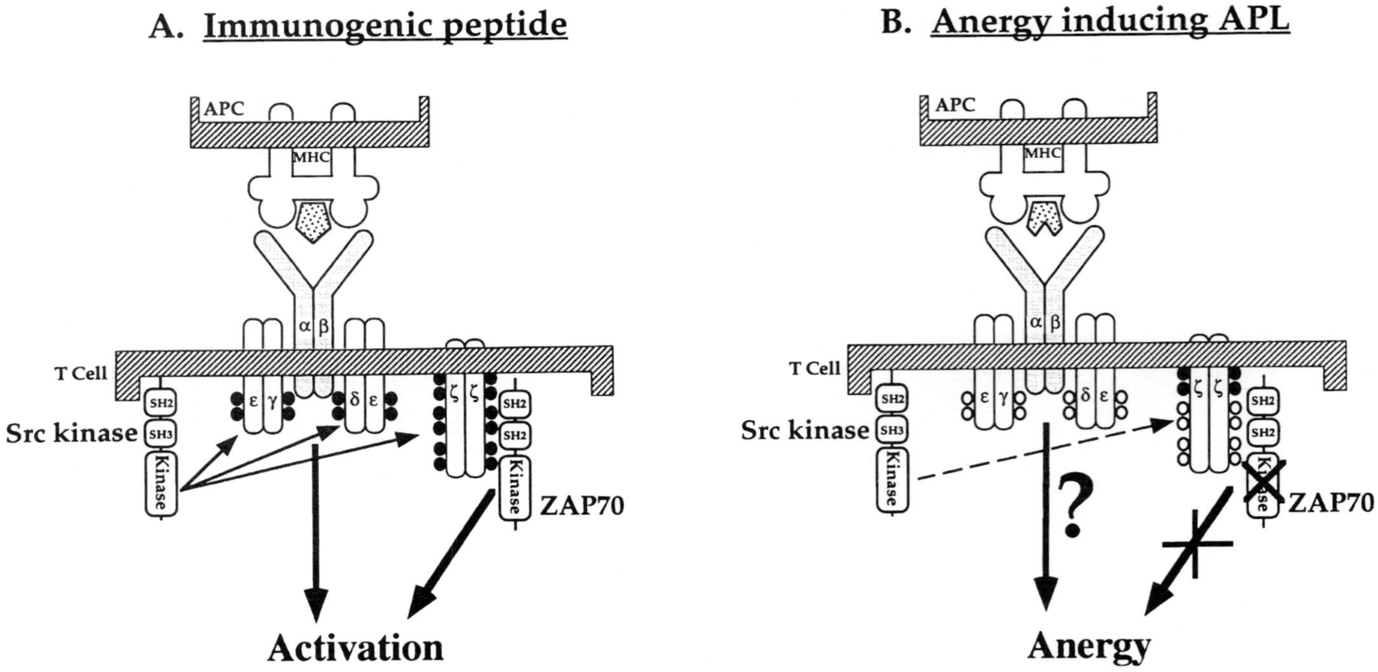

A. Immunogenic peptide

APC
MHC
α β
ε γ
δ ε
ζ ζ
T Cell
Src kinase
SH2
SH3
Kinase
ZAP70
SH2
SH2
Kinase
Activation

B. Anergy inducing APL

APC
MHC
α β
ε γ
δ ε
ζ ζ
T Cell
Src kinase
SH2
SH3
Kinase
ZAP70
SH2
SH2
Kinase
?
Anergy
Anergy

APL can specifically activate membrane proximal signal transduction events leading to phosphorylation of the TCR ζ chain.

Altered peptide ligands induce a qualitatively different signal

It was important to distinguish whether this altered phosphorylation of the TCR ζ chain was simply a reflection of a lower level of stimulation by the Ser70 APL which should, therefore, be mimicked by activation with lower doses of Hb(64–76), or whether this APL was in fact inducing a unique signal in the T cells. PL.17 cells were stimulated with various concentrations of Hb(64–76) and the TCR ζ chain phosphorylation patterns were compared to those stimulated with an anergy-inducing dose of the Ser70 APL. There was no concentration of Hb(64–76) at which the pattern of ζ chain tyrosine phosphorylation resembled that seen after stimulation with the Ser70 APL. Thus, the Ser70 APL appeared to deliver a unique signal to the T cell.

The altered phospho-ζ pattern also correlated precisely with the ability of different APLs to induce T cell anergy. We generated a panel of APLs in which we substituted the Ala residue at position 70 with all of the other 19 amino acids. Each of these peptides was then tested for its ability to induce anergy of PL.17 cells and for the phospho-ζ pattern induced. There was a direct correlation between the ability to induce anergy and the stimulation of an altered phospho-ζ pattern. These findings taken together clearly indicated that the anergy inducing APLs were stimulating a qualitatively different signal to the T cells, as reflected by the unique pattern of TCR ζ chain phosphorylation.

FIG. 1. A proposed model of signal transduction events involved in altered peptide ligand (APL)-induced anergy. (A) Possible activation events occurring after engagement of a T helper 1 clone by a fully stimulatory ligand are illustrated. Recognition of the immunogenic ligand results in the recruitment and activation of Src family kinases ($p56^{lck}$ and $p59^{fyn}$). These kinases then phosphorylate both tyrosine residues in the immunoreceptor tyrosine activation motifs (ITAMs) of the CD3 γ, δ and ε chains, and the T cell receptor (TCR) ζ chains (filled symbols). The phosphorylated ITAMs then bind various Src homology 2 (SH2) domain-containing proteins, e.g. ZAP-70, which are subsequently activated. Together, the activated Src kinases, ZAP-70 and other recruited signalling molecules lead to downstream signalling events, resulting in full T cell activation. (B) Recognition of the APL results in a less than optimal interaction, leading to less efficient recruitment and activation of the Src family kinases. Little or no phosphorylation of the CD3 ITAMs occurs (open symbols), whereas an altered TCR ζ chain phosphorylation pattern is detectable (filled and open symbols). The altered ζ chain phosphorylation results in a failure of ZAP-70 to be activated, preventing activation of downstream signalling events that depend upon ZAP-70 activation. Other as yet unidentified signalling events do occur (?) and the integration of these in the absence of the ZAP-70 pathway results in anergy, instead of clonal expansion. APC, antigen-presenting cell; MHC, major histocompatibility complex.

The altered ζ chain phosphorylation pattern results in the failure of ZAP-70 to be activated

Since the TCR phospho-ζ chain associates with downstream signalling molecules, we then assessed the tyrosine kinase activity associated with the ζ chain after T cell stimulation with the Ser70 APL compared with stimulation by the immunogenic ligand. Stimulation with Hb(64–76) resulted in the activation of the ZAP-70 tyrosine kinase. In contrast, stimulation with the Ser70 APL resulted in a complete lack of ZAP-70 activation. Thus, the altered phospho-ζ pattern stimulated by the anergy-inducing APL dramatically affected downstream signalling events by preventing the activation of the ZAP-70 tyrosine kinase.

Model for anergy induction by an altered peptide ligand

Figure 1 shows a proposed model, based upon the current model of T cell activation, of how an APL could induce anergy. Recognition of an immunogenic peptide by the TCR (Fig. 1A) results in the activation of the Src family protein tyrosine kinases, $p56^{lck}$ and $p59^{fyn}$. Once activated, these kinases then phosphorylate tyrosines in many different proteins, including those of the critically important immunoreceptor tyrosine activation motifs (ITAMs) of the CD3 and ζ chains. These phosphorylation events result in the initiation of multiple signalling pathways, including the activation of the ZAP-70 kinase. The integration of the constellation of the intracellular signals delivered by these different pathways then leads to full T cell activation.

Recognition of an APL by a T cell (Fig. 1B) results in a lower level of activation of $p56^{lck}$ and $p59^{fyn}$. Subsequently, there is a reduced level of tyrosine phosphorylation, including a lack of phosphorylation of the ITAM of the CD3 chains and an altered phosphorylation pattern of the ζ chain ITAMs. This altered ζ phosphorylation pattern results in the failure to activate the ZAP-70 pathway, although other signalling pathways are still activated. The integration of these different signals leads to the induction of anergy in the T cell.

The ability of a T cell to interact productively with less than optimal ligands has profound implications for much of T cell biology. These types of interactions are critically involved in the positive and negative selection of T cells in the thymus (Hogquist et al 1994, Ashton-Rickardt et al 1994, Jameson et al 1994, Hsu et al 1995). For peripheral T cells, these types of interactions could be involved in TCR antagonism, cross-reactive and autoreactive immune responses, and maintenance of memory T cells in the absence of antigen (Evavold et al 1993b, Sette et al 1994).

Summary

These studies show definitively that the TCR–CD3 complex can engage in selective signalling pathways, depending upon the nature of the interaction with its ligand. We have begun to establish a biochemical basis for the induction of anergy by APLs by showing that they can stimulate tyrosine phosphorylation pathways. The signal delivered in response to such APLs leads to a different pattern of tyrosine phosphorylation from that delivered by the stimulatory ligand. This results in a phenotypic outcome of anergy as opposed to activation. These findings demonstrate that anergy results from an active and selective signalling process directly via the TCR.

Acknowledgements

This work was supported by grants from the National Institutes of Health and the American Cancer Society.

References

Ashton-Rickardt PG, Bandeira A, Delaney JR et al 1994 Evidence for a differential avidity model of T cell selection in the thymus. Cell 76:651–663

Babbitt BP, Allen PM, Matsueda G, Haber E, Unanue ER 1985 Binding of immunogenic peptides to Ia histocompatibility molecules. Nature 317:359–361

Bjorkman PJ, Saper MA, Samraoui B, Bennett WS, Strominger JL, Wiley DC 1987 Structure of the human class I histocompatibility antigen, HLA-A2. Nature 329: 506–512

De Magistris MT, Alexander J, Coggeshall M et al 1992 Antigen analog–major histocompatibility complexes act as antagonists of the cell receptor. Cell 68: 625–634

Evavold BD, Allen PM 1991 Separation of IL-4 production from Th cell proliferation by an altered T cell receptor ligand. Science 252:1308–1310

Evavold BD, Williams SG, Hsu BL, Buus S, Allen PM 1992 Complete dissection of the Hb(64–76) determinant using T helper 1, T helper 2 clones, and T cell hybridomas. J Immunol 148:347–353

Evavold BD, Sloan-Lancaster J, Hsu BL, Allen PM 1993a Separation of T helper 1 clone cytolysis from proliferation and lymphokine production using analog peptides. J Immunol 150:3131–3140

Evavold BD, Sloan-Lancaster J, Allen PM 1993b Tickling the TCR: selective T cell functions stimulated by altered peptide ligands. Immunol Today 14:602–609

Germain RN 1994 MHC-dependent antigen processing and peptide presentation: providing ligands for T lymphocyte activation. Cell 76:287–299

Hedrick SM, Cohen DI, Nielsen EA, Davis MM 1984 Isolation of cDNA clones encoding T cell-specific membrane-associated proteins. Nature 308:149–153

Hogquist KA, Jameson SC, Heath WR, Howard JL, Bevan MJ, Carbone FR 1994 T cell receptor antagonist peptides induce positive selection. Cell 76:17–27

Hsu BL, Evavold BD, Allen PM 1995 Modulation of T cell development by an endogenous altered peptide ligand. J Exp Med 185:805–810

Jameson SC, Carbone FR, Bevan MJ 1993 Clone-specific T cell receptor antagonists of major histocompatibility complex class I-restricted cytotoxic T cells. J Exp Med 177:1541–1550

Jameson SC, Hogquist KA, Bevan MJ 1994 Specificity and flexibility in thymic selection. Nature 369:750–752

Jenkins MK, Schwartz RH 1987 Antigen presentation by chemically modified splenocytes induces antigen-specific T cell unresponsiveness *in vitro* and *in vivo*. J Exp Med 165:302–319

Lorenz RG, Allen PM 1988 Direct evidence for functional self protein/Ia-molecule complexes *in vivo*. Proc Natl Acad Sci USA 85:5220–5223

Racioppi L, Ronchese F, Matis LA, Germain RN 1993 Peptide–major histocompatibility complex class II complexes with mixed agonist/antagonist properties provide evidence for ligand-related differences in T cell receptor-dependent intracellular signaling. J Exp Med 177:1047–1060

Sette A, Alexander J, Ruppert J et al 1994 Antigen analogs/MHC complexes as specific T cell receptor antagonists. Annu Rev Immunol 12:413–431

Sloan-Lancaster J, Evavold BD, Allen PM 1993 Induction of T-cell anergy by altered T-cell-receptor ligand on live antigen-presenting cells. Nature 363:156–159

Sloan-Lancaster J, Evavold BD, Allen PM 1994a Th2 cell clonal anergy as a consequence of partial activation. J Exp Med 180:1195–1205

Sloan-Lancaster J, Shaw AS, Rothbard JB, Allen PM 1994b Partial T cell signaling: altered phospho-ζ and lack of Zap70 recruitment in APL-induced T cell anergy. Cell 79:913–922

Spain LM, Jorgensen JL, Davis MM, Berg LJ 1994 A peptide antigen antagonist prevents the differentiation of T cell receptor transgenic thymocytes. J Immunol 152:1709–1717

Weiss A, Littman DR 1994 Signal transduction by lymphocyte antigen receptors. Cell 76:263–274

DISCUSSION

Fitch: I would like to mention some relevant observations that we have made (Lancki et al 1995) using T cell clones derived from the Fyn knockout mice generated by Stein et al (1992). We've been able to derive T helper 1 (Th1) and Th2 clones, as well as conventional $CD8^+$ cytotoxic T lymphocyte (CTL) clones, from these mice. Signals generated via the T cell receptor (TCR) in these clones appear to be identical to those observed in clones derived from wild-type mice. However, there is one exception. In the wild-type clones, signalling can occur via the CDw90 (Thy-1) cell surface molecule, and some anti-CDw90 monoclonal antibodies can transmit signals in the absence of TPA (12-*0*-tetradecanoylphorbol 13-acetate). However, the Fyn knockout clones are defective in signalling through CDw90 and also probably through Ly6. It should be noted that we derived these clones using antigen-presenting cells (APCs) from normal mice. This raises the question as to whether Fyn is involved on the APC side of the continuing dialogue. Thus, we conclude that Fyn does not seem to be required for signalling via the TCR, although it may be required for signalling via CDw90.

Allen: It may even be more complicated than that because Jurkat cells have a large amount of Fyn but they do not apparently require it, and 2B4 clones also have a large amount of Fyn and basically no Lck (lymphocyte-specific tyrosine kinase). T cell clones may be selected to deal with whatever kinases they express.

Fitch: But it suggests that the CDw90 receptor may be linked to particular Src kinases.

Allen: And that the kinases are not completely redundant.

Mitchison: The most conservative view is that differential oligomerization is dependent on different affinities. In principle, this is easy to test because there should be a direct relationship between affinity and the propensity to behave like that.

Allen: I'm collaborating with Mark Davis to obtain affinity measurements of different ligands for the soluble TCR. His studies show that the different ligands have a decreased affinity (Matsui et al 1994). These affinity measurements are in the micromolar range, and the altered peptide ligands (APLs) have about a threefold lower affinity. Affinity plays a role but the net avidity is also important.

Mitchison: In principle, it wouldn't be difficult to map your interpretation onto the double signalling model because this may also affect polymerization.

Allen: I would like to put the two ideas together, i.e. that you need both oligomerization and conformational changes in normal T cell signalling, and that APLs affect both processes.

Mitchison: Are the differences that you've detected the same as those detected by Quill et al (1992) in anergized clones?

Allen: Helen Quill has studied the signalling properties of anergized cells when they are reactivated. She has found that the levels of Fyn are increased. We're looking at the initial events of anergy induction, and she has examined cells that are already anergized.

Mitchison: Are the two sets of results contradictory?

Allen: There are not enough results to answer this.

Fitch: Jenkins and colleagues (Mueller et al 1989) have studied the defect in anergized cells, and you've been looking at the defect that may lead to anergy. These are quite different defects.

Allen: One would expect the T cell to reset after a while. The kinase activity of Fyn after 13 days may be different to looking at it after 10 min.

Dutton: If T cells are being produced in the thymus all the time, and we pick out the ones that will meet the peptide that triggered them, what's the probability of the average T cell first encountering a related peptide that will anergize it?

Allen: If one believes that peptides are involved in positive selection, the re-recognition of these peptides in the periphery won't normally lead anywhere because the avidity of the receptor for that complex has been tuned by the level of co-receptors and other molecules. Therefore, the activity is increased in the thymus to allow positive selection. It then decreases in the periphery so that the peptide is not re-recognized. We've manipulated haemoglobin in a transgenic system so that it becomes an endogenous TCR antagonist. In the thymus this has an effect in that it lowers the avidity of the T cells which develop but it does not have any effects in the periphery and it is ignored by those T cells. When we add exogenous haemoglobin at a final concentration of about 10 mg/ml, we see a small amount of antagonism. Therefore, the T cells in the periphery may not become anergized because they are conditioned to ignore the concentrations of self-antigens that are normally present. However, during a viral infection, a large amount of antigen that could act as an antagonist is expressed, and the T cells could then interact productively with these higher levels of antagonist.

Mitchison: Ed Palmer supports the conformational change hypothesis (DiGiusto & Palmer 1994, and personal communication). The oligomerization hypothesis is difficult to test, whereas the conformational change hypothesis has numerous possibilities. For example, it may be possible to make small structural changes in the TCR, by mutating TCR residues and searching for alterations that might push the cells towards either an anergic or an activation response. A logical way to screen for these is to look for alterations that will affect positive but not negative selection, or vice versa. Have you tried to manipulate the mouse TCR in this way?

Allen: We have made a TCR transgenic mouse specific for haemoglobin. It is useful to look at clones but it is also important to look at Th precursor (Thp), Th1 and Th2 cells with the same receptor. If we give them the same APL, we may see different signalling pathways being activated.

Cantrell: The phosphotyrosine Western blots that you published (Sloan-Lancaster et al 1994) contained a band, which may have been Fyn, that was hyperphosphorylated. However, Fyn kinase activity was decreased.

Allen: It is possible that the band at 56–59 kDa is Fyn. The hyperphosphorylation would be consistent with the requirement for the removal of the regulatory phosphate for kinase activation.

Mitchison: The gels that you published looked quite different from one another (Sloan-Lancaster et al 1994). Could these be simply kinetic differences?

Cantrell: It is possible. I suspect that the difference between the gels represents scratching at the tip of an iceberg because in the positive control peptide there were only two or three tyrosine phosphorylated proteins, and there were about 100 tyrosine phosphorylated proteins in the activated cell. Kinetic differences could also be quite important because there are kinetic differences in the activation of kinases.

Lotze: Is it possible that a differential interaction with the α and β chain of the TCR is involved, producing partial agonism with one but not the other. There may be enough cross-linking to cause stimulation, and full activation of both chains in association with their CD3 chains may be required. One way of testing this is to use some of the different peptide ligands that you've tested, which would prevent full activation. If you take one of your APLs and introduce additional changes, does this kill activation altogether?

Allen: We first need to determine the structure of the TCR before we can guess how it interacts with the peptide/major histocompatibility complex (MHC) ligand.

Lotze: We have also studied melanoma-derived peptides. We have made amino acid substitutions in the class I (HLA-AZ) antigen-presenting molecule, which would presumably cause similar changes. We have observed a spectrum of activation responses, depending on the amino acid substitutions that are made within the same clone or different clones expressing different TCRs. Activation is measured by cytokine release and cytotoxicity, and we observe cytokine release when there is a loss of cytotoxicity (Castelli et al 1995).

Allen: The avidity of the interaction would be affected by changing either the peptide, MHC ligand or TCR.

Mitchison: Many cellular immunologists are desperate for a way of getting a signal out of an anergized cell. If this could be achieved, one would be in a better position to ask whether anergy is a component of autoimmunity and self-tolerance.

Allen: I prefer to work with the TCR transgenic mice rather than clones because I can look at primary T cells and if anergy does exist, I can try to determine the factors that reverse it.

Cantrell: In relation to this work, Narin Osman in my group has taken a purely biochemical approach to determine which proteins bind to the different immunoreceptor tyrosine activation motifs (ITAMs) (unpublished results). He synthesized these different ITAMs from the TCR ζ chain and CD3 chains as double-phosphorylated tyrosine phosphopeptides. He coupled these peptides to an affinity matrix, and he used them to fish out proteins from normal peripheral blood, [^{35}S]methionine-labelled T cell lysates. He found that ITAMs bind to a limited subset of proteins from these cell lysates. Titration of the peptides showed that these ITAMs are different in terms of their affinities for ZAP-70. The γ and δ ITAMs are always indistinguishable, predictably perhaps because there's only one amino acid difference in the intervening sequence between the two tyrosines. The ζ2 ITAM is not particularly good at binding to ZAP-70, although the ε ITAM binds the weakest. Therefore, they all bind to ZAP-70 but not equally. When we started to look at other proteins that had the potential to interact with ITAMs, we saw more pronounced differences.

Fyn also bind to ITAMs. In contrast to ZAP-70, it binds most strongly to $\zeta 2$ ITAM and poorly to the γ and δ ITAMs. It also binds poorly to the ε ITAM.

In contrast to Fyn, Shc (SH2 domain-containing adaptor protein) does not bind to the $\zeta 2$ ITAM. Shc binds most strongly to the γ and δ ITAMs, and it binds poorly to the ε ITAM.

Grb2 (growth factor receptor-bound protein 2) is a small adaptor protein containing an SH2 domain which is important in a number of regulatory pathways. It binds to $\zeta 1$, γ and δ, but not to $\zeta 2$, z3 or ε.

These binding analyses were performed with doubly phosphorylated ITAMs, and they show that these ITAMs are not equivalent. The biggest differences are not in the binding of the tyrosine kinases but in the binding of the adaptor proteins.

We have also studied how the phosphorylation state of the ITAMs effects binding. We made a number of variants of the $\zeta 1$ ITAM, so that it was either non-phosphorylated or phosphorylated at one or other of the tyrosine residues (Osman et al 1995). We found that that ZAP-70 binds preferentially to the doubly phosphorylated $\zeta 1$ ITAM. It will bind to $\zeta 1$ ITAM phosphorylated at only one tyrosine, with a preference for the N-terminally phosphorylated peptide, and it does not bind to the non-phosphorylated $\zeta 1$ ITAM.

A different pattern is observed for Fyn binding to different phosphoforms of the 3′ ITAM. Fyn doesn't bind to the unphosphorylated peptide but it binds preferentially to the C-terminally phosphorylated $\zeta 1$ ITAM. Also, in contrast to ZAP-70, it binds least strongly to the doubly phosphorylated $\zeta 1$ ITAM.

There are also differences in Shc and Grb2 binding. Grb2 will bind to the doubly phosphorylated $\zeta 1$ ITAM, but not to $\zeta 1$ ITAM with only one phosphotyrosine residue. In contrast, Shc prefers the C-terminally phosphorylated $\zeta 1$ ITAM.

Therefore, there are many different ways of generating signal heterogeneity from a TCR, for example by phosphorylating different ITAMs and by varying the phosphorylation state of a single ITAM.

Paul Allen's results demonstrate *in vivo* variable phosphorylation of the TCR. Therefore, variable phosphorylation must have important functional consequences on the cells. What controls this variable phosphorylation? What kinases are involved? We need to answer these questions.

Also, what is known of the biochemical responses in anergic cells as opposed to the biochemical responses that induce anergy?

Mitchison: Research in that area has been rather unprofitable (see Kang et al 1992).

Allen: Doreen Cantrell, have you tried to do the same experiment in Jurkat cells? Because they apparently only require Lck.

Cantrell: Narin Osman in my laboratory is doing this at the moment in Jurkat cells. Lck doesn't bind to ITAM peptides in these cells. There is probably another Src kinase binding in these cells because there is a kinase at about 58 kDa associated with the ITAM peptides in the Jurkat cells. This kinase cannot be Fyn because it's difficult to find Fyn in these cells. It also cannot be Lck because the molecular weight is too high. We've done this experiment in B cells and we have found that the ITAM peptides bind to Syk and Lyn, which is analogous to their binding to ZAP-70 and Fyn.

Abbas: Is there any evidence that Syk can take over the role of ZAP-70 in thymocyte populations or patients with ZAP-70-deficient T cells?

Cantrell: The TCR does not lose all function in the ZAP-70-deficient patients, which could mean that Syk takes over the ZAP-70 role.

Mitchison: Reth sequences, which bind tyrosine kinases and are homologous to sequences found in the cytoplasmic tail of diverse transmembrane signal-transducing proteins, are potential targets for medicinal chemistry. If there are differences in these sequences, they will become more attractive as targets. It's possible that medicinal chemists will make probes that can be used to answer some of these questions. Doreen Cantrell, what is your opinion of this?

Cantrell: This is a possible approach. I would predict that there would be some selectivity.

Mitchison: There are better compounds than phosphorylated peptides for antagonizing binding to Reth sequences; or at least, we can expect better drugs with this mode of action to emerge in the future.

Locksley: When are we going to find the ligand for CD45? This step is critical and probably mediates the earliest dephosphorylation after TCR stimulation. Paul Allen could also be seeing activation of phosphatases in his system assaying responses to altered ligands. I don't know if this has been looked at. These various CD45 isoforms all have different extracellular domains, but no one seems to have identified a ligand.

Mitchison: People have tried unsuccessfully to identify a ligand.

Cantrell: This may be because one does not exist. CD45 is essential for TCR signalling but it's possible that the initial signal from the TCR goes to CD45. I'm not sure what this signal could be but if it exists, it would generate an inside-out feedback pathway that does not require a ligand. The CD45 isoforms are a puzzle and may not be explained by this idea.

References

Castelli C, Storkus WJ, Maeurer MJ et al 1995 Mass spectrometric identification of a naturally processed melanoma peptide recognized by $CD8^+$ cytotoxic T lymphocytes. J Exp Med 181:363–368

DiGiusto DL, Palmer E 1994 An analysis of sequence variation in the beta chain framework and complementarity determining regions of an allo-reactive T cell receptor. Mol Immunol 31:693–699

Kang S-M, Beverly B, Tran A-C, Brorson K, Schwartz RH, Lenardo MJ 1992 Transactivation by AP-1 is a molecular target of T cell clonal anergy. Science 257:1134–1138

Lancki DW, Qian D, Fields P, Gajewski T, Fitch FW 1995 Differential requirement for protein tyrosine kinase Fyn in the functional activation of antigen specific T lymphocyte clones through the TCR or Thy-1. J Immunol 154:4363–4370
Matsui K, Boniface JJ, Steffner P, Reay PA, Davis MM 1994 Kinetics of T cell receptor binding to peptide/I-E^k complexes. Correlation of the dissociation rate with T cell responsiveness. Proc Natl Acad Sci USA 91:12862–12866
Mueller DL, Jenkins MK, Schwartz RH 1989 Clonal expansion versus functional clonal inactivation: a costimulatory signalling pathway determines the outcome of T cell antigen receptor occupancy. Annu Rev Immunol 7:445–480
Osman N, Lucas SC, Turner H, Cantrell D 1995 A comparison of the interaction of Shc and the tyrosine kinase ZAP-70 with the T cell antigen receptor ζ chain tyrosine-based activation motif. J Biol Chem 270:13981–13986
Quill H, Riley MP, Cho EA, Casnellie JE, Reed JC, Torigoe T 1992 Anergic Th1 cells express altered levels of the protein kinases $p56^{lck}$ and $p59^{fyn}$. J Immunol 149:2887–2893
Sloan-Lancaster J, Shaw AS, Rothbard JB, Allen PM 1994 Partial T cell signaling: altered phospho-ζ and lack of Zap70 recruitment in APL-induced T cell anergy. Cell 79:913–922
Stein PL, Lee HM, Rich S, Soriano P 1992 $pp59^{fyn}$ mutant mice display differential signaling in thymocytes and peripheral T cells. Cell 70:741–750

The two faces of interleukin 12: a pro-inflammatory cytokine and a key immunoregulatory molecule produced by antigen-presenting cells

Giorgio Trinchieri

The Wistar Institute for Anatomy and Biology, 3601 Spruce Street, Philadelphia, PA 19104, USA

Abstract. Interleukin 12 (IL-12) is produced by phagocytic cells, antigen-presenting cells and B lymphocytes in response to bacteria or intracellular parasites. IL-12 acts on T and natural killer (NK) cells inducing: production of cytokines, particularly γ-interferon (IFN-γ); proliferation; and enhancement of cell-mediated cytotoxicity. Early in infection, IL-12 acts as a proinflammatory cytokine and induces IFN-γ production by NK and T cells. IFN-γ activates the phagocytes and increases their ability to produce IL-12. Unlike IFN-γ, IL-10, IL-4, IL-13 and transforming growth factor β are negative regulators of the production and activity of IL-12. IL-12 sets the stage for the ensuing adaptive immune response by stimulating the generation of T helper 1 (Th1) cells. It is likely that the balance between IL-12 (favouring a Th1 response) and IL-4 (favouring a Th2 response) determines the eventual outcome of the Th1/Th2 dichotomy during an immune response. HIV-infected patients have a deficient production of IL-12, even at early stages of the disease. However, exogenous IL-12 can improve the deficient immune responsiveness of these patients' T and NK cells *in vitro*, suggesting a possible role of the IL-12 deficiency in HIV disease pathogenesis and a potential therapeutic role of IL-12 both against opportunistic pathogens and HIV infection itself.

1995 T cell subsets in infectious and autoimmune diseases. Wiley, Chichester (Ciba Foundation Symposium 195) p 203–220

Interleukin 12 (IL-12) is a heterodimeric cytokine, composed of a heavy chain of 40 kDa and a light chain of 35 kDa. It was orginally described with the name of natural killer stimulatory factor (NKSF) as a product of Epstein–Barr virus (EBV)-transformed human B lymphoblastoid cell lines (Kobayashi et al 1989). A cytotoxic lymphocyte maturation factor (CLMF) was later described, also as a product of B cell lines (Stern et al 1990). Cloning of the two genes coding for the heavy and light chains of NKSF and CLMF revealed that the two cytokines were identical (Wolf et al 1991), and the name of IL-12 is now commonly used.

IL-12 structure, receptor, production and functions

The two genes encoding the two chains of IL-12 are separate and unrelated. The gene encoding the p35 light chain has limited homology with other single chain cytokines, whereas the gene encoding the p40 heavy chain is homologous to the extracellular domain of the haemopoietic cytokine receptor family (Merberg et al 1992). The p35 and the p40 chains are covalently linked to form a biologically active heterodimer (p70) (Kobayashi et al 1989). The biologically active IL-12 heterodimer resembles a cytokine covalently linked to a soluble form of its receptor. An analogous but not identical situation is observed for IL-6 and ciliary neurotrophic factor (CNTF), two cytokines that can bind in solution to the soluble form of their specific receptor chain. These complexes bind to other transmembrane chains of their receptor (including the shared gp130 chain) inducing signal transduction and biological functions. Interestingly, the p40 heavy chain of IL-12 is most homologous to the CNTF receptor and to the IL-6 receptor (α chain), whereas the p35 light chain has some homology with IL-6 itself (Merberg et al 1992). Thus, it is likely that IL-12 is evolutionarily derived from a primordial cytokine similar to IL-6/CNTF and from a chain of its original receptor.

The structure of the cellular receptor for IL-12 is not completely understood. Two or more binding affinities are observed on IL-12-responsive cells. The receptors with the highest affinity, in the picomolar range, are probably responsible for IL-12 biological activity (Chua et al 1994). The chain of the human receptor that binds IL-12 with low affinity (2–5 nM, Chua et al 1994) has been cloned and is most homologous to the gp130 chain of IL-6 and other receptors.

Although IL-12 was discovered as a product of B cell lines, B lymphocytes do not appear to be the most important physiological producers of IL-12. IL-12 appears to be produced both *in vitro* and *in vivo* mainly by phagocytic cells (monocytes, macrophages and neutrophils) (D'Andrea et al 1992, Cassatella et al 1995) and other cell types with antigen-presenting capabilities, as discussed below. The expression of the IL-12 p40 gene is highly regulated. It is expressed only in the cell types producing IL-12 biologically active heterodimers, whereas the message for the p35 gene is expressed, often constitutively, in almost every cell type. However, p35 alone, in the absence of p40, does not appear to be secreted. In IL-12-producing cells p35 gene expression is regulated by mechanisms similar, but not identical, to those regulating p40 gene expression (Cassatella et al 1995). Surprisingly, in all cell types producing IL-12, p40 mRNA is always much more abundant than p35 mRNA, and a large excess of p40 protein (10- to 100-fold) is produced compared with the p35 protein. Thus, only 1 : 10 to 1 : 100 of the p40 protein is secreted in the form of biologically active heterodimer (D'Andrea et al 1992, Cassatella et al 1995). The secretion of a large excess of the free p40 chain raises the possibility that p40 competes with the biologically active heterodimer for binding to the IL-12 receptor, and thus serves as a natural antagonist. Crude murine preparations of recombinant p40 were shown to antagonize the biological activity of IL-12 (Mattner et al 1993); however, these results could not be confirmed in humans, where only p40 homodimers, but not the free p40 chains, inhibit the IL-12 heterodimer at a high molar ratio (Ling et al 1995). These apparently contrasting results are most likely explained by the observation that the cloned IL-12 receptor chain binds to p40 (in the homodimeric or heterodimeric form) and that the murine IL-12 receptor has an affinity for p40 10-fold higher than the human receptor (Chua et al 1994, M. Gately, personal communication). Thus, the mouse p40 homodimer is an efficient antagonist for IL-12 binding on mouse cells, whereas the p40 homodimer is not an efficient antagonist on human cells because of the low affinity binding to the human receptors. These results suggest the possibility of using the mouse p40 homodimer as an IL-12 antagonist *in vivo* in murine systems. However, the absence of results showing the natural production of p40 homodimer means that a physiological role of p40 homodimers as a natural antagonist is yet to be demonstrated.

The major target cells for IL-12 biological functions are T cells and NK cells. IL-12 mediates three major effects on these cells: (1) induction of cytokine production; (2) proliferative effects; and (3) enhancement of cytotoxic functions (Kobayashi et al 1989, Gately et al 1992). γ-Interferon (IFN-γ) is the predominant cytokine induced by IL-12. IL-12 achieves this by inducing IFN-γ gene transcription itself and synergizing with other inducers (e.g. IL-2, antigen stimulation, mitogens, CD28 or FcRIII/CD16 triggering) to induce both IFN-γ transcription and production (Chan et al 1991, 1992). Other cytokines, including tumour necrosis factor α (TNF-α), granulocyte macrophage colony-stimulating factor (GM-CSF) and IL-2, are also induced by IL-12. IL-12 activity can be easily detected on preactivated T or NK cells, possibly because of up-regulation of the IL-12 receptor on these cells (Gately et al 1992, Perussia et al 1992). IL-12 can also act as a cofactor that mediates the proliferation of T cells in conjunction with other mitogens or antigens (Kobayashi et al 1989, Perussia et al 1992). In addition, it is often required for optimal proliferation of T cells, e.g. for the antigen-induced proliferation of T cell clones (Murphy et al 1994). The enhancing effect of IL-12 on cell-mediated cytotoxicity can be demonstrated by a short-term enhancement of NK cell-mediated cytotoxicity and by its ability to enhance the generation of lymphokine-activated killer (LAK) cells and cytotoxic T lymphocytes (CTLs) (Kobayashi et al 1989, Chehimi et al 1993, Stern et al 1990).

IL-12 also has an enhancing effect on colony formation by early pluripotent and committed haematopoietic progenitor cells induced by other colony stimulatory factors. This stimulatory effect of IL-12 is mediated by a direct effect on the progenitor cells, and it may be masked *in vitro* and *in vivo* by an inhibitory effect on haematopoiesis. This inhibition is mediated by inhibitory cytokines (particularly IFN-γ and TNF-α) produced by NK and T cells in response to IL-12 (Jacobsen et al 1993, Bellone & Trinchieri 1994, Eng et al 1995).

IL-12: resistance to pathogens and endotoxic shock

Early *in vitro* studies of IL-12 production from phagocytic cells showed that bacteria, bacterial products and intracellular parasites were among the most powerful inducers of IL-12 production (D'Andrea et al 1995). Subsequently, several *in vivo* models showed that IL-12 is responsible for the early IFN-γ production, primarily by NK cells but also by T cells, observed during infection with *Listeria monocytogenes* and *Toxoplasma gondii* (Tripp et al 1993, Gazzinelli et al 1993). IFN-γ induced by IL-12 not only acts as an activator of macrophages in the inflamed tissue, increasing their phagocytic and bacteriocidal activity, but also increases the ability of the macrophages to produce IL-12, in a powerful positive feedback loop (Kubin et al 1994a). GM-CSF is another cytokine that enhances macrophage production of IL-12 (Kubin et al 1994a). The synergistic activity of IFN-γ and lipopolysaccharide (LPS) or other bacterial stimuli can be easily demonstrated *in vitro* in monocyte/macrophage cultures. LPS has little IL-12-inducing effect in these cultures unless the cells have been preincubated with IFN-γ for several hours. This results in IL-12 production that can be easily detected at the level of transcription (Ma et al 1995). The positive feedback loop induced by IL-12-induced IFN-γ, if not interrupted, would lead to uncontrolled life-threatening cytokine production. However, a negative feedback loop intervenes rapidly to prevent an uncontrolled production of proinflammatory cytokines. Negative regulators of IL-12 are primarily IL-10, IL-4, IL-13 and TGF-β (D'Andrea et al 1993, 1995). These cytokines are produced by various cell types—monocyte/macrophages, lymphocytes and others at later stages of the inflammatory response—and they act on the IL-12 system at various levels and with different mechanisms. IL-10 is a potent inhibitor of IL-12 production by phagocytic cells, and it prevents T cell activation primarily by blocking the expression of IL-12 and other co-stimulatory molecules (e.g. B7 antigens) in accessory cells (D'Andrea et al 1993, Kubin et al 1994b, Murphy et al 1994). IL-4 and IL-13 inhibit IL-12 production from phagocytic cells when present simultaneously with the activating stimulus. However, when present for 24 h or longer, they induce the differentiation of monocyte/macrophages to an activation state in which they are able to produce several-fold higher levels of IL-12 (D'Andrea et al 1995). Furthermore, IL-4 but not IL-13 prevents the activation of T cells by IL-12, probably by affecting co-stimulatory signals rather than IL-12 itself. TGF-β is likely the most effective inhibitor of the IL-12 system because it inhibits both IL-12 production and the effect of IL-12 on T and NK cells (D'Andrea et al 1995).

Although these proinflammatory effects of IL-12 are of primary importance in the innate resistance of the organism against infectious organisms, they may also lead to pathological effects, best exemplified in the mouse by the generalized Shwartzman-like reaction (Ozmen et al 1994) and by LPS-induced endotoxic shock (Wysocka et al 1995). In the Shwartzman reaction, a local LPS injection in the footpad sensitizes the mouse to a subsequent intravenous LPS injection 24 h later, resulting in the death of the mouse. This sensitizing effect is mediated by the production of IFN-γ in response to the LPS-induced IL-12. IL-12 and IFN-γ probably play a key role in the pathological events leading to the death of the mice, and they are induced by the challenging dose of LPS (Ozmen et al 1994).

A direct involvement of IL-12 in the pathogenic mechanisms is demonstrated more directly by the LPS-induced endotoxic model in *Mycobacterium bovis* bacillus Calmette–Guérin (BCG)-primed mice. In these mice small doses of LPS induce a rapid production of TNF-α one to two hours after LPS injection, followed by a peak of serum IL-12 three to four hours after LPS injection (Wysocka et al 1995). Both TNF-α and IL-12 are required for production of IFN-γ, which peaks around five to seven hours (Wysocka et al 1995) after LPS injection. The activity of each of these three cytokines is required for inducing the endotoxic shock that leads to the death of the mice within 18 h because neutralizing antibodies against each of these cytokines can protect the mice from death (Wysocka et al 1995). The endotoxic shock pathology is due to a dysregulation of cytokine production, representing a pathological exaggeration of the physiological response to infection. However, it appears that the physiological negative feedback loops are still operative because of the temporal succession of transient increases in expression of the various cytokines.

Immunoregulatory functions of IL-12

The production of IL-12 during the early inflammatory response to pathogens participates in setting the stage and influences profoundly the characteristics of the ensuing adaptive immune response against the pathogen. This activity of IL-12 is mediated by a double stimulatory/inhibitory action. IL-12 induces the generation of T helper 1 (Th1) cells, which produce IFN-γ and IL-2, and it favours cell-mediated immunity, macrophage activation and production of the opsonizing IgG2a isotype. It also inhibits the generation of Th2 cells, which produce IL-4, IL-5 and IL-10, and it favours humoral immunity and production of IgG1, IgE and IgA isotypes. These activities of IL-12 have been clearly documented both *in vitro* in humans (Manetti et al 1993, 1994) and mice, (Hsieh et al 1993) and *in vivo* in mice (Heinzel et al 1993). It is likely that the Th1/Th2 dichotomy in the immune response is regulated mostly by the balance, early during an immune response, of IL-12 that favours Th1 responses and IL-4 that favours Th2 responses (Trinchieri 1993).

In vitro studies on the cytokine production of human T cell clones generated by limiting dilution in the presence or absence of IL-12 have clearly demonstrated that IL-12 can prime virtually all peripheral blood $CD4^+$ or $CD8^+$ T cells for high IFN-γ production, provided that IL-12 is present during the first one or two weeks of clonal expansion. After that time, the priming for high IFN-γ production is irreversible even if the T cell clones are maintained in the absence of IL-12 (Manetti et al 1994). Thus, IL-12 can endow T cells with the capacity to produce IFN-γ, which results in a direct differentiation-inducing effect on each T cell. However, the inhibition of cells producing IL-4 and other Th2 cytokines (Manetti et al 1993, D'Andrea et al 1993, Heinzel et al 1993), which is required for the generation of a complete Th1 response, does not appear to be due to a direct effect on T cell maturation/differentiation (Manetti et al 1994), but rather on a selective mechanism (preferential growth of Th1 cells or inhibition/cytotoxicity of Th2 cells) which, at least in part, may be mediated by IL-12-induced cytokines such as IFN-γ.

The early requirement for IL-12 during an immune response for the effective induction of a Th1 immune response is best exemplified by the *Leishmania major* infection model in mice. In this model the susceptible BALB/c mice, which have an ineffective Th2 response to the parasite, can be 'cured' with IL-12 by inducing a protective Th1 response. This response occurs only if IL-2 is given within the first week of infection (Heinzel et al 1993). However, a significant proportion of BALB/c mice that have an established infection can still be shifted to a Th1 response and cured if they are simultaneously treated with IL-12 and antimonial anti-parasite drugs (Nabors et al 1995). This demonstrates that in the appropriate experimental or therapeutic conditions the immune response still has a certain plasticity, even during an established infection.

Optimal cytokine production and proliferation of differentiated Th1 clones *in vitro* in response to antigen often requires co-stimulation by IL-12 and B7 antigens (activating the CD28 receptor on the T cells). Both of these co-stimulatory molecules are expressed by antigen-presenting cells (APCs) (Murphy et al 1994). In the murine system (Murphy et al 1994) and on human T cells (Kubin et al 1994b), B7 and IL-12 synergize strongly to induce cytokine production and proliferation of T cells. This shows that both a surface-bound product (B7) and a secreted product (IL-12) of APCs synergize both with T cell receptor (TCR) stimulation and amongst themselves for optimal T cell stimulation. The requirement of differentiated Th1 cells for IL-12 raises the question of whether IL-12 is required *in vivo* for both initiation and maintenance of a Th1 response. However, during *T. gondii* infection in mice, anti-IL-12 antibody suppresses the induction of a protective Th1 response when administered during the first week of infection, but the antibody is ineffective at suppressing an established Th1 response and IFN-γ production when administered to mice with chronic infection (Gazzinelli et al 1994). This suggests that the presence of IL-12 is not an obligatory requirement for maintenance of a Th1 response *in vivo*.

The ability of IL-12 to bias the immune system to a Th1 response can be practically utilized by using IL-12 as an adjuvant in the vaccinations for those pathogens against which cellular immunity or macrophage activation are effective (Afonso et al 1994). This concept was originally demonstrated for *L. major* infection in mice, in which vaccination with *Leishmania* antigen and IL-12 as adjuvant induces a persistent Th1-type immunological memory that results in a protective response to subsequent challenge with live *L. major* promastigotes (Afonso et al 1994). The effectiveness of this use of IL-12 as an adjuvant has now been demonstrated with several other pathogens, nominal antigens, viruses and tumour antigens.

The production of IL-12 by phagocytic cells during bacterial infection results in high circulating levels of IL-12 that persist with a relatively long half-life of three to four hours. This massive production of IL-12 probably results in a systemic induction of Th1 responses in the organism, and indeed bacterial infections are normally characterized by a Th1 type of immune response. The response of the mice to bacterial infection, and probably also to other antigens concomitantly present, is biased towards the Th1 response. However, increasing evidence shows that in addition to macrophages, professional APCs, such as dendritic cells, are also able to produce IL-12. For example, spleen dendritic cells can induce anti-ovalbumin TCR transgenic T cells to develop into Th1 cells if endogenously produced IL-4 is inhibited. This effect is mediated by the IL-12 produced by dendritic cells (Macatonia et al 1995). It appears that activated T cells themselves, during antigen presentation, induce the dendritic cells to produce IL-12, although the mechanisms of these interactions need to be clarified. Unlike phagocytic cells, which produce high levels of IL-12, dendritic cells most likely produce locally limited amounts of IL-12 that affect only the contiguous T cells. In this situation, the maturation of T cells is sensitive to the delicate balance between IL-12 production and IL-4 production, as clearly suggested by the *in vitro* model with TCR transgenic mice (Macatonia et al 1995). The exact role of dendritic cell-produced IL-12, the interplay of positive and negative signals during the interaction between APCs and T cells, the regulation of the production of IL-12 by the APCs, and the responsiveness of the T cells to Th1- and Th2-inducing cytokines are probably all central in the regulation of T cell activation during an immune response, and they all remain to be fully investigated.

Role of IL-12 in the pathogenesis and therapy of HIV disease

IL-12 enhances the cytotoxic activity of NK cells from HIV-infected patients, and at least partially corrects their deficient activities observed at a late stage of the disease (Chehimi et al 1992). Also, IL-12, alone or in synergy with IL-2, stimulates IFN-γ production from HIV-infected patients' peripheral blood lymphocytes (PBL) (Chehimi et al 1992, Clerici et al 1993). More recently, IL-12 was found to enhance and almost completely correct the deficient T cell proliferation of HIV-infected patients in response to recall antigens, including the opportunistic pathogen *Mycobacterium avium*, HIV antigens, alloantigens and mitogens (Clerici et al 1993, Newman et al 1994). The enhancement of antigen-specific proliferation with PBL from HIV patients was similar but more marked than that observed with PBL from healthy donors (Perussia et al 1992, Newman et al 1994). These results suggest that IL-12 production is deficient in patients with HIV. This deficiency was confirmed by a study of 73 HIV-infected patients which showed that in response to *in vitro* stimulation with *Staphylococcus aureus*, the peripheral blood mononuclear cells (PBMC) of the patients produced 10–20-fold less p40 and fivefold less p70, on average, than PBMC from healthy donors (Chehimi et al 1994). The deficiency in IL-12 production was also observed in patients at the early stage of the disease who are asymptomatic and have a $CD4^+$ T cell count close to normal. This early defect in IL-12 production is concomitant with the early loss of T cell proliferative ability in HIV-infected patients (Miedema et al 1988). It is conceivable that the IL-12 defect is at least in part responsible for the decreased T cell responsiveness in the patients, and that the deterioration of other immunological parameters, including $CD4^+$ cell number, may contribute to the failure of the patients to resist opportunistic infections.

Although *in vitro* IL-12 improves some of the deficient peripheral blood T cell responses such as cell-mediated cytotoxicity, IFN-γ production and antigen-dependent proliferation, these acute effects of IL-12 would be of limited therapeutic value unless continuous chronic treatment is used. It was, therefore, important to test whether IL-12 could induce a permanent effect on patients' T cells. Indeed, it was observed that $CD4^+$ and $CD8^+$ T cell clones from the HIV-infected patients generated *in vitro* in the presence of IL-12 produce approximately fivefold more IFN-γ in response to stimulation than clones generated in the absence of IL-12 (Paganin et al 1995). This priming for high IFN-γ production was irreversible in the continued culture in the absence of IL-12, and it was observed with T cells from patients at every stage of disease, from asymptomatic to full-blown AIDS. Even in patients with almost no $CD4^+$ T cells, and from whom no $CD4^+$ clones were obtained, a striking priming for IFN-γ production was observed in $CD8^+$ T cells. This priming is probably the key mechanism by which IL-12 induces Th1 cell generation, which suggests that IL-12, used in HIV-infected patients as an adjuvant in vaccinations or in the treatment of opportunistic infections, may result in the generation of long-lasting, Th1-biased memory T cells. The production of these cells may be able to protect the patients from opportunistic infections even after termination of the treatment. However, the possibility that IL-12 treatment may improve the cell-mediated immune response against HIV itself, remains to be investigated.

Conclusions

The central role of IL-12 in inflammation and immunoregulation is becoming apparent within a few years of its discovery. Although not discussed in this paper, there is also the possibility that IL-12 plays important roles in the induction of autoimmune diseases and in the immune response against tumours. Indeed, the powerful anti-tumour activity of IL-12 in experimental models has generated much interest in ongoing clinical trials in cancer patients. Phase I clinical trials are also ongoing in HIV-infected patients, even though the most rational modalities for treatment of this complex disease still remain to be evaluated. In many other infectious diseases, IL-12 offers possibilities of therapeutic treatment, or it may be used as an adjuvant in prophylactic vaccination. From an immunological point of view, the research on IL-12 has focused on the interplay between innate resistance and adaptive immunity, and on the production of immunoregulatory factors by APCs. These are two important aspects of the immune response that, until recently, deserved more attention than that devoted to them. Undoubtedly, the research on IL-12 will provide new and important information for our understanding of immune mechanisms and will hopefully also create new therapeutic treatments.

Acknowledgments

I thank Marion Kaplan for typing the manuscript. The experimental work described in this paper was supported by United States Public Health Service grants CA 10815, CA 20833, CA 32898 and AI 34412, by a grant from the W. W. Smith Charitable Trust and by a contribution for AIDS research from the Commonwealth of Pennsylvania.

References

Afonso LCC, Scharton TM, Vieira LQ, Wysocka M, Trinchieri G, Scott P 1994 The adjuvant effect of interleukin-12 in a vaccine against *Leishmania major*. Science 263:235–237

Bellone G, Trinchieri G 1994 Dual stimulatory and inhibitory effect of NK cell stimulatory factor/IL-12 on human hematopoiesis. J Immunol 153:930–937

Cassatella MA, Meda L, Gasperini S, D'Andrea A, Ma X, Trinchieri G 1995 Interleukin-12 production by human polymorphonuclear leukocytes. Eur J Immunol 25:1–5

Chan SH, Perussia B, Gupta JW et al 1991 Induction of interferon-γ production by natural killer cell stimulatory factor: characterization of the responder cells and synergy with other inducers. J Exp Med 173:869–879

Chan SH, Kobayashi M, Santoli D, Perussia B, Trinchieri G 1992 Mechanisms of IFN-γ induction by natural killer cell stimulatory factor (NKSF/IL-12): role of transcription

and messenger RNA stability in the synergistic interaction between NKSF and IL-2. J Immunol 148:92–98
Chehimi J, Starr SE, Frank I et al 1992 Natural killer (NK) cell stimulatory factor increases the cytotoxic activity of NK cells from both healthy donors and human immunodeficiency virus-infected patients. J Exp Med 175:789–796
Chehimi J, Valiante NM, D'Andrea A et al 1993 Enhancing effect of natural killer cell stimulatory factor (NKSF/interleukin-12) on cell-mediated cytotoxicity against tumor-derived and virus-infected cells. Eur J Immunol 23:1826–1830
Chehimi J, Starr SE, Frank I et al 1994 Impaired interleukin-12 production in human immunodeficiency virus-infected patients. J Exp Med 179:1361–1366
Chua AO, Chizzonite R, Desai BB et al 1994 Expression cloning of a human IL-12 receptor component: a new member of the cytokine receptor superfamily with strong homology to gp130. J Immunol 153:128–136
Clerici M, Lucey DR, Berzofsky JA et al 1993 Restoration of HIV-specific cell-mediated immune responses by interleukin-12 *in vitro*. Science 262:1721–1724
D'Andrea A, Rengaraju M, Valiante NM et al 1992 Production of natural killer cell stimulatory factor (interleukin-12) by peripheral blood mononuclear cells. J Exp Med 176:1387–1398
D'Andrea A, Aste-Amezaga M, Valiante NM, Ma X, Kubin M, Trinchieri G 1993 Interleukin-10 (IL-10) inhibits human lymphocyte interferon-γ production by suppressing natural killer cell stimulatory factor/IL-12 synthesis in accessory cells. J Exp Med 178:1041–1048
D'Andrea A, Ma XJ, Aste-Amezaga M, Paganin C, Trinchieri G 1995 Stimulatory and inhibitory effects of interleukin (IL)-4 and IL-13 on the production of cytokines by human peripheral blood mononuclear cells: priming for IL-12 and tumor necrosis factor-α production. J Exp Med 181:537–546
Eng VM, Car BD, Schnyder B et al 1995 The stimulatory effects of IL-12 on hematopoiesis are antagonized by IL-12-induced IFN-γ *in vivo*. J Exp Med 181: 1893–1898
Gately MK, Wolitzky AG, Quinn PM, Chizzonite R 1992 Regulation of human cytolytic lymphocyte responses by interleukin-12. Cell Immunol 143:127–142
Gazzinelli RT, Hieny S, Wynn TA, Wolf S, Sher A 1993 Interleukin 12 is required for the T-lymphocyte-independent induction of interferon-γ by an intracellular parasite and induces resistance in T-cell-deficient hosts. Proc Natl Acad Sci USA 90: 6115–6119
Gazzinelli RT, Wysocka M, Hayashi S et al 1994 Parasite-induced IL-12 stimulates early IFN-γ synthesis and resistance during acute infection with *Toxoplasma gondii*. J Immunol 153:2533–2543
Heinzel FP, Schoenhaut DS, Rerko RM, Rosser LE, Gately MK 1993 Recombinant interleukin-12 cures mice infected with *Leishmania major*. J Exp Med 177:1505–1509
Hsieh C-S, Macatonia SE, Tripp CS, Wolf SF, O'Garra A, Murphy KM 1993 Development of T_H1 $CD4^+$ T cells through IL-12 produced by *Listeria*-induced macrophages. Science 260:547–549
Jacobsen SEW, Veiby OP, Smeland EB 1993 Cytotoxic lymphocyte maturation factor (interleukin-12) is a synergistic growth factor for hematopoietic stem cells. J Exp Med 178:413–418
Kobayashi M, Fitz L, Ryan M et al 1989 Identification and purification of natural killer cell stimulatory factor (NKSF), a cytokine with multiple biologic effects on human lymphocytes. J Exp Med 170:827–845
Kubin M, Chow JM, Trinchieri G 1994a Differential regulation of interleukin-12

(IL-12) tumor necrosis factor-α, and IL-1β production in human myeloid leukemia cell lines and peripheral blood mononuclear cells. Blood 83:1847–1855

Kubin M, Kamoun M, Trinchieri G 1994b Interleukin-12 synergizes with B7/CD28 interaction in inducing efficient proliferation and cytokine production of human T cells. J Exp Med 180:211–222

Ling P, Gately MK, Gubler U et al 1995 Human IL-12 p40 homodimer binds to the IL-12 receptor but does not mediate biologic activity. J Immunol 154:116–127

Ma X, Chow JM, Carra G et al 1995 The interleukin 12 p40 gene promoter is primed by interferon-γ in monocytic cells. J Exp Med, in press

Macatonia SE, Hosken NA, Litton M et al 1995 Dendritic cells produce IL-12 and direct the development of Th1 cells from naive $CD4^+$ cells. J Immunol 154:5071–5079

Manetti R, Parronchi P, Giudizi MG et al 1993 Natural killer cell stimulatory factor [interleukin-12 (IL-12)] induces T helper type 1 (Th1)-specific immune responses and inhibits the development of IL-4-producing Th cells. J Exp Med 177: 1199–1204

Manetti R, Gerosa F, Giudizi MG et al 1994 Interleukin-12 induces stable priming for interferon-γ (IFN-γ) production during differentiation of human T helper (Th) cells and transient IFN-γ production in established Th2 cell clones. J Exp Med 179: 1273–1283

Mattner F, Fischer S, Guckes S et al 1993 The interleukin-12 subunit p40 specifically inhibits effects of the interleukin-12 heterodimer. Eur J Immunol 23:2202–2208

Merberg DM, Wolf SF, Clark SC 1992 Sequence similarity between NKSF and the IL-6/ G-CSF family. Immunol Today 13:77–78

Miedema F, Petit AJC, Terpstra FG et al 1988 Immunological abnormalities in human immunodeficiency virus (HIV)-infected asymptomatic homosexual men: HIV affects the immuune system before $CD4^+$ T helper cell depletion occurs. J Clin Invest 82:1908–1914

Murphy EE, Terres G, Macatonia SE et al 1994 B7 and interleukin-12 cooperate for proliferation and interferon-γ production by mouse T helper clones that are unresponsive to B7 costimulation. J Exp Med 180:223–231

Nabors GS, Afonso LCC, Farrell JP, Scott P 1995 Switch from a type 2 to a type 1 T helper cell response and cure of established *Leishmania major* infection in mice is induced by combined therapy with interleukin-12 and pentostan. Proc Natl Acad Sci USA 92:3142–3146

Newman GW, Guarnaccia JR, Vance EA 3rd, Wu J, Remold HG, Kazanjian PH Jr 1994 Interleukin-12 enhances antigen-specific proliferation of peripheral blood mononuclear cells from HIV-positive and negative donors in response to *Mycobacterium avium.* AIDS 8:1413–1419

Ozmen L, Pericin M, Hakimi J et al 1994 Interleukin-12, interferon-γ and tumor necrosis factor-α are the key cytokines of the generalized Schwartzman reaction. J Exp Med 180:907–915

Paganin C, Frank I, Trinchieri G 1995 Priming for high interferon-γ production induced by interleukin-12 in both $CD4^+$ and $CD8^+$ T cell clones from HIV-infected patients. J Clin Invest 96:1677–1682

Perussia B, Chan SH, D'Andrea A et al 1992 Natural killer (NK) cell stimulatory factor or IL-12 has differential effects on the proliferation of $TCR\alpha\beta^+$, $TCR\gamma\delta^+$ lymphocytes T, and NK cells. J Immunol 149:3495–3502

Stern AS, Podlaski FJ, Hulmes JD et al 1990 Purification to homogeneity and partial characterization of cytotoxic lymphocyte maturation factor from human B-lymphoblastoid cells. Proc Natl Acad Sci USA 87:6808–6812

Trinchieri G 1993 Interleukin-12 and its role in the generation of T_H1 cells. Immunol Today 14:335–338
Tripp CS, Wolf SF, Unanue ER 1993 Interleukin 12 and tumor necrosis factor alpha are costimulators of interferon gamma production by natural killer cells in severe combined immunodeficiency mice with listeriosis, and interleukin 10 is a physiological antagonist. Proc Natl Acad Sci USA 90:3725–3729
Wolf SF, Temple PA, Kobayashi M et al 1991 Cloning of cDNA for natural killer cell stimulatory factor, a heterodimeric cytokine with multiple biologic effects on T cells and natural killer cells. J Immunol 146:3074–3081
Wysocka M, Kubin M, Vieira LQ et al 1995 Interleukin-12 is required for interferon-γ production and lethality in LPS-induced shock in mice. Eur J Immunol 25:672–676

DISCUSSION

Romagnani: When you generated clones from HIV^+ patients in the presence or absence of interleukin 12 (IL-12), did you find that the clones cultured in the absence of IL-12 replicated the virus more efficiently than those cultured in the presence of IL-12?

Trinchieri: We did not do that experiment with clones from HIV^+ patients. The experiment was performed with $CD4^+$ clones from normal individuals. When we infected them with HIV *in vitro*, the clones generated in the presence of IL-12 replicated the virus much less efficiently and produced a much higher level of γ-interferon (IFN-γ) than the clones generated in the absence of IL-12.

Romagnani: Does IFN-γ have a role in this inhibition of replication?

Trinchieri: I have not determined this yet.

Mitchison: Were you surprised to observe high levels of IL-4 in the clones cultured in the presence of IL-12?

Trinchieri: We concluded that the priming for high IFN-γ production is a differentiation effect at the single-cell level, and that the ability of IL-12 to inhibit IL-4 production is either a selective effect or an indirect effect via another cytokine. However, if we keep a polyclonal T cell culture for a week or longer in the presence of IL-12 before cloning, then we observe a decrease in the level of IL-4. This suggests that the mechanism of selection is present in the polyclonal culture, and that's probably also true of the situation *in vivo*.

Dutton: Ann Kelsoe cloned concanavalin A-stimulated T cells after three days and determined the cytokines produced by panels of $CD4^+$ and $CD8^+$ clones (Kelsoe et al 1991). She found every combination and no clear division between the CD4 and CD8 patterns.

Coffman: We have some unpublished results (S. Mocci & R. L. Coffman) which suggest that mouse T helper 2 (Th2) cells, once they have differentiated into strong IL-4 producers, are relatively insensitive to IL-12. This situation may be different in humans but it is consistent with results that both Abul Abbas and I presented earlier on trying to shift the Th1/Th2 balance *in vitro* (this volume: Abbas et al 1995, Coffman et al 1995). Is it possible that there are differences in IL-12 responsiveness between the mouse and human Th clones?

Trinchieri: It is possible that these differences exist. For example, compared with mouse clones, human clones are not as polarized towards Th1 and Th2 responses. However, as far as I know, there is no definitive evidence that mouse Th2 clones do not respond to IL-12. We have some preliminary evidence showing that mouse Th2 clones express the IL-12 receptor (IL-12R) (L. Showe, C. Son & G. Trinchieri, unpublished results), although I understand that Ken Murphy has found that they do not express some of the late transcription factors (Jacobson et al 1995). It is possible that only part of the signal transduction mechanism is affected, so that the downstream effects are different.

Locksley: Is there a defect in dendritic or other antigen-presenting cells (APCs) in the IL-4 knockout mouse, or is the loss of IL-4 compensated by IL-13?

Trinchieri: As far as I know, IL-2R γ chains are present in IL-4R but not in IL-13R. One would expect IL-4R γ chain knockout mice to be responsive to IL-13.

Flavell: In the experiment where you took T cells and differentiated them, could you exclude the possibility that the cells were originally derived from two different progenitor cell types? Or are they both from a single cell type?

Trinchieri: Although we are also dealing with purified $CD34^+$ progenitor cells, these experiments were performed using peripheral blood monocytes (M. Kubin & G. Trinchieri, unpublished observations, Koch et al 1995). These cells were cultured in the presence of granulocyte macrophage colony-stimulating factor (GM-CSF) and tumour necrosis factor α (TNF-α), or GM-CSF and either IL-4 or IL-13. The cells cultured in the presence of GM-CSF and IL-4 or IL-13 lost some of the characteristics of monocytes (e.g. CD14) and acquired functional and phenotypic characteristics of Langerhans cells (e.g. CD1). During the period in culture, there were times in which the majority of the cells expressed both CD14 and CD1, suggesting that monocytes in these cultures differentiate and acquire some of the characteristics of Langerhans cells, although they may not necessarily become the *in vitro* equivalent to skin Langerhans cells. However, it is interesting that these cells produce high levels of IL-12 in response to lipopolysaccharide (LPS) and are potent APCs. These activities are present to a much lesser extent in monocytes cultured in GM-CSF and TNF-α.

Lotze: We have also cultured $CD1^+$ $CD14^-$, macrophage-depleted cells with IL-4 plus GM-CSF, and we found that they proliferate relatively well but they do not make IL-12 (W. J. Storkus & M. T. Lotze, unpublished results). We have not treated these cells with LPS or *Staphylococcus aureus* Cowan (SAC) but if we add exogenous IL-12, they change from a classic diffuse dendritic morphology to a rounded up morphology and they also start to proliferate. If we add IL-12 plus stimulated allo-peripheral blood lymphocytes, the development of the cytolytic response is enhanced markedly. Therefore, it is conceivable, based also on your results, that you have a mixture of both macrophages and conventional dendritic cells in your culture, and that if you depleted all of the macrophages, you might not observe IL-12 production.

Trinchieri: I do not want to make any assumptions on the nature of the cells in our cultures, which I agree might be heterogeneous. However, their antigen-presenting activity was, in part, dependent on their ability to produce IL-12 and we have evidence that both mouse and human dendritic cells and Langerhans cells produce IL-12, which is required for their ability to generate Th1 responses (Koch et al 1995, Murphy et al 1994, Kang et al 1995).

Lotze: But you don't observe IL-12 production in the absence of stimulation.

Trinchieri: No, the production of IL-12 was only inducible.

Lotze: We have stimulated the same cells, derived from $CD34^+$ cells in the bone marrow, with IL-4 plus GM-CSF. We then added peptide and used them as antigen-stimulating cells *in vivo*. We found that mice with Day 14 tumours induced by human papillomavirus 16 (HPV-16) E7 are cured by injecting these peptide-pulsed dendritic cells intravenously (Mayordomo et al 1995). Have you tested whether these cells are responsive to other cytokines?

Trinchieri: I can only tell you that these cells are not responsive to IFN-γ. They appear to be already primed for IL-12 production and if IFN-γ is added, IL-12 production does not increase significantly, in contrast to monocytes/macrophages. We do not know if these cells proliferate in the presence of IL-12. They are derived from terminally differentiated monocytes and they have a slow, if any, rate of proliferation in culture. This contrasts with cultured $CD34^+$ precursor cells, which proliferate and differentiate.

Lotze: Most people believe that IL-4 suppresses monocyte differentiation into a macrophage phenotype. CD14 is a phosphoinositol glycan-linked molecule. It can be released from macrophages following IL-4 stimulation.

Trinchieri: LPS, which promotes maturation of the dendritic cells, also induces the production of IL-12. Therefore, it is possible that IL-12 has an endogenous effect in the maturation process.

Dutton: What is the significance of using retinoic acid to stimulate the Langerhans cells?

Trinchieri: Retinoic acid has been shown to activate dendritic and Langerhans cells, and it is widely used in dermatology (Meunier et al 1994). Our dermatologist collaborators in these studies, A. Rook and K. Cooper, are interested in a possible effect on the immunological activity of skin Langerhans cells.

Lotze: We used the same culture system as Giorgio Trinchieri, except that we depleted the macrophages from peripheral blood mononuclear cells with colloidal-S. We then removed the non-adherent cells and cultured the adherent cells with GM-CSF and IL-4 for four or five days. We either used these cells to stimulate a primary response *in vitro* with peptide or we incubated them for another day or two with other cytokines. If the cells were incubated with IL-12 without SAC or LPS, we observed that they developed a more rounded shape and they started proliferating, i.e. IL-12 is a dendritic cell growth factor.

We were also interested in the differences between murine IL-10 and viral IL-10 in terms of their ability to alter biological effects *in vivo*. We tested for the enhanced expression of various co-stimulatory molecules using their corresponding antibodies. We found that CD80 and CD86 are both expressed by cells treated with murine or viral IL-10. However, the expression of both B7.1 and B7.2 is decreased in cells treated with murine IL-10 compared to those treated with viral IL-10. We also have mouse models which suggest that there are major differences in terms of the biology of viral and cellular IL-10 (Suzuki et al 1995).

Our results on the IL-12 anti-tumour effects (Nastala et al 1993, 1994, Tahara et al 1994) are similar to those presented by Giorgio Trinchieri. IL-12 was originally cloned from an Epstein–Barr virus (EBV)-infected immortalized B cell line. It seems strange that this cell line makes IL-12 because IL-12 is a potent stimulator of IFN-γ production. It is also curious that these cells produce large amounts of IL-10, far more than normal B cells, and that, at the end of the viral life cycle, the cells make viral IL-10, which presumably occurs during lysogeny. Both forms of IL-10 will cause a decrease in IFN-γ. One of the reasons why IL-12 is made by B cells is that IL-12 comes back and acts on the B cell as a signal for proliferation and differentiation (Jelinek & Braaten 1995). In addition, Eliot Kieff and Mark Birkenbach (unpublished results) have shown that the EBV-induced gene EBI3 has about 30–40% homology with the p40 chain of IL-12 and exists as a non-covalently associated heterodimer with the p97 molecule. If EBI3 is added to cultures either alone or with IL-12, nothing really happens. We're now making retroviral constructs expressing EBI3 and we expect that EBI3, like the p40 homodimer, will block IL-12R when presented appropriately by the p97 molecule.

There are more differences between viral IL-10 and murine IL-10. Sixty percent of wild-type mice will develop a tumour if 10^5 B16 cells are injected. If a retroviral vector containing the viral IL-10 gene is transfected into the tumour that is injected into the mice, the mice all develop tumours and die. However, if murine IL-10 is similarly transfected into the tumour, all the mice develop tumours that then regress. In murine models if high doses of murine IL-10 are administered systemically, tumour regression is observed. There is also a synergistic effect with IL-12, which is a paradox because IL-10 inhibits IL-12 production by macrophages (Suzuki et al 1995).

Ramshaw: What is IL-10 doing in the tumour model?

Lotze: IL-10 has a variety of stimulatory effects as well as being a chemoattractant, e.g. it affects APCs, including dendritic cells. It is possible that it regulates the expression of B7.1 and B7.2. Its effects can be blocked with irradiation and are, therefore, presumably mediated by immunological effects.

Lachmann: Are the effects you observe on tumour growth transferable to untreated mice and if so, by which cells?

Lotze: The best transfer experiments were performed by Forni's group (Pericle et al 1994). In most cytokine transfection or systemic cytokine administration studies, T cells are predominantly responsible but it depends on which models you look at. In the IL-12 tumour models either CD8 or CD4 are involved.

Allen: It is confusing that as dendritic cells get better at driving Th1 responses, they get further away from being able to phagocytose antigens. Are you inducing an intermediate stage of dendritic cell development? Is it possible that the fully differentiated cells aren't the real effectors *in vivo*.

Trinchieri: The cells that are induced with IL-4 and IL-13 are endocytic, even though they may not be phagocytic. When they are induced to differentiate with LPS or TNF-α, they lose their phagocytic and endocytic abilities, and they become better APCs. At the same time, they produce IL-12. It makes sense that an antigen can be endocytosed and induce maturation of dendritic cells. If there is a particulate antigen, then macrophages may be necessary to process the antigen.

Mitchison: Is there any evidence that dendritic cells are phagocytic *in vivo*?

Trinchieri: I am not aware of any, but they can be endocytic.

Allen: If Th1 cells are required for immunity to intracellular pathogens, the APCs have to be, by definition, phagocytic. It is possible that there's cooperation between APCs, so that macrophages might start the processing by releasing soluble antigens which are then processed and presented by the dendritic cells.

Trinchieri: Phagocytic cells respond to bacterial infection by producing high levels of IL-12. In acute infections, IL-12 may be detectable at effective concentrations in the serum and it probably induces a Th1 response. During more limited immune responses, professional APCs, such as dendritic cells, probably produce only low, local levels of IL-12. Thus, in this situation, the balance between IL-12 and IL-4 in the local environment in which the antigen presentation takes place (e.g. in the draining lymph nodes) might determine the balance of Th1 and Th2 responses.

Röllinghoff: Christine Blank et al (1993) used electron microscopy to show that Langerhans cells can take up *Leishmania in vivo* in the skin. The view is that after the Langerhans cells take up *Leishmania*, they are triggered to travel to the lymph node where they present the antigen.

Lotze: We've taken the same dendritic cells that are cultured in GM-CSF plus IL-4, and we've shown that they will take up gold beads that are used to carry plasmid DNA. This suggests that they are actively phagocytic (W. J. Storkus & M. T. Lotze, unpublished results). If they are stimulated with TNF or IL-12, their phagocytic capability is decreased. This *in vitro* observation may represent the difference between what is observed within the tissue and what they are supposed to do at the lymph node site, where they may not be as phagocytic.

Abbas: Ken Rock (personal communication) has evidence that protein antigens covalently coupled to beads are presented by purified Langerhans cells and by B lymphocytes. The important point is that a negative result in a conventional phagocyte assay may not be meaningful because APCs may take up all of the small amount of antigen that is required for presentation.

Mitchison: But dendritic cells aren't part of the classical reticular endothelial system. If a mouse is injected with indian ink, the ink does not end up in dendritic cells.

Sher: Is IL-12 absolutely necessary for the initiation of a Th1 response? What are the results from the IL-12 knockout mice experiments?

Trinchieri: In the IL-12 knockout mice, IFN-γ production in response to LPS is reduced by about 80%. The Th1 cells in these mice produce low levels of IFN-γ, suggesting that IL-12 is required for optimal Th1 cell differentiation, but it may not be absolutely required for the development of Th1 functions, other than for IFN-γ production.

Abbas: Andrew McKnight has published that anti-IL-12 antibody inhibits the Th1 response to conventional hapten–protein conjugates in complete Freund's adjuvant (McKnight et al 1994).

References

Abbas AK, Perez VL, van Parijs L, Wong RCK 1995 Differentiation and tolerance of $CD4^+$ T lymphocytes. In: T cell subsets in infectious and autoimmune diseases. Wiley, Chichester (Ciba Found Symp 195) p 7–19

Blank C, Fuchs H, Rappersberger K, Röllinghoff M, Moll H 1993 Parasitism of epidermal Langerhans cells in experimental cutaneous leishmaniasis with *Leishmania major*. J Infect Dis 167:418–425

Coffman RL, Correa-Oliviera R, Mocci S 1995 Reversal of polarized T helper 1 and T helper 2 cell populations in murine leishmaniasis. Wiley, Chichester (Ciba Found Symp 195) p 20–33

Jacobson NG, Szabo SJ, Weber-Nordt RM et al 1995 Interleukin 12 signaling on T helper type 1 (Th1) cells involves tyrosine phosphorylation of signal transducer and activator of transcription (Stat)3 and Stat4. J Exp Med 181:1755–1762

Jelinek DF, Braaten JK 1995 Role of IL-12 in human B lymphocyte proliferation and differentiation. J Immunol 154:1606–1613

Kang K, Kubin M, Cooper KD, Lessin SR, Trinchieri G, Rook AH 1995 Constitutive and inducible IL-12 synthesis by human Langerhans cells: preferential production relative to epidermal keratinocytes, submitted

Kelsoe A, Troutt AB, Maraskovsky E et al 1991 Heterogeneity in lymphokine profiles of $CD4^+$ and $CD8^+$ T cells and clones activated *in vivo* and *in vitro*. Immunol Rev 123:85–114

Koch F, Heufler C, Stanzi U et al 1995 Interleukin-12 is produced by dendritic cells and mediates Th2 development as well as IFN-γ production by Th1 cells, submitted

Mayordomo JI, Zorina T, Storkus WJ et al 1995 Bone marrow-derived dendritic cells pulsed with tumor peptides effectively treat established murine tumors. Nature, submitted

McKnight AJ, Zimmer G, Fogelman I, Wolf SF, Abbas AK 1994 Effects of interleukin-12 on helper T cell-dependent immune responses *in vivo*. J Immunol 152:2172–2179

Meunier L, Bohianen K, Yoorhees JJ, Cooper KD 1994 Retinoic acid upregulates human Langerhans cell antigen presentation and surface expression of HLA-DR and CD11c, a beta integrin critically involved in T-cell activation. J Invest Dermatol 103:775–779

Murphy EE, Terres G, Macatonia SE et al 1994 B7 and IL-12 cooperate for proliferation and IFN-γ production by mouse T helper clones that are unresponsive to B7 costimulation. J Exp Med 180:223–231

Nastala CL, Edington H, Storkus WJ, Lotze MT 1993 Recombinant interleukin-12 (rmIL-12) mediates regression of both subcutaneous and metastatic murine tumors. Surg Forum 44:518–521

Nastala CL, Edington H, Storkus WJ et al 1994 Recombinant interleukin-12 induces tumor regression in murine models: interferon-gamma but not nitric oxide dependent effects. J Immunol 153:1697–1706

Pericle F, Giovarelli M, Colombo MP et al 1994 An efficient Th2-type memory follows $CD8^+$ lymphocyte-driven and eosinophil-mediated rejection of spontaneous mouse mammary adenocarcinoma engineered to release IL-4. J Immunol 153:5659–5673

Suzuki T, Tahara H, Robbins P, Narula S, Moore K, Lotze MT 1995 The human herpes virus 4 cIL-12 homologue, vIL-10, induces local anergy to tumor allo and autografts. J Exp Med, in press

Tahara H, Zeh H, Storkus WJ et al 1994 Fibroblasts genetically engineered to secrete interleukin-12 can suppress tumor growth *in vivo* and induce antitumor immunity to a murine melanoma. Cancer Res 54:182–189

Interleukin 15 and its receptor

David Cosman, Satoru Kumaki, Minoo Ahdieh, June Eisenman, Kenneth H. Grabstein, Ray Paxton, Robert DuBose, Della Friend, Linda S. Park, Dirk Anderson and Judith G. Giri

Immunex Research and Development Corporation, 51 University Street, Seattle, WA 98101, USA

Abstract. Interleukin 15 (IL-15) is a member of the four-helix bundle cytokine family that shares many *in vitro* biological activities with IL-2. Previous work demonstrated that IL-15 utilizes the β and γ chains of the IL-2 receptor (IL-2R), and that these are essential for IL-15-mediated signal transduction. However, several lines of evidence indicated the existence of an additional, IL-15-specific receptor component. An IL-15 binding chain was identified on a murine T cell clone, and direct expression cloning was used to isolate the corresponding cDNA. The predicted structure of this protein shows sequence similarity to the IL-2R α chain. Transfection of this cDNA into a murine, IL-3-dependent myeloid cell line, 32D-01, conferred IL-15 binding and, together with transfection of the IL-2R β chain, rendered the cells responsive to IL-15 stimulation. This experiment confirmed that the IL-15 binding chain is part of the IL-15 receptor, and it is designated as the IL-15Rα subunit. The expression pattern of the IL-15Rα mRNA is distinct from that of IL-2Rα mRNA. Recombinant expression of a soluble form of IL-15Rα demonstrated that it is a potent inhibitor of IL-15 biological activity.

1995 T cell subsets in infectious and autoimmune diseases. Wiley, Chichester (Ciba Foundation Symposium 195) p 221–233

Interleukin 15 (IL-15) is a recently discovered addition to the subfamily of cytokines that mediates biological responses in T cells. It was initially discovered as an activity, produced at low constitutive levels by a simian epithelial cell line, that stimulates the proliferation of an IL-2-responsive, murine T cell line, CTLL-2 (Grabstein et al 1994). This paper will review: the purification and cloning of IL-15; the similarities between the biological activities of IL-15 and IL-2; the shared components of the IL-2 receptor (IL-2R) and the IL-15 receptor (IL-15R); and the current progress in the characterization of an IL-15-specific component of the IL-15R. Similarities and differences between the IL-2/IL-2R and IL-15/IL-15R systems will be highlighted.

Purification and cloning of IL-15

Simian IL-15 was purified from the African Green monkey kidney epithelial cell line CV-1/EBNA by a combination of hydrophobic interaction, anion-exchange chromatography and high performance liquid chromatography (HPLC). Fractions from HPLC were analysed on SDS-polyacrylamide gels followed by gel slicing and elution of proteins for biological assay (Grabstein et al 1994). Results indicated a 14–15 kD band as a likely candidate for IL-15. This band was transferred to a membrane filter and sequenced directly. From the resulting N-terminal amino acid sequence, degenerate oligonucleotides were designed and used in PCRs to clone a 92 base-pair cDNA fragment from CV-1/EBNA cDNA. This fragment served as a probe to isolate full-length simian IL-15 cDNA clones, which in turn were used to obtain human and murine IL-15 cDNAs. Recombinant expression of biologically active protein confirmed the identity of the IL-15 cDNAs.

Structure of IL-15

Comparison of the predicted amino acid sequence of IL-15 from cDNA clones with the experimentally determined N-terminus of the purified protein showed that IL-15 is synthesized as a 162 amino acid precursor with an unusually long 48 amino acid leader sequence that precedes the mature N-terminus. The mature protein contains four cysteine residues that form two disulfide bonds. There are two potential N-linked glycosylation sites in simian IL-15, three in the human and four in the mouse homologues. Human and simian IL-15 share 97% amino acid identity; and human and murine IL-15 share 73% identity (Grabstein et al 1994, Anderson et al 1995). The sequence of IL-15 shows no similarity to any protein in current sequence databases at the primary amino acid level. However, molecular modelling of IL-15 was used to predict that it is a member of the four-helix bundle family of cytokines with a particularly close structural relationship to IL-2 (Bazan 1992, Grabstein et al 1994). The intron/exon structure of the IL-15 gene also shows similarities with that of IL-2 and other helical cytokines (Anderson et al 1995).

Expression of IL-15

The range of cell and tissue types that express IL-15 has been assessed primarily by Northern blotting. This analysis shows that IL-15 mRNA is present both in a variety of cell lines of epithelial, fibroblast and stromal origin, and in peripheral blood mononuclear cells. Many tissues also contain IL-15 mRNA, with the highest levels found in the placenta and skeletal muscle. However, the biological activity of IL-15 has been harder to detect, and more rigorous studies are needed to establish precisely which cell types make IL-15 *in vivo*. Nevertheless, a striking observation was that IL-15 mRNA could not be detected in activated peripheral blood T cells, which are the major source of IL-2 and express many other cytokines. This suggests a different biological role for IL-15 *in vivo* than that of IL-2, despite the similarities in their *in vitro* activities that are discussed below.

Biological activities of IL-15

Recombinant IL-15 was used in a variety of biological assays to determine its spectrum of activity. This analysis revealed a striking similarity between the actions of IL-15 and IL-2. Both molecules are involved in the proliferation of a variety of murine T cell clones; the co-stimulation of primary human $CD4^+$ and $CD8^+$ T cell subsets with phytohaemagglutinin (PHA); the induction of alloantigen-specific cytotoxic T cell function; the generation of lymphokine-activated killer cells; the proliferation and cytolytic activation of natural killer (NK) cells; and the production of cytokines from NK cells (Carson et al 1994, Grabstein et al 1994). Additionally, IL-2 and IL-15 are both involved in the co-stimulation (with anti-IgM) of B cell proliferation, and the co-stimulation (with CD40 ligand) of IgM, IgG1 and IgA production (Armitage et al 1995, Matthews et al 1995). These results suggested that IL-15 might function by utilization of IL-2R components or by inducing IL-2 production. The latter possibility was eliminated when neutralizing antibodies to IL-2 failed to inhibit IL-15 function (Grabstein et al 1994), which led to the examination of the role of the IL-2R in IL-15 signalling.

Interaction of IL-15 with IL-2 receptor components

The IL-2R has been studied in great detail for a number of years (reviewed in Taniguchi & Minami 1993). The IL-2R α chain binds IL-2 with low affinity but appears to play no role in signal transduction. The β and γ subunits are both essential for signal transduction and cooperate with the α chain to increase the affinity of IL-2 binding. In the human system the β and γ chains together can bind IL-2 with an affinity intermediate between that of α alone and the $\alpha\beta\gamma$ complex, and they can signal in response to a sufficiently high concentration of IL-2. Murine $\beta\gamma$ complexes, however, are not able to bind IL-2 with measurable affinity, and the α chain is required for receptor function (Kumaki et al 1993). The β and γ chains are members of the haematopoietin receptor family, whereas the α chain is not. The γ chain has been shown to participate in the receptors for IL-4, IL-7 and IL-9 in addition to IL-2R (Kondo et al 1993, 1994, Noguchi et al 1993a, Russell et al 1993, 1994). Mutations in the gene encoding the γ chain have been shown to be the cause of X-linked severe combined immunodeficiency (X-SCID) (Noguchi et al 1993b).

Participation of the IL-2R α, β and γ chains in the IL-15R was examined in several ways. Antibodies to the IL-2Rα, which effectively neutralized IL-2 biological activity, had no effect on IL-15-mediated proliferation of PHA-activated human peripheral blood mononuclear cells or on binding of IL-15 to these cells (Giri et al 1994, Grabstein et al 1994). However, antibodies to the β chain were capable of inhibiting IL-15 binding and biological activity in several cell types (Carson et al 1994, Giri et al 1994, Grabstein et al 1994). The participation of the β chain in the IL-15R was confirmed by using a reconstitution assay that was previously developed for the IL-2R system (Hatakeyama et al 1989). Transfection of the murine, IL-3-dependent pro-B lymphoblast cell line, BAF/BO3, with the human IL-2R β chain had been shown to confer IL-2 responsiveness by reconstitution of the full IL-2R in combination with the endogenously expressed α and γ chains. Similarly, transfection of the IL-15 non-responsive BAF/BO3 cells with the human IL-2R β chain allowed the cells to proliferate in the presence of IL-15, and this proliferation was blocked by a monoclonal antibody to the β chain (Giri et al 1994).

Participation of the IL-2R γ chain in the IL-15R was examined using a different reconstitution system that had previously examined the role of the γ chain in IL-2-mediated signalling (Asao et al 1993). Mouse L fibroblast cells were stably transfected with either the human IL-2R α and β chains or the α, β and γ chains and tested for *c-myc* and *c-fos* mRNA induction in response to IL-2 or IL-15. As previously demonstrated for IL-2 (Asao et al 1993), the presence of the γ chain was required for IL-15-mediated signalling (Giri et al 1994).

Finally, the interaction of IL-15 with the IL-2Rβ and IL-2Rγ subunits was measured directly by the binding of IL-15 to African green monkey COS kidney cells transfected with CDNAs encoding human IL-2Rβ and IL-2Rγ, and in co-immunoprecipitation assays with recombinantly expressed extracellular domains of IL-2Rβ and IL-2Rγ. In both cases, no detectable binding of IL-15 to β or γ alone was seen, whereas the combination of β and γ bound IL-15 (Giri et al 1994).

Taken together, these results show clearly that the IL-2R β and γ chains are essential components of the IL-15R, but that the IL-2R α chain appears not to participate in IL-15 binding or signalling.

Evidence for IL-15-specific receptor components

Several lines of evidence pointed to the existence of IL-15R subunit(s) in addition to IL-2Rβ and IL-2Rγ. Not all murine IL-2-responsive cell lines responded to IL-15 (Giri et al 1994, Grabstein et al 1994). L cells transfected with human IL-2Rα and IL-2Rβ bound IL-15 with the same affinity as L cells expressing IL-2Rα, IL-2Rβ and IL-2Rγ despite the inability of IL-2Rβ alone to bind IL-15 (Giri et al 1994). Similarly, IL-15 bound well to Epstein–Barr virus

(EBV)-transformed B cells from patients with X-SCID in which IL-2Rγ expression was undetectable (Giri et al 1994, Kumaki et al 1995). A search was undertaken to find a cell line suitable for molecular characterization of additional components of the IL-15R.

Characterization of a novel IL-15-binding protein

Most murine T cell lines tested showed a similar dose response to IL-2 and IL-15 for proliferation. However, one T helper 2 (Th2) clone manifested a proliferative response to concentrations of IL-15 that were 10-fold lower than those of IL-2 necessary to give the same response. The enhanced proliferation to IL-15 correlated with a large number of IL-15 binding sites on these cells (7000–25 000), which was in vast excess of the number of high affinity IL-2 binding sites that could be measured (a few hundred). Affinity cross-linking of radiolabelled IL-15 to these cells showed that IL-15 bound to a protein of 58–60 kDa and that unlabelled IL-15, but not unlabelled IL-2, could compete for that binding. Accordingly, these cells appeared to be an ideal source from which to attempt the expression cloning of this IL-15-specific binding protein. Transfection of pools of cDNA clones from an expression library constructed with mRNA from this cell line into COS cells followed by binding of radiolabelled IL-15 and autoradiography allowed the isolation of a single cDNA clone that expressed high levels of an IL-15-binding protein. The cDNA encoded a type I membrane protein of 263 amino acids including a 175 amino acid extracellular domain and a 37 amino acid cytoplasmic domain. Sequence comparisons showed that this protein was not a member of the haematopoietin receptor family but instead showed some sequence similarity to IL-2Rα. Specifically, at the mature N-terminus of the IL-15-binding protein and IL-2Rα is a region of about 50–60 amino acids that show about 25% amino acid identity between the IL-15-binding protein and IL-2Rα subunits from different species. Similar conserved amino acid motifs have been found in a variety of other proteins, including complement components, complement receptors, coagulation factors and adhesion molecules, where they have been called 'sushi' domains, GP-1 motifs, or short consensus repeats (reviewed in Ichinose et al 1990). In IL-2Rα, but not in the IL-15-binding protein, there is a second sushi domain, and in other proteins the sushi domains are often repeated multiple times. The rest of the IL-15-binding protein sequence shows a lower degree of similarity with IL-2Rα.

Reconstitution of IL-15 binding and signalling by transfection of the IL-15-binding protein

In order to prove that the IL-15-binding protein was a functional component of the IL-15R, reconstitution experiments were undertaken using a subline of the murine IL-3-dependent myeloid cell line 32D. Although 32D cells were reported to proliferate weakly in response to IL-2 and not at all to IL-15 (Grabstein et al 1994), the subline, 32D-01, was unable to proliferate in response to IL-2 or IL-15. By flow cytometry, 32D-01 expressed undetectable amounts of IL-2Rβ but modest levels of IL-2Rα and IL-2Rγ. IL-15 binding, as detected with Flag epitope-tagged IL-15 followed by a monoclonal antibody against the Flag epitope, was undetectable. Transfection of 32D-01 with the IL-15-binding protein gave a high level expression of IL-15 binding, but the cells were unable to proliferate in response to IL-15. Transfection of 32D-01 with murine IL-2Rβ allowed the cells to proliferate in response to IL-2 but not to IL-15. Co-transfection of IL-2Rβ with the IL-15-binding protein reconstituted both IL-15 binding and IL-15 responsiveness. In view of these results we have designated the IL-15-binding protein as the α chain of the IL-15 receptor (IL-15Rα).

Expression of the IL-15 receptor α chain

Preliminary Northern blotting experiments show a wide distribution for IL-15Rα transcripts. IL-15Rα mRNA is expressed in a variety of tissues, with highest levels in the liver and skeletal muscle. It is expressed in epithelial and stromal cell lines as well as T, B and macrophage cell lines. IL-15 binding is also detected on a wide variety of cell types (Giri et al 1994) and can be up-regulated on T cell clones following activation by antibodies to CD3 or phorbol ester and on macrophage cell lines by γ-interferon (IFN-γ). Comparison of IL-15Rα mRNA expression with IL-2Rα mRNA shows several striking differences. Liver expresses very low levels of IL-2Rα mRNA, and epithelial and stromal cell lines express undetectable levels of IL-2Rα mRNA, implying that there may be cell types that respond to IL-15 but not to IL-2. Further work is required to delineate the nature of the cells within tissues that express IL-15Rα and to establish if IL-15 is able to signal in non-lymphoid cell types.

A recombinant soluble form of the IL-15 receptor α chain

The extracellular domain of IL-15Rα was expressed by transient transfection. Metabolic labelling experiments showed that the soluble receptor was secreted as a protein of about 55 kDa. Supernatants containing soluble IL-15Rα were highly effective at neutralizing the biological activity of IL-15 but they did not affect the activity of IL-2. IL-15Rα thus appears to have a much higher affinity for IL-15 than IL-2Rα has for IL-2. Soluble IL-2R α chains are shed naturally from activated T cells, and both natural and recombinant soluble forms of IL-2Rα have been shown to be relatively poor antagonists of IL-2 because the binding of IL-2 to IL-2Rα has a K_d of about 10 nM (Robb &

Kutny 1987). Further analysis is required to determine the precise affinity of binding of IL-15 to IL-15Rα and to examine the utility of soluble IL-15Rα as an IL-15 antagonist.

Summary

The work described in this paper has defined a novel cytokine, IL-15, and delineated the structure of its receptor. The IL-15R consists of the β and γ chains of the IL-2R and an IL-15-specific α chain that shows some sequence homology to the α chain of the IL-2R. It is possible that these two α chains are the first members of a new receptor family. In the murine system it is clear that IL-15 responsiveness requires the presence of the IL-15Rα subunit in addition to the β and γ chains, the situation in human cells needs further investigation as it remains possible that IL-15, like IL-2, could signal through a $\beta\gamma$ complex. Despite the overall similarity in the IL-2 and IL-15 receptors, the α chains appear to play different roles. IL-2Rα binds IL-2 with low affinity, and the β and γ chains contribute significantly to the high affinity of the $\alpha\beta\gamma$ complex. In contrast, IL-15Rα, at least in the murine system, appears to contribute the majority of the binding energy.

Although IL-15 and IL-2 show very similar *in vitro* biological activities, the different expression patterns of the cytokines and their respective receptor α chains suggest different *in vivo* roles. At the present time it is unclear what those roles are. Naturally occurring mutations in human diseases and targeted disruption of genes in mice have started to increase our understanding of IL-2 biology, but much remains to be determined. The sharing of subunits between cytokine receptors complicates the analysis. In mice with disrupted IL-2 or IL-4 genes (or both), normal numbers of mature T and B cells are found (Kühn et al 1991, Schorle et al 1991, Sadlack et al 1994). In mice with a disrupted IL-7R gene, there is a major decrease in both T and B cell numbers (Peschon et al 1994). Mutation of the murine gene encoding the IL-2R γ chain also has profound effects on lymphocyte development and NK cell development (DiSanto et al 1995). However, the corresponding disease in humans, X-SCID, results in normal to elevated levels of B cells with depressed T and NK cell development (Conley 1992). It is possible that IL-15 could be important for T or NK cell differentiation and function (or both). Targeted disruption of the IL-15 gene and comparison of an IL-2Rβ knockout mouse with an IL-2 knockout mouse may help to resolve these questions. The cloning of the IL-15Rα cDNA should allow the generation of reagents that will facilitate analysis of the IL-15/IL-15R system.

Acknowledgment

We thank Anne C. Bannister for excellent editorial assistance.

References

Anderson DM, Johnson L, Glaccum MB et al 1995 Chromosomal assignment and genomic structure of Il15. Genomics 25:701–706

Armitage RJ, Macduff BM, Eisenman J, Paxton R, Grabstein KH 1995 IL-15 has stimulatory activity for the induction of B cell proliferation and differentiation. J Immunol 154:483–490

Asao H, Takeshita T, Ishii N, Kumaki S, Nakamura M, Sugamura K 1993 Reconstitution of functional interleukin 2 receptor complexes on fibroblastoid cells: involvement of the cytoplasmic domain of the γ chain in two distinct signaling pathways. Proc Natl Acad Sci USA 90:4127–4131

Bazan JF 1992 Unraveling the structure of IL-2. Science 257:410–413

Carson WE, Giri JG, Lindemann MJ et al 1994 Interleukin (IL)-15 is a novel cytokine that activates human natural killer cells via components of the IL-2 receptor. J Exp Med 180:1395–1403

Conley ME 1992 Molecular approaches to analysis of X-linked immunodeficiences. Annu Rev Immunol 10:215–238

DiSanto JP, Müller W, Guy-Grand D, Fischer A, Rajewsky K 1995 Lymphoid development in mice with a targeted deletion of the interleukin 2 receptor γ chain. Proc Natl Acad Sci USA 92:377–381

Giri JG, Ahdieh M, Eisenman J et al 1994 Utilization of the β and γ chains of the IL-2 receptor by the novel cytokine IL-15. EMBO J 13:2822–2830

Grabstein KH, Eisenman J, Shanebeck K et al 1994 Cloning of a T cell growth factor that interacts with the β chain of the interleukin-2 receptor. Science 264:965–968

Hatakeyama M, Mori H, Doi T, Taniguchi T 1989 A restricted cytoplasmic region of IL-2 receptor β chain is essential for growth signal transduction but not for ligand binding and internalization. Cell 59:837–845

Ichinose A, Bottenus RE, Davie EW 1990 Structure of transglutaminases. J Biol Chem 265:13411–13414

Kondo M, Takeshita T, Ishii N et al 1993 Sharing of the interleukin-2 (IL-2) receptor γ chain between receptors for IL-2 and IL-4. Science 262:1874–1877

Kondo M, Takeshita T, Higuchi M et al 1994 Functional participation of the IL-2 receptor γ chain in IL-7 receptor complexes. Science 263:1453–1454

Kühn R, Rajewsky K, Müller W 1991 Generation and analysis of interleukin-4 deficient mice. Science 254:707–710

Kumaki S, Kondo M, Takeshita T, Asao H, Nakamura M, Sugamura K 1993 Cloning of the mouse interleukin 2 receptor γ chain: demonstration of functional differences between the mouse and human receptors. Biochem Biophys Res Commun 193: 356–363

Kumaki S, Ochs HD, Timour M et al 1995 Characterization of B cell lines established from two X-linked severe combined immunodeficiency patients: interleukin-15 binds to the B cells but is not internalized efficiently. Blood 86:1428–1436

Matthews DJ, Clark PA, Herbert J et al 1995 Function of the interleukin-2 (IL-2) receptor γ chain in biologic responses of X-linked severe combined immunodeficient B cells to IL-2, IL-4, IL-13, and IL-15. Blood 85:38–42

Noguchi M, Nakamura Y, Russell SM et al 1993a Interleukin-2 receptor γ chain: a functional component of the interleukin-7 receptor. Science 262:1877–1880

Noguchi M, Yi H, Rosenblatt HM et al 1993b Interleukin-2 receptor γ chain mutation results in X-linked severe combined immunodeficiency in humans. Cell 73:147–157

Peschon JJ, Morrissey PJ, Grabstein KH et al 1994 Early lymphocyte expansion is severely impaired in interleukin 7 receptor-deficient mice. J Exp Med 180:1955–1960

Robb RJ, Kutny RM 1987 Structure-function relationships for the IL 2-receptor system. 4. Analysis of the sequence and ligand-binding properties of soluble Tac protein. J Immunol 139:855–862

Russell SM, Keegan AD, Harada N et al 1993 Interleukin-2 receptor γ chain: a functional component of the interleukin-4 receptor. Science 262:1880–1883

Russell SM, Johnston JA, Noguchi M et al 1994 Interaction of IL-2Rβ and γ_c chains with Jak1 and Jak3: implications for XSCID and XCID. Science 266:1042–1045

Sadlack B, Kühn R, Schorle H, Rajewsky K, Müller W, Horak I 1994 Development and proliferation of lymphocytes in mice deficient for both interleukins-2 and -4. Eur J Immunol 24:281–284

Schorle H, Holtschke T, Hünig T, Schimpl A, Horak I 1991 Development and function of T cells in mice rendered interleukin-2 deficient by gene targeting. Nature 352: 621–624

Taniguchi T, Minami Y 1993 The IL-2/IL-2 receptor system: a current overview. Cell 73:5–8

DISCUSSION

Abbas: Have you added the soluble interleukin 15 receptor (IL-15R) to cultures stimulated with anything other than IL-15?

Cosman: Not yet.

Lotze: Is differential splicing involved?

Cosman: We have used PCR analysis and we have looked at additional cDNA clones, but we have not observed differential splicing in the mouse system. We will need to develop more sensitive assays to determine if the receptor is shed, but it probably will be because most membrane proteins are shed to one degree or another.

Lotze: Do you have an antibody to IL-15R?

Cosman: No.

Mitchison: Do you have purified soluble IL-15R?

Cosman: We have supernatants of COS African Green monkey kidney cells that contain IL-15R.

Cantrell: IL-15 mRNA was present in various tissues, some of which were presumably not stimulated. Does this mean that there is a high serum level of IL-15?

Cosman: We have not yet been able to detect this. We have not studied the IL-15 protein. Most of our results were obtained using Northern blotting or PCR analysis.

Cantrell: You showed that the IL-15R and IL-2R α chains were expressed differentially in the liver, for example. Does the liver contain IL-2R β and γ chains? Because a functional IL-2R is present in the liver.

Cosman: It is possible. There is a broad distribution of the IL-15R α chain, so we're trying to determine if the β and γ chains are present in non-lymphoid cells. So far, we have found that they are present in some cases and absent in others. Also, we are investigating whether a different signal can be generated in those cases where the β and γ chains are not present.

Cantrell: Does the cytoplasmic domain of IL-15Rα have a particular moiety which suggests that it may be involved in signalling?

Cosman: It is probably not involved in proliferation. We have removed the cytoplasmic domain and transfected the murine IL-3-dependent myeloid 32D cell line and we have found that this does not affect proliferation. However, it may be involved in other types of signals.

Liew: We have found IL-15 in human peripheral blood serum. We have also found that patients with active rheumatoid arthritis have high levels of IL-15 in the synovium (I. B. McInnes & F. Y. Liew, unpublished results).

Lachmann: Is transcription of the gene encoding IL-15 inhibited by cyclosporin? If it is not, IL-15 would represent an IL-2-like cytokine that is resistant to cyclosporin.

Cosman: We haven't looked at that.

Sher: We have looked at whether IL-15 has a similar role to IL-12 in the initiation of immune responses (M. Doherty, R. Seder & A. Sher, unpublished results). We have studied the expression of IL-15 mRNA using PCR analysis. Unfortunately, we do not have an antibody to IL-15 itself, but we can measure its biological activity as an additional parameter. We have found that it is a product of activated macrophages and that its expression is enhanced by pretreatment with γ-interferon (IFN-γ). It is peculiar in that, unlike IL-12 and tumour necrosis factor, expression of IL-15 is not down-regulated by cytokines such as IL-10, IL-4, IL-13 and transforming growth factor β.

Lamb: We have followed some of Nadler's work (Boussiotis et al 1993), and we have found that if human T cell clones are exposed to a high concentration of peptide in the absence of antigen-presenting cells (APCs), they are refractory to re-challenge with the ligand (Lamb et al 1983). If high concentrations of IL-2 are added, that induction of anergy is prevented. We have also observed the same effects with IL-15, whereas the addition of IL-4 leads to a partial inhibition of the induction of anergy. Interestingly, IL-7 has no effect (J. R. Lamb, E. Liew & H. Yssel, unpublished observations).

Coffman: Once anergy is induced, it can usually be overcome by adding IL-2. Has anyone looked at whether IL-15 has a similar effect?

Lamb: I haven't looked at the reversal of anergy. Hans Yssel has investigated the factors that induce the proliferation of refractory T helper 1 (Th1) and Th2 cells. He has observed the same pattern, i.e. that the cells respond to IL-4, IL-2 and IL-15 but not to IL-7.

Romagnani: We have compared the effects of IL-15, IL-2 or both on the differentiation of antigen-specific T cell clones *in vitro* and we have found no significant differences in the type of T cell clones that were generated under these three different experimental conditions (R. Manetti, F. Annunziato, E. Maggi & S. Romagnani, unpublished results).

Coffman: If one form of anergy results in the inhibition of IL-2 production by Th1 cells, how does IL-15 fit in with its entirely different pattern of regulation?

Romagnani: This is a good point, particularly with regard to the possible role of IL-15 in forming the T cell response to autoantigens that are presented by non-professional APCs. IL-15 may bypass the requirement for co-stimulatory signals and favour the activation of autoreactive T cells.

Cantrell: IL-2 and IL-15 clearly have similar effects *in vitro* but IL-15 is not able to substitute for IL-2 in IL-2 knockout mice. The crucial question is what actually controls the production of IL-15? Until we know the answer to this, we cannot understand its function *in vivo*.

Cosman: It may have an essential role for natural killer (NK) cells.

Mitchison: What phenotype do the IL-2 knockout mice have?

Cantrell: They are born normally and they have T cells. However, they don't have normal T cell function because they die after about 8–12 weeks of autoimmune diseases.

Abbas: They get inflammatory bowel disease and autoimmune haemolytic anaemia; and they have multiple autoantibodies, including anti-DNA antibodies (Sadlack et al 1993). This leads to an interesting hypothesis that an important function of IL-2 is to programme T cells for apoptosis and peripheral tolerance.

Cosman: IL-2 knockout mice don't have all of the defects that one would have predicted, based on the *in vitro* experiments. For example, one would expect them to have defective T cells.

Cantrell: I would not have expected that. There is no evidence that IL-2 has a role in thymus development, only that a small subset of thymocytes have the α chain. However, the α chain is not functional by itself, and IL-2 does not affect thymocyte proliferation.

Lotze: IL-15 has a potential role in the gut. Steve Gillis (Immunex, Seattle, personal communication) has suggested that IL-15 protects against gut toxicity associated with chemotherapeutic agents or irradiation. However, when we tried to repeat these studies with gut toxicity associated with chemotherapeutic agents, we found that death was enhanced. Are these results suggesting a protective role for IL-15 solid?

Cosman: I would say they're preliminary. The indications we have are that the mice have to be pre-treated with IL-15 before they are irradiated or exposed to the drug in order to get a protective effect. These experiments use recombinant cytokine and I'm not sure what the relevance is to the normal function of IL-15.

Lotze: Ken Grabstein has some results with IL-15 (personal communication). In terms of systemic effects, it induces much less of a vascular leak syndrome than IL-2, suggesting that it has a different biology to IL-2.

Cosman: But the caveat is that he's using a human or simian cytokine in mice.

Liew: Do NK cells have a high level of IL-15R?

Cosman: No. But NK cells are depressed in the IL-2R γ chain knockout mouse, and if they were normal in IL-2 knockout mice, then that would point to a role for IL-15 in NK cell development. At the moment, our best suggestion for the function of IL-15 is in the development of NK cells and also, in combination with IL-12 produced by macrophages, in stimulating NK cells to make cytokines.

Flavell: Have you looked at the receptor expression on thymic subsets.

Cosman: No, we haven't.

Mitchison: You have observed that IL-15 is produced in tissues like muscle, for example. Are you sure that IL-15 is not being produced by other cells that are present in muscle tissue?

Cosman: We're not absolutely sure. We obtained the tissue blots from Clontech (Palo Alto, CA), so we had no control over how the tissue was prepared. We're currently looking at any possible effects of IL-15 on muscle.

Swain: There seemed to be a high level of constitutive expression of IL-15R in those Northern blots. Is this correct?

Cosman: It's not a high level. The typical IL-15R levels are in the range of hundreds per cell with the cell types we've looked at. The only places where thousands of receptors per cell are observed are on certain T cell clones.

Swain: What is the level of expression of IL-15?

Cosman: The level of expression of IL-15 is low in the cells from which we isolated cDNA clones.

Lotze: Is it possible to stimulate the expression of IL-15 in these cells?

Cosman: Not in these cells. So far, we have found that macrophages are the only cell types in which IL-15 expression can be induced (Grabstein et al 1994).

Lotze: Do you observe IL-15 expression if normal cell lines are treated with IFN-γ?

Cosman: We have not looked at that.

Mosmann: IL-15 may act as a maintenance factor for the survival of T cells in a general way because IL-2 is a survival factor for resting T cells as well as a growth factor for activated ones. During an immune response, when T cells are proliferating, T cells use large amounts of IL-2. The standard IL-2 assay doesn't measure the concentration of IL-2, it measures the amount of IL-2. Therefore, it is possible that the background levels of IL-15 may maintain small numbers of unactivated T cells but there is not enough IL-15 to sustain a focused local immune response, so the inducible synthesis of IL-2 is required.

Sher: Mark Doherty has had problems in finding a suitable class of macrophage in which to study IL-15 expression. Most macrophage populations, such as inflammatory peritoneal cells and even resident cells, produce IL-15 mRNA spontaneously. Therefore, he used granulocyte macrophage colony-stimulating factor-stimulated bone marrow-derived cells because they have a relatively low level of constitutive IL-15 mRNA expression. I like Tim Mosmann's idea that IL-15 may be a growth component for the immune system, and that it may be up-regulated under certain inflammatory situations, but that it differs fundamentally from IL-2 because it's not produced by T cells.

Liew: So the general conclusion is that there is some degree of functional overlap between IL-2 and IL-15, which suggests some degree of evolutionary redundancy.

Swain: There is little redundancy in terms of the immune response. A responding T cell produces a high level of IL-2 directly at the site of T cell proliferation. In contrast, only background levels of IL-15 are produced by non-responding cells, and such levels are not going to have the same effects.

References

Boussiotis VA, Freeman GJ, Gray G, Gribben J, Nadler LM 1993 B7 but not intracellular adhesion molecule 1 costimulation prevents the induction of human alloantigen-specific tolerance. J Exp Med 178:1753–1763

Grabstein KH, Eisenman J, Shanebeck K et al 1994 Cloning of a T cell growth factor that interacts with the β chain of the interleukin-2 receptor. Science 264:965–968

Lamb JR, Skidmore BJ, Green N, Chiller J, Feldmann M 1983 Induction of tolerance in influenza virus immune T lymphocyte clones with synthetic peptides of influenza haemagglutinin. J Exp Med 157:1434–1447

Sadlack B, Merz H, Schosle H, Schimpl A, Feller AC, Horak I 1993 Ulcerative colitis-like disease in mice with a disrupted interleukin-2 gene. Cell 75:253–261

Nitric oxide in infectious and autoimmune diseases

F. Y. Liew

Department of Immunology, University of Glasgow, Western Infirmary, Glasgow G11 6NT, UK

Abstract. Nitric oxide (NO) is a critical mediator of a variety of biological functions. A range of micro-organisms, including viruses, bacteria, protozoa and helminths, is sensitive to NO produced by macrophages activated with γ-interferon (IFN-γ) and lipopolysaccharide. In contrast, NO is involved in a number of important immunopathologies, including diabetes, graft-vs-host reaction, rheumatoid arthritis, systemic lupus erythematosus, experimental autoimmune encephalomyelitis and multiple sclerosis. Thus, it is crucial that the synthesis of NO is under tight regulation. This is achieved, in part, through the opposing cytokines produced by T helper 1 (Th1) and Th2 cells. Th1 cells produce IFN-γ, which is the most powerful inducer of inducible NO synthase (iNOS). In contrast, interleukin 4 is produced by Th2 cells and inhibits the induction of iNOS at the level of transcription. Furthermore, NO is also produced by Th1 cells, whose proliferation can be inhibited by high concentrations of NO. Thus, apart from being a mediator of Th1/Th2 interaction, NO may also be an important self-regulatory molecule that prevents the over-expansion of Th1 cells which are implicated in a range of severe immunopathologies.

1995 T cell subsets in infectious and autoimmune diseases. Wiley, Chichester (Ciba Foundation Symposium 195) p 234–244

There has been a considerable interest in nitric oxide (NO) in recent years. This is because NO is a critical mediator of a variety of biological functions, including vascular and muscle relaxation, platelet aggregation, neuronal cell function, microbicidal and tumoricidal activity, and a range of immunopathologies (reviewed in Moncada et al 1991, Liew 1993, Nathan & Xie 1994, Bredt & Snyder 1994). NO is derived from L-arginine and molecular oxygen in a reaction catalyzed by NO synthase (NOS). There are at least two types of NOS (Marletta 1994). The Ca^{2+}-dependent form is present constitutively in a variety of tissues, and it produces the physiological concentration of NO needed for house-keeping. The Ca^{2+}-independent form is inducible in a number of cell types (including macrophages, hepatocytes, neutrophils, muscle and endothelium) by a variety of immunological stimuli such as γ-interferon (IFN-γ), tumour necrosis factor α (TNF-α) and bacterial lipopolysaccharide (LPS). Once induced, these cells produce a large amount of NO, which is cytotoxic. Thus, NO has a normal physiological function and is also required in large amounts to combat infectious agents and tumours. The production of excessive amounts of NO could lead to a different range of pathological outcomes and important pathologies; therefore, the expression of inducible NOS (iNOS) is necessarily under tight regulation.

This article summarizes the apparent paradoxical roles of NO in infectious and autoimmune diseases, and discusses the possible mechanism for the regulation of NO synthesis. This is based, in some cases, on our unpublished data.

Antimicrobial activity

Antimicrobial activity is one of the major functions of NO. This can be clearly illustrated with leishmaniasis, which is caused by the intracellular parasite *Leishmania* transmitted by sand flies. In mammalian hosts *Leishmania* only infects macrophages. Currently, there are over 12 million cases of leishmaniasis worldwide, some of which will be fatal.

When CBA mice were infected in the footpad with 2×10^6 *Leishmania major* promastigotes they developed a localized lesion starting six days after infection. These mice developed significantly larger lesions and four orders of magnitude higher parasite loads when the lesion was injected with 5 mg/mouse per day of L-*N*-monomethyl arginine (L-NMMA) than when controls were injected with the inert D-enantiomer, D-NMMA, or saline (Liew et al 1990).

The role of NO in resistance against leishmaniasis is formally shown in iNOS knockout mice. Mice with a disruption in the gene encoding iNOS are not able to resist *L. major* infection in contrast to heterozygote littermates of the MF1 background, which are highly resistant to the same infection (Wei et al 1995).

Leishmania is one of an increasing range of micro-organisms that are susceptible to NO synthesized by cytokine-activated macrophages. These include *Cryptococcus neoformans*, *Schistosoma mansoni*, *Plasmodium* and *Toxoplasma gondii*. These studies have been carried out in murine models, and there is considerable controversy as to whether human macrophages are capable of producing NO. I have shown that human monophage THP-1 cell line can produce large amounts of NO when cultured with the superantigen, staphylococcal enterotoxin B (SEB) (F. Y. Liew, unpublished results). This NO production is inconsistent suggesting that the exact conditions needed for the optimal activation of human macrophages, and consequent production of NO, are distinct from those of murine macrophages. These conditions need to be defined.

Synthesis of nitric oxide by human macrophages

The production of NO by human macrophages can also be demonstrated by using freshly isolated human tissues. When the synovial tissues of patients with active rheumatoid arthritis were cultured *in vitro* as single cell suspensions in the presence of SEB, high levels of nitrite (up to 50 μM) were detected in the culture supernatants after 24 h. When the tissues were fixed and double stained with anti-iNOS and anti-CD68 antibodies, many cells (20–30%) stained positively, suggesting that they express both iNOS and CD68, a specific marker for human macrophages. Therefore, these results demonstrate that human macrophages can express iNOS, and that NO may be closely associated with rheumatoid arthritis. The latter finding is supported by results from the murine collagen-induced arthritis model.

The pathological effect of nitric oxide

DBA/1 mice develop an arthritis-like disease when injected with collagen in Freund's complete adjuvant. Symptoms of the disease start at about 28 days after challenge. The progression of the disease can be stopped by a daily i.p. (intraperitoneal) injection of L-NMMA, starting at Day 28 after induction, at a standard dose of 250 mg/kg per day for 10 days. More impressively, when the mice were similarly treated with L-NMMA at the beginning of collagen injection, the disease was markedly reduced throughout the experiment, even though the treatment was only for a short period, i.e. for the first 10 days (I. B. McInnes & F. Y. Liew, unpublished results). These results demonstrate that NO is not only involved in the manifestation of arthritis, but may also be involved in the induction of arthritis.

The synovium also contains a high level of interleukin 15 (IL-15), a recently discovered cytokine produced by macrophages (but not T cells) that shares many of the functional properties of IL-2. We have recently shown that IL-15, like IL-2, is a potent attractant of T cells (Wilkinson & Liew 1995). Therefore, we postulate that the following sequence of events occurs: activated macrophages in the synovium produce IL-15, which attracts T cells into the synovium; the migrant T cells produce cytokines, such as IFN-γ, which stimulate macrophages and other synovial cells to produce NO; NO then manifests cartilage destruction either directly or by inducing the synthesis of metalloproteinases.

Another example of the pathological effect of NO is the murine lupus model. MRL-lpr/lpr (lymphoproliferation) mice develop, among other symptoms, excessive proteinuria starting at about 100 days old. This and other symptoms were significantly reduced if the mice were treated with L-NMMA suggesting an important role for NO in the pathogenesis of systemic lupus erythematosus (SLE). Indeed, MRL-lpr/lpr mice produce significantly higher levels of NO, as evident from the level of nitrate in the serum than control MRL- + / + mice or BALB/c mice. In addition, spleen and peritoneal cells from the MRL-lpr/lpr mice produce significantly more NO when cultured with LPS and IFN-γ *in vitro* than those of the MRL- + / + mice. The culture supernatants from the MRL-lpr/lpr mice cells also contained significantly higher levels of IL-12 than those from the control MRL- + / + mice cells. When spleen cells from the MRL-lpr/lpr mice were cultured with IL-12 in the presence of LPS, they produced NO in a dose-dependent manner. These levels are significantly higher than those from the control MRL- + / + mice. The production of NO can be markedly, but not completely, inhibited by anti-IFN-γ antibody, and can be modestly inhibited by anti-TNF-α antibody. IL-12 does not directly induce macrophages to produce NO. It appears to act through natural killer (NK) cells and T cells. Thus, it is possible that macrophages of MRL-lpr/lpr mice have an exaggerated capacity to produce IL-12, which may stimulate NK cells and T cells to produce both IFN-γ and perhaps an additional undefined factor. Together, these factors may activate macrophages to produce high levels of NO, which may mediate at least some of the lupus disease symptoms. The relationship between this scheme and the gene encoding Fas ligand, a defect of which has been implicated in SLE, is at present unclear.

The human body, therefore, faces an important dilemma. To combat infectious pathogens and possibly also tumours, it has to produce significant levels of NO. However, an excessive production of NO will lead to serious pathology. Therefore, an intrinsic homeostatic mechanism that maintains this important balance must exist. An understanding of the regulatory mechanism of NO synthesis may lead to an effective therapy for some of the major diseases.

Inhibition of nitric oxide synthesis by cytokines

One of the first natural mechanisms that was recognized for regulating NO synthesis was through IL-4. When macrophages were activated with IFN-γ and a low dose of LPS, they produced significant amounts of NO and expressed high levels of NOS. Both production of NO and expression of NOS were inhibited in a dose-dependent manner by preincubating the cells with IL-4 (Liew et al 1991). PCR analysis using NOS-specific primers demonstrated that this inhibition was at the level of transcription. Cells that were preincubated with IL-4 then activated with IFN-γ and LPS expressed markedly reduced levels of NOS mRNA than control cells activated only with IFN-γ and LPS (Sands et al 1994, Bogdan et al 1994).

Nitric oxide inhibits its own synthesis by a feedback mechanism

Macrophages start producing detectable levels of NO about 6 h after activation with IFN-γ and LPS, reaching a plateau by 24 h. NOS activity occurs earlier, peaking at 12 h and declining rapidly thereafter. These kinetics suggest that NO inhibits its own production by inactivating NOS. Indeed, the activity of NOS from the brain or macrophages can be inhibited, in a dose-dependent manner, by *S*-nitroso-*N*-acetyl-penicillamine (SNAP), which is a producer of NO (Rogers & Ignarro 1992, Assreuy et al 1993, Rengasamy & Johns 1993, Park et al 1994). This feedback control mechanism, whereby a potentially harmful molecule prevents its own over-production, is characteristic of an important biological system.

Another effective way by which NO can regulate its own synthesis is to inhibit the production of IFN-γ by T helper 1 (Th1) cells. A cloned malaria-specific Th1 cell line produced a modest level of IFN-γ when stimulated with 5 μg/ml of concanavalin A. This is significantly enhanced in a dose-dependent manner by the presence of the NOS inhibitor, L-NMMA. Conversely, the production of IFN-γ by the cells was completely abolished by the addition of SNAP (Taylor-Robinson et al 1994). These results therefore demonstrate that T cells can produce NO which, in turn, can inhibit IFN-γ production by T cells.

NO also inhibits the production of IL-2 by Th1 cells and the proliferation of these cells. When cloned Th1 cells specific for malaria antigens were cultured with malaria-parasitized erythrocytes, they produced a vigorous proliferative response that was completely inhibited, in a dose-dependent manner, by the addition of SNAP. In contrast, the proliferation of the malaria-specific Th2 clones was not affected by the presence of similar concentrations of SNAP. Thus, the differential susceptibility to NO may be an important functional parameter to differentiate these two subsets of $CD4^+$ T cells. It is presently unclear if this finding is generally applicable to other T cell clones or to uncloned primary Th1 or Th2 cells.

Conclusions

These are a few examples of the mechanisms by which NO synthesis is regulated. The details of the signalling pathways and the transcription factors involved in the induction and expression of NOS are, at present, unknown. The possibility that T cells produce a large amount of NO which also modulates the secretion of IL-2 and IFN-γ by the T cells may have important implications. These cytokines are products of Th1-like cells that have been implicated in a range of autoimmune diseases. Therefore, NO may be a self-regulatory molecule that prevents the potentially deleterious effect of over-expansion of this subset of T cells. Evidence supporting this notion is accumulating. Understanding these molecular interactions will not only shed light on this major biological system, but will also lead to effective therapy of some of the major diseases.

Acknowledgements

I would like to thank William Sands, Iain McIness, Xiao-qing Wei, Fang ping Huang, Andrew Taylor-Robinson and Stephen Phillips for their important contributions. The work described here was supported by the Wellcome Trust, The Robertson Trust and the Medical Research Council.

References

Assreuy J, Cunha FQ, Liew FY, Moncada S 1993 Feedback inhibition of nitric oxide synthase activity by nitric oxide. Br J Pharmacol 108:833–837

Bogdan C, Vodovotz Y, Paik J, Xie Q-W, Nathan C 1994 Mechanism of suppression of nitric oxide synthase expression by IL-4 in primary mouse macrophages. J Leukocyte Biol 55:227–233

Bredt DS, Snyder SH 1994 Nitric oxide: a physiologic messenger molecule. Annu Rev Biochem 63:175–195

Liew FY 1993 The role of nitric oxide in parasitic disease. Ann Trop Med Parasitol 87:637–642

Liew FY, Millott S, Parkinson C, Palmer RMJ, Moncada S 1990 Macrophage killing of *Leishmania* parasite *in vivo* is mediated by nitric oxide from L-arginine. J Immunol 144:4794–4797

Liew FY, Li Y, Severn A et al 1991 A possible novel pathway of regulation by murine T helper type-2 (T_h2) cells of a T_h1 cell activity via the modulation of the induction of nitric oxide synthase on macrophages. Eur J Immunol 21:2489–2494

Marletta MA 1994 Nitric oxide synthase: aspects concerning structure and catalysis. Cell 78:927–930

Moncada S, Palmer RMJ, Higgs EA 1991 Nitric oxide: physiology, pathophysiology and pharmacology. Pharmacol Rev 43:109–142

Nathan C, Xie Q-W 1994 Regulation of biosynthesis of nitric oxide. J Biol Chem 269:13725–13728

Park SK, Lin HL, Murphy N 1994 Nitric oxide limits transcriptional induction of NO synthase in CNS glial cells. Biochem Biophys Res Commun 201:762–768

Rengasamy A, Johns RA 1993 Regulation of nitric oxide synthase by nitric oxide. Mol Pharmacol 44:124–128

Rogers NE, Ignarro LJ 1992 Constitutive nitric oxide synthase from endothelium is reversibly inhibited by nitric oxide formed from L-arginine. Biochem Biophys Res Commun 189:242–249

Sands WA, Bulut V, Severn A, Xu DM, Liew FY 1994 Inhibition of nitric oxide synthesis by interleukin-4 may involve inhibiting the activation of protein kinase C epsilon. Eur J Immunol 24:2345–2350

Taylor-Robinson AW, Liew FY, Severn A et al 1994 Regulation of the immune response by nitric oxide differentially produced by T helper type 1 and T helper type 2 cells. Eur J Immunol 24:980–984

Wei X-Q, Charles I, Smith A et al 1995 Altered immune response in mice lacking inducible nitric oxide synthase. Nature 375:408–411

Wilkinson PC, Liew FY 1995 Chemoattraction of human blood T lymphocytes by interleukin-15. J Exp Med 181:1255–1260

DISCUSSION

Kaufmann: Production of nitric oxide (NO) by human macrophages is an important observation. How does staphylococcal enterotoxin B (SEB) cause NO production in more conventional macrophages, e.g. blood monocytes?

Liew: We studied NO production in cells from the synovial tissue of patients with rheumatoid arthritis. We have not analysed NO production in peripheral blood mononuclear cells (PBMC). I suspect that PBMC will not produce NO because there are few mature macrophages in the blood. There's no reason for circulating monocytes to produce NO. This would only lead to unnecessary pathology.

Kaufmann: Is NO an important effector molecule in defence against intracellular bacteria or protozoa infections?

Liew: I think NO is an important effector molecule against both bacteria and protozoa.

Ramshaw: With many viruses, γ-interferon (IFN-γ) induces an antiviral effect that is mediated through NO. This is easily demonstrated *in vitro* (Karupiah & Harris 1995). Some viruses that we have looked at are good inducers of NO, and NO is important in the immune-mediated control of their growth *in vivo* (Karupiah et al 1994). If we make recombinant viruses that express interleukin 4 (IL-4), a cytokine that down-regulates T helper 1 (Th1) responses, both responses mediated by cytotoxic T lymphocytes and the production of NO are inhibited. Such viruses have increased pathogenicity (Sharma et al 1995).

When did you inject L-NMMA (L-*N*-monomethyl arginine)?

Liew: At the time of induction, during the first 10 days.

Ramshaw: There may be problems with how you administer it. Injecting L-NMMA twice a day may not be sufficient if the half-life of L-NMMA activity is only an hour or so.

Liew: It's sufficient in our *Leishmania* system.

Ramshaw: But it may not be sufficient in all systems, and it may depend on the life cycle of the virus that you use. Some viruses, for example the pox viruses, are highly susceptible to NO, whereas retroviruses are not. I don't know how susceptible the LCMV (lymphocytic choriomeningitis virus) is.

Liew: Retroviruses may be less susceptible to NO because they pick up certain functions from the host cells.

Mitchison: Is the LCMV inflammation inhibitable by L-NMMA?

Liew: We have not looked at this.

Swain: You mentioned that NO may be involved in inducing apoptosis in the Th1 cell population. Is this process Fas and Fas ligand dependent?

Liew: These are purely speculative ideas. We are currently investigating the details, so at the moment I do not know whether the process is Fas and Fas ligand dependent.

Lotze: You showed that the iNOS (inducible NO synthase) knockout mice have an increased response to concanavalin A. Presumably, this was using normal splenocytes, and it could be recapitulated with L-NMMA. This suggests that there's something different about the cells *in vivo*. I understand that L-NMMA can inhibit NO and can augment a mixed lymphocyte response in the rat, where there is significant NO production; but that its effects are minimal in the mouse. Therefore, how can you explain the dramatic effects in the iNOS knockout mice?

Liew: We have obtained these iNOS knockout mice only recently, so many of our experiments with them are in progress. However, it is possible to explain this apparent contradiction. I believe that there is a constitutive level of NO that is required for the maintenance of T cell development, but that too much NO can prevent the overexpansion of Th1 cells. We recently investigated the effect of increasing concentrations of *S*-nitrosyl-*N*-acetyl-penicillamine (SNAP), which is an NO donor, in a particular Th1 cell clone. We found that in the absence of SNAP, there was a normal level of proliferation in the presence of concanavalin A, and that the proliferation decreased with increasing concentrations of SNAP. This suggests that the low level of NO, which is produced by the constitutive form of NOS, is required to maintain the normal proliferation of Th1 cells. However, when iNOS is induced, there is an increase in the levels of NO and proliferation ceases. This explains why there is an overrepresentation of Th1 cells in iNOS knockout mice.

Allen: Have you looked at T cell development in the iNOS knockout mice?

Liew: Not yet.

Flavell: Could you elaborate on what you think the role of the NO effector mechanisms is in experimental autoimmune encephalomyelitis models?

Liew: We have three possible explanations for this. (1) NO expands the cardiovascular system so that there is an influx of inflammatory cells into the lesions. (2) Free radical-induced damage is involved. (3) NO induces the inflammatory cells to produce other inflammatory cytokines such as IL-1β and tumour necrosis factor α (TNF-α).

Flavell: Is there any evidence for the level at which it is acting? Is it acting at the level of cell induction or in the brain?

Liew: There is no evidence for this.

Sher: I believe that NO has an immunosuppressive role. In *Toxoplasma* infections it has been shown to be an important suppressor of early lymphocyte proliferation *in vitro* (Candolfi et al 1995). Thus, it may be important in turning off the inflammatory response. I have some difficulty accepting the concept that NO is selectively inhibiting Th1 cells because NO is a non-specific molecule. It is membrane permeable and it inhibits many enzyme functions in cells. How could a non-specific molecule inhibit the production of specific cytokines?

Mitchison: Eddy Liew did say that he hasn't been able to produce differential effects on fresh cells. It's probably too early to discuss specific effects until that issue has been clarified.

Please could you tell us about the *lsh* gene? What is its relationship with NO?

Liew: The *lsh* gene was cloned by Philip Gros' lab (Vidal et al 1993), and it is homologous but not identical to the gene controlling the transport of nitrites. It maps to a different location to the gene encoding iNOS, and it is unclear how these two gene products are related. It is possible that the activation of iNOS requires a backwards and forwards transport of nitrite, so that interruption of the gene encoding Nramp (natural resistance-associated macrophage protein) reduces the production of iNOS and the subsequent survival of intracellular pathogens.

Mason: Does IL-15 rescue the T helper 1 cells from NO-mediated cell death?

Liew: We have not looked at this.

Allen: Is the effect of IL-15 on T cells mediated by chemotaxis (i.e. the directed movement of T cells) or chemokinesis (i.e. the random movement of T cells)?

Liew: This is a specific interaction. If a collagen gel is used to make a cytokine gradient, one can see the cells boring through gel towards higher concentrations of IL-15.

Romagnani: Is IL-15 expressed in normal synovial tissue?

Liew: It is ethically difficult to obtain normal synovial tissue.

Mitchison: Do you have an antibody to IL-15?

Cosman: We have antibodies to human IL-15 but not to murine IL-15 or to the specific receptor α chains.

Lotze: I would like to ask a couple of questions concerning the role of IL-15 in early inflammatory events. If systemic IL-2 is given intravenously to a patient, the natural killer (NK) cells disappear and then within 4 h all the T cells disappear. However, it is possible that the cytokine causing some of the early recruitment of T cells is IL-15. Did you only look at the migration of activated T cells, or did you also look at naive T cells and memory T cells? Also, have you examined the ability of IL-15 to cause the migration of NK cells?

Liew: I do not know whether the cells that we obtain from peripheral blood are activated or not. They are probably a mixed population of cells.

Lotze: But you could separate them.

Liew: We haven't done that. But if we culture the cells overnight, they're more responsive to IL-15.

In answer to your second question, NK cells do not respond to IL-15 in our cell polarization experiments.

Lachmann: In human lupus, and probably also in mouse lupus, the inflammation is caused predominantly by immune complexes and is essentially mediated by polymorphs. I am concerned that you are giving a paramount role to monocyte-derived products rather than to products of the polymorphs. Do polymorphs make NO?

Liew: Yes, they do.

Lachmann: Therefore, it is possible that your results in lupus are due to an effect on the polymorphs, rather than on monocytes.

Romagnani: I agree. There are also different models of lupus. For example, lupus can be induced by chemicals (such as $HgCl_2$) or by allogeneic reactions. These are Th2-mediated disorders (Goldman et al 1991) and, therefore, NO is not produced.

Lachmann: You said that NO is produced by both polymorphs and macrophages. When you look at the synovial fluid, which is full of polymorphs, how much NO is polymorph derived and how much is macrophage derived? Complex interactions between the myeloid system and the antibody complement system may be involved.

Liew: I gather that polymorphs do not live for long. We have found that macrophages are not the only cells that produce NO in the synovium; for example, neutrophils may also produce NO. It varies among individuals, but about 20% of the cells expressing iNOS are $CD68^+$, and are therefore likely to be macrophages.

Lachmann: That is true but the many cells seen in the fluid have presumably arrived in the last 24–48 h. Numerous polymorph products are therefore likely to be present.

Navikas: Can NO induce myelin damage directly?

Liew: There is a report that NO can damage myelin directly (Cross et al 1994).

Sher: Isabelle Oswald and Stephanie James demonstrated that mouse endothelial cells can be up-regulated to produce iNOS (Oswald et al 1994). Did you look for up-regulation of iNOS *in situ* in the endothelium?

Liew: Yes, the expression of iNOS was up-regulated in the endothelium.

Mitchison: Could you give an overview on the state of NO inhibitory drugs and clinical trials?

Liew: From an industrial pharmaceutical point of view, there are two major targets for NO inhibitory drugs: one that targets shock and the other that targets the cardiovascular system. The aim is to produce a compound that blocks iNOS but not the constitutive form of NOS. I believe that a compound may have been developed which is 10–100 times more selective against iNOS than the constitutive form of NOS.

Some companies have developed a NO-releasing patch that is given to pregnant women undergoing premature labour. The result is that the muscles relax and labour stops.

Lotze: Kilbourn has shown that L-NMMA reverses IL-2-and TNF-induced hypotension in dogs and humans, and it does not impact on other manifestations of IL-2 toxicity (Kilbourn & Belloni 1990, Kilbourn et al 1990).

Liew: NO has also been implicated for penile erection. Therefore, the possibility of using NO to treat impotence is being investigated.

Mitchison: Is it likely that NO, as a major inflammatory mediator, will be used in the treatment of shock?

Liew: Yes. There are a few clinical trials in progress for shock. A clinical trial is also about to go ahead for rheumatoid arthritis. This is unlikely to involve direct injection into the joints because that's too dangerous. It may be possible to use some sort of patch that releases an inhibitor of NO.

Mitchison: Is L-NMMA toxic?

Liew: No, it's not toxic. It's been used in humans on quite a few occasions.

Lotze: Aminoguanidine has also been used for the treatment of diabetes and to inhibit NO production.

Mitchison: Why are they so hesitant to test aminoguanidine in, for example, rheumatoid arthritis?

Lachmann: It causes a substantial rise in blood pressure.

Abbas: It constricts blood vessels, so there would be tremendous cardiovascular complications.

References

Candolfi E, Hunter CA, Remington JS 1994 Mitogen- and antigen-specific proliferation of T cells in murine toxoplasmosis is inhibited by reactive nitrogen intermediates. Infect Immun 62:1995–2001

Cross AH, Misko TP, Lin RF, Hickey WF, Trotter JL, Tilton RG 1994 Aminoguanidine, an inhibitor of inducible nitric oxide synthase, ameliorates experimental autoimmune encephalomyelitis in SJL mice. J Clin Invest 93:2684–2690

Goldman M, Druet P, Gleichmann E 1991 Th2 cells in systemic autoimmunity: insights from allogeneic diseases and chemically induced autoimmunity. Immunol Today 12:223–227

Karupiah G, Xie Q-w, Buller ML, Nathan C, Duarte C, MacMicking JD 1994 Inhibition of viral replication by interferon-γ-induced nitric oxide synthase. Science 261:1445–1448

Karupiah G, Harris N 1995 Inhibition of viral replication by nitric oxide and its reversal by ferrous sulfate and tricarboxylic acid cycle metabolites. J Exp Med 181:2171–2179

Kilbourn RG, Belloni P 1990 Endothelial cell production of nitrogen oxides in response to interferon γ in combination with tumor necrosis factor, interleukin-1, or endotoxin. J Natl Cancer Inst 82:772–776

Kilbourn RG, Gross SS, Jurban A et al 1990 N^G-methyl-L-arginine inhibits tumor necrosis factor-induced hypotension: implications for the involvement of nitric oxide. Proc Natl Acad Sci USA 87:3629–3632

Oswald IP, Eltoum I, Wynn TA et al 1994 Endothelial cells are activated by cytokine treatment to kill an intravascular parasite, *Schistosoma mansoni*, through the production of nitric oxide. Proc Natl Acad Sci USA 91:999–1003

Sharma DP, Ramsay AJ, Magure D, Rolph M, Ramshaw IA 1995 Interleukin 4 expression enhances the pathogenicity of vaccinia virus and suppresses cytotoxic T cell responses, submitted

Vidal SM, Malo D, Vogan K, Skamene E, Gros P 1993 Natural resistance to infection with intracellular parasites: isolation of a candidate for Bcg. Cell 73:469–475

Final discussion

Mitchison: I would like to raise a general point relating to Giorgio Trinchieri's presentation. When you grow T cells from HIV^+ patients in culture and switch their phenotype so that they produce γ-interferon (IFN-γ), why don't you put them back into the patients? In other words, what is your opinion of autoadoptive therapy of the style that Phil Greenberg has been initiating (Riddell et al 1992)?

Trinchieri: It's certainly possible to use these cells for autoadoptive therapy. We know that *in vitro* these cells do not replicate the virus very well. However, they are generated by stimulation with either anti-CD3 antibody or phytohaemagglutinin, so they may not be antigen-specific clones. It is possible that because they make more IFN-γ, they will provide at least some of the non-specific early inflammatory responses and they will be effective in improving the resistance to infections. Alternatively, interleukin 12 (IL-12) may have the same effect *in vivo* if it's present during opportunistic infection. Clinical trials with IL-12 are currently underway. Single low doses were used in phase I trials to avoid any side effects. Hopefully, these doses may be used as systemic outpatient treatment. If IL-12 is given at the time of infection, it may provide a prolonged immunization against those infections. IL-12 stimulates the bone marrow *in vivo* to produce more lymphocytes. Therefore, in the treatment of patients with AIDS it will probably cause an increased number of $CD4^+$ cells, although it is not known whether this will be accompanied by increased viral replication. Also, we need to find out the extent of possible side effects to determine if it is feasible to treat asymptomatic patients with IL-12.

Cantrell: I was surprised at your results which suggested that IL-12 is necessary for IFN-γ production. How much of the IFN-γ produced in those cultures is by a direct effect of IL-12 and how much is produced by an indirect effect through the production of IL-2? Because you showed that anti-IL-12 antibody blocks the proliferation of the cells, and this implies that there is also a defect in IL-2 production.

Trinchieri: IL-12 has a direct effect on the transcription of the gene encoding IFN-γ.

Cantrell: Does it also have a direct effect on the transcription of the gene encoding IL-2?

Trinchieri: There is a small effect but it is not significant. However, there's a strong synergistic effect of IL-2 and IL-12 in inducing IFN-γ production. This is due mostly to a post-transcriptional mechanism because it affects the stability of IFN-γ mRNA. The addition of anti-IL-2 antibody results in a decrease in IFN-γ production because of this synergism.

Ramshaw: I would like to raise another issue. The immune system involves cell-mediated immunity (CMI) and the production of antibodies. These are inversely related, so that a CMI response results in the suppression of an antibody response, and vice versa. Why has the immune system evolved such a sophisticated mechanism? One suggestion is that the immune system selects the most appropriate type of immune response for a particular infectious agent (Bretscher et al 1992). However, it also has to suppress the other response because, for example, if an infectious agent stimulated both CMI and antibody production, the antibodies may not only be ineffective but they may also block CMI.

Mitchison: The immune system expands and contracts locally but as a whole it stays pretty much the same, so it is under homeostatic control. Presumably, this is achieved by the control of lymphocyte apoptosis by survival factors, although none of these survival factors are known. These survival factors may be members of the interleukin family.

Lachmann: Thymectomized chickens produce a fairly normal humoral response in the absence of T cell control mechanisms. This implies that the plasticity of the immune response is greater than you suggest.

Mosmann: The immune system is definitely regulated in a number of different ways. The suggestion that some of the survival factors may be lurking amongst existing cytokines is very likely. For example, IL-2 is probably a maintenance factor for resting T cells, and granulocyte macrophage colony-stimulating factor keeps neutrophils alive (Lee et al 1993).

The suppression of inappropriate immune responses may also play a part in the regulation of the immune system. In many diseases much of the tissue damage is due to immunopathology. Therefore, the immune system can be extremely damaging to tissues. It makes sense to limit any response which isn't absolutely required. Damage to the tissue must be an important consideration.

Sher: But that's a self-regulatory mechanism. Ian Ramshaw was talking about cross-regulation.

Ramshaw: The immune system does not induce all types of immune responses in the hope that one of them works. It is highly selective for this reason, i.e. it does not cause damage to self.

Mosmann: That situation would be very damaging.

Romagnani: Immune responses can also be damaging if they are prolonged. For example, I believe that during infection with *Mycobacteria*, if the bacteria are not removed, the prolonged Th1 response may also cause tissue damage.

Liew: Perhaps it is simply just uneconomical to switch on all the immune responses.

Mosmann: That would explain the striking cross-regulation. Alternatively, it may be just inherent in what responses are required for different organisms, and what responses might interfere.

Lotze: Are we just looking at extreme models? It is possible that *in vivo* both a little bit of CMI and a little bit of antibody production occur at the same time, and that there is an interplay between the two that is driven by the amount of antigen.

Ramshaw: I disagree. The appearance of both antibody and CMI may be due to recognition of different antigen determinants or it may be temporally related, i.e. antibody production may occur after a CMI response.

Shearer: It is probable that both CMI and antibody production occur during a particular immune response. CMI responses often decrease after immunization or infection. This may reflect the immune system's attempt to prevent an autoimmune or immunopathological condition mediated by CMI. The production of antibodies may then increase. This whole process may be regulated by cytokines that induce cross-regulation of Th1-like and Th2-like responses, and this may be the way that the system is self-regulated. As antibody production increases, one often sees a decrease in the magnitude of the cellular response. This phenomenon is illustrated in the figure shown in our paper on vaccination against HIV (Salk et al 1993).

Lotze: Can you introduce antigen load into that model?

Shearer: One way to induce a dominant CMI or Th1-like response is to immunize with low doses of antigen. In the early 1970s, several people (including Eddy Liew) demonstrated that low-dose immunization of mice induced a cellular response, but not an antibody response (reviewed in Parish 1972).

During an acute HIV infection, there is an increased HIV-specific cytotoxic T lymphocyte (CTL) activity concomitant with the drop in viral load. Neutralizing antibody does not appear until after the virus load in the blood is reduced. It is noteworthy that antibody access is increased at the same time that CTL activity is declining (R. L. Koup, unpublished observations). It has been argued that CMI and CTLs cannot be protective against initial HIV infection because people become infected and die of AIDS despite a potent initial CMI response (Miedema et al 1993). However, investigators forget that the role of prophylactic immunization is actually to give a jump start to the immune system. It is important that CMI is strong, but it actually decreases after a while and antibody production increases. In my opinion, if one wanted to achieve persistent protection and a continuous reduction in virus load in HIV-infected individuals, a potent CMI response must be maintained and antibody levels should remain relatively low. Antibodies are probably present in a strong CMI setting and may be protective. People developing a vaccine for AIDS currently want the best of both worlds, and they would probably not accept a vaccine that didn't produce a high level of antibody. I doubt that IL-12 or some of the other adjuvants being considered will give the best of both worlds and cancel out the effects of immune cross-regulation. If the current vaccine strategy doesn't work, despite plenty of antibody and some CMI activity, I would like to see someone develop an approach that maintained strong CMI and disregarded the level of antibody.

Mason: Everyone is forgetting the adrenal gland. If a rat is immunized with myelin basic protein in Freund's adjuvant it becomes paralysed but then it recovers spontaneously. However, if the adrenal glands are removed before the signs of paralysis appear, the subsequent paralysis is progressive and the rat finally dies. If the loss of corticosterone that follows adrenalectomy is made good by steroid-replacement therapy, one sees the same time course of disease and the same recovery as that observed in rats whose adrenal glands are intact (MacPhee et al 1989). More strikingly, mice (or rats) that are adrenalectomized before immunisation with complete Freund's adjuvant die within 48 h because they cannot regulate the tumour necrosis factor production induced by the adjuvant (Bertini et al 1988). It has been shown that IL-1 acts on the hypothalamic pituitary adrenal axis to induce corticosteroid hormone release from the adrenal gland (Berkenbosch et al 1987). This is a potent feedback mechanism that controls the overall inflammatory response and in experimental autoimmune encephalomyelitis it is the curative mechanism. We have recently shown that if cells are activated *in vitro* in the presence of dexamethasone, a Th2-type cytokine response is favoured and the Th1 cells undergo apoptosis (F. Ramirez, unpublished results). These results indicate that the neuroendocrine system can influence cytokine synthesis in a selective way. This regulatory mechanism is independent of cytokines *per se* except that it is switched on by IL-1.

Mitchison: That's true, and Daynes & Araneo (1994) have also worked on the balanced regulation by steroids.

Liew: I would like to clarify some historical points. In the paper that Chris Parish and I published in 1972 (Parish & Liew 1972), we obtained a mirror image between delayed-type hypersensitivity and antibody responses only at a certain time point, i.e. at Day 28 after the injection. This relationship was not as clear-cut at other time points.

Mosmann: The location of the immune response may also be important. A powerful systemic response may regulate systemically, so that it's easier to see reciprocal regulation. In more localized responses it may be possible to have different responses generated simultaneously in different locations. There may be moderate systemic crosstalk but it may be not complete. Therefore, it may be possible to get a mixture of responses by immunizing in different places.

Ramshaw: Can you give us an example of that?

Mosmann: We did some experiments in collaboration with Rich Locksley (Sadick et al 1991). We studied *Leishmania* and *Nippostrongylus* infections in the same mouse, and we expected them to interfere more than they actually did. The mouse seemed to handle them separately, so we thought that the two responses were in two separate locations. A similar experiment was done later by Jay Berzofsky and collaborators (Actor et al 1994) with vaccinia virus and *Schistosoma mansoni*. In this case the immune responses did interfere.

Ramshaw: These are different responses to different agents. This is distinct from different responses to the same agent in different sites of the body. I don't know of an example where that might occur.

Mosmann: I'm suggesting a vaccine strategy whereby two different immune response types might be induced in the same person by vaccinating simultaneously in different locations.

Swain: In mucosa there is a different response than in lymph nodes, in terms of cytokine production by T cells as well as antibody production. This suggests the existence of specialized environments.

Lotze: Thierry Boon and colleagues have done some similar experiments. They have immunized mice on two different limbs, and they have found that there are qualitatively and quantitatively different results in the two sets of regional lymph nodes, in terms of response to individual antigens (Van den Eynde et al 1991). However, these were CMI responses and not antibody responses.

Ramshaw: That's not what Tim Mosmann is addressing. He is talking about antibody and CMI to the same antigen at different sites in the body.

Lotze: But the starting conditions (i.e. which cells get recruited, which endothelia gets activated and which pro-inflammatory cells are present) may be critical for determining the type of response.

References

Actor JK, Marshall MA, Eltoum IA, Buller RM, Berzofsky JA, Sher A 1994 Increased susceptibility of mice infected with *Schistosoma mansoni* to recombinant vaccinia virus: association of viral persistence with egg granuloma formation. Eur J Immunol 24:3050–3056

Berkenbosch F, Oers JV, Del Rey A, Tilders F, Besedovsky H 1987 Corticotropin-releasing factor-producing neurons in the rat activated by interleukin-1. Science 238:524–526

Bertini R, Bianchi M, Chezzi P 1988 Adrenalectomy sensitizes mice to the lethal effects of interleukin 1 and tumour necrosis factor. J Exp Med 167:1708–1712

Bretscher PA, Wei G, Menon JN, Bielefeldt-Ohmann H 1992 Establishment of stable, cell-mediated immunity that makes 'susceptible' mice resistant to *Leishmania major*. Science 257:539–542

Daynes RA, Araneo BA 1994 The development of effective vaccine adjuvants employing natural regulators of T-cell lymphokine production *in vivo*. Ann N Y Acad Sci 730:144–161

Lee A, Whyte MK, Haslett C 1993 Inhibition of apoptosis and prolongation of neutrophil functional longevity by inflammatory mediators. J Leukocyte Biol 54: 283–288

MacPhee IAM, Antoni FA, Mason DW 1989 Spontaneous recovery of rats from experimental allergic encephalomyelitis is dependent on regulation of the immune system by endogenous adrenal corticosteroids. J Exp Med 169:431–445

Miedema F, Meyaard L, Klein MR 1993 Protection from HIV infection or AIDS? Science 262:1074–1075

Parish CR 1972 Relationship between humoral and cell-mediated immunity. Transplant Rev 13:35–66

Parish CR, Liew FY 1972 Immune responses to chemically modified flagellin. 3. Enhanced cell-mediated immunity during high and low zone antibody tolerance to flagellin. J Exp Med 135:298–311

Riddell SR, Watanabe KS, Goodrich JM, Li CR, Agha ME, Greenberg PD 1992 Restoration of viral immunity in immunodeficient humans by the adoptive transfer of T-cell clones. Science 257:238–241

Sadick MD, Street N, Mosmann TR, Locksley RM 1991 Cytokine regulation of murine leishmaniasis: interleukin 4 is not sufficient to mediate progressive disease in resistant C57BL/6 mice. Infect Immun 59:4710–4714

Salk J, Bretscher PA, Salk PL, Clerici M, Shearer GM 1993 A strategy for prophylactic vaccination against HIV. Science 260:1270–1272

Van den Eynde B, Lethe B, Van Pel A, De Plaen E, Boon T 1991 The gene encoding for a major tumor rejection antigen of tumor P815 is identical to the normal gene of syngeneic DBA-2 mice. J Exp Med 173:1373–1384

Index of contributors

Non-participating co-authors are indicated by asterisks. Entries in bold type indicate papers; other entries refer to discussion contributions.

Indexes compiled by Liza Weinkove.

Subject index